GENERAL TOPOLOGY AND MODERN ANALYSIS

F. Burton Jones

GENERAL TOPOLOGY AND MODERN ANALYSIS

Edited by

L. F. McAULEY
Department of Mathematics
State University of New York
Binghamton, New York

M. M. RAO
Department of Mathematics
University of California
Riverside, California

ACADEMIC PRESS
A Subsidiary of Harcourt Brace Jovanovich, Publishers
New York London Toronto Sydney San Francisco 1981

Academic Press Rapid Manuscript Reproduction

Proceedings of a Conference on General Topology and Modern Analysis
Held at the University of California, Riverside, May 28-31, 1980,
in honor of F. Burton Jones.

ACADEMIC PRESS, INC.
111 Fifth Avenue, New York, New York 10003

United Kingdom Edition published by
ACADEMIC PRESS, INC. (LONDON) LTD.
24/28 Oval Road, London NW1 7DX

Library of Congress Cataloging in Publication Data
Main entry under title:

General topology and modern analysis.

"Proceedings of the Conference on General Topology and Modern Analysis held in May 1980 at the University of California, Riverside, in honor of the retirement of Professor F. Burton Jones"--Pref.
Bibliography: p.
1. Topology--Congresses. 2.Mathematical analysis--Congresses. 3. Jones, F. Burton. I. McAuley, L. F. (Louis F.), Date. II. Rao, M. M. (Malempati Madhusudana), Date. III. Jones, F. Burton. IV. Conference on General Topology and Modern Analysis (1980: University of California, Riverside)
QA611.A1G454 514 81-2249
ISBN 0-12-481820-X

PRINTED IN THE UNITED STATES OF AMERICA

81 82 83 84 9 8 7 6 5 4 3 2 1

CONTENTS

CONTRIBUTORS

Numbers in parentheses indicate the pages on which the authors' contributions begin.

Leonard A. Asimow (365), Department of Mathematics, University of Wyoming, Laramie, Wyoming 82071

David P. Bellamy (31, 143), Department of Mathematics, University of Delaware, Newark, Delaware 19711

Donald E. Bennett (39), Department of Mathematics, Murray State University, Murray, Kentucky 42071

R. H. Bing (3), Department of Mathematics, University of Texas, Austin, Texas 78712

Karol Borsuk (147), Mathematics Institute, Śniadeckich 8, 00-950 Warszawa, Poland

Beverly Brechner (151), Department of Mathematics, University of Florida, Gainesville, Florida 32611

C. Edmund Burgess (169), Department of Mathematics, University of Utah, Salt Lake City, Utah 84108

Bruce L. Chalmers (373), Department of Mathematics, University of California, Riverside, California 92521

John E. de Pillis (383), Department of Mathematics, University of California, Riverside, California 92521

Nicolae Dinculeanu (391), Department of Mathematics, University of Florida, Gainesville, Florida 32611

James Dugundji (177), Department of Mathematics, University of Southern California, Los Angeles, California 90007

Eldon Dyer (185), Graduate School and University Center, City University of New York, New York, New York 10036

C. A. Eberhart (209), Department of Mathematics, University of Kentucky, Lexington, Kentucky 40506

Edward G. Effros (217), Department of Mathematics, University of California, Los Angeles, California 90024

Joseph Brauch Fugate (209), Department of Mathematics, University of Kentucky, Lexington, Kentucky 40506

B. D. Garrett (229), Texas A & M University, Texas Transportation Institute, College Station, Texas, 77843

Jack T. Goodykoontz, Jr. (43), West Virginia University, Morgantown, West Virginia 26506

George R. Gordh, Jr. (53, 65), Department of Mathematics, California State University, 6000 J Street, Sacramento, California 95819

Edward E. Grace (71, 493), Department of Mathematics, Arizona State University, Tempe, Arizona 85281

Charles L. Hagopian (83, 239), Department of Mathematics, California State University, Sacramento, California 95819

Roger W. Hansell (405), Department of Mathematics, University of Connecticut, Storrs, Connecticut 06268

Robert P. Hunter (89), Department of Mathematics, Pennsylvania State University, McAllister Building, University Park, Pennsylvania 16802

Robert C. James (347), Department of Mathematics, Claremont Graduate School, Claremont, California 91711

F. Burton Jones (19), Department of Mathematics, University of California, Riverside, California 92521

V. Kannan (241), School of Mathematics & CIS, University of Hyderabad, Nampally Station Road, Hyderabad-500 001 India

Lewis Lum (143), Department of Mathematics, Salem College, Winston-Salem, North Carolina 27108

Garr S. Lystad (247), 1213 Woodbine Street, Lewisville, Texas 75028

Byron L. McAllister (255), Department of Mathematics, Montana State University, Bozeman, Montana 59717

Louis F. McAuley (117, 265), Department of Mathematics, State University of New York, Binghamton, New York 13901

Donald A. Martin (417), Department of Mathematics, University of California, Los Angeles, California 90024

John C. Mayer (151), Department of Mathematics, University of Florida, Gainesville, Florida 32608

Austin C. Melton (281), Department of Mathematics, Marshall University, Huntington, West Virginia 25701

Mark Michael (291), Department of Mathematics, Southeast Missouri State University, Cape Girardean, Missouri 63701

Deane Montgomery (295), School of Mathematics, Institute for Advanced Study, Princeton, New Jersey 08540

Jan Mycielski (431), Department of Mathematics, University of Colorado, Boulder, Colorado 80309

Issac Namioka (437), Department of Mathematics, University of Washington, Seattle, Washington 98195

Peter J. Nyikos (441), Department of Mathematics, University of South Carolina, Columbia, South Carolina 29208

Roman Pol (451), Department of Mathematics, University of Washington, Seattle, Washington 98195

M. M. Rao (457), Department of Mathematics, University of California, Riverside, California 92521

James T. Rogers, Jr. (97), Department of Mathematics, Tulane University, New Orleans, Louisiana 70118

Leland E. Rogers (101), Cook, Washington 98605

Mary E. Rudin (305), Department of Mathematics, University of Wisconsin, Madison, Wisconsin 53706

Elias Saab (475), Department of Mathematics, The University of British Columbia, #121-1984 Mathematics Road, Vancouver, British Columbia V6T1Y4

Paulette Saab (475, 485), Department of Mathematics, The University of British Columbia, #121-1984 Mathematics Road, Vancouver, British Columbia V6T1Y4

Albert R. Stralka (247), Department of Mathematics, University of California, Riverside, California 92521

Franklin D. Tall (309), Department of Mathematics, University of Toronto, Toronto, Canada M551A1

Eric K. van Douwen (43, 399), Department of Mathematics, Ohio University, Athens, Ohio 45701

Eldon J. Vought (105), Department of Mathematics, California State University, Chico, California 95929

John J. Walsh (317), Department of Mathematics, University of Tennessee, Knoxville, Tennessee 37917

Lewis E. Ward, Jr. (327), Department of Mathematics, University of Oregon, Eugene, Oregon 97403

David C. Wilson (337, 341), Department of Mathematics, University of Florida, Gainesville, Florida 32611

Edythe P. Woodruff (265), Department of Mathematics, Trenton State College, Trenton, New Jersey 08625

C. T. Yang (295), Department of Mathematics, University of Pennsylvania, Philadelphia, Pennsylvania 19104

PREFACE

This volume contains the proceedings of the Conference on General Topology and Modern Analysis held in May 1980 at the University of California, Riverside, in honor of the retirement of Professor F. Burton Jones. The variety of topics covered included set theory as well as some applications, and reflected Professor Jones' wide-ranging interest in mathematics.

Among Professor Jones' many contributions to topology, perhaps his idea of aposyndesis and his creation of its theory have generated the most research activity. So, we have made a special effort to present the current status of work in this area and have devoted one of the three major sections of the book to this topic. Each of these sections starts with a survey article, followed by other articles in alphabetical order by author. Professor E. E. Grace, who has provided able assistance for the special section, has prepared a nearly exhaustive annotated bibliography of aposyndesis that is included at the end of the volume. We warmly appreciate his work. For delivering the inaugural address of the conference, Professor R. H. Bing, a fellow creator of "circles of pseudoarcs" along with Professor Jones, also deserves applause.

The idea of honoring Burton Jones with such a broadly based conference originated with his colleagues, former students, and many friends. The organizing committee consisted of J. de Pillis, L. F. McAuley, M. M. Rao, P. Roy, and A. R. Stralka, with Professors McAuley and Rao as cochairmen and later as coeditors of the Proceedings. The responses to our invitations were quite enthusiastic; the sessions were well attended with a good geographic representation. Many people even paid their own way to attend the conference.

The seventieth birthday of Professor F. Burton Jones was celebrated on November 22, 1980. As a token of respect and affection for his mathematical and other contributions, as well as his humane qualities, we present this volume to him, and wish him many years of fruitful mathematical activity.

ACKNOWLEDGMENTS

The necessary financial assistance for the conference was graciously provided by Dr. W. Mack Dugger, Dean of the College of Natural and Agricultural Sciences, and Dr. Michael D. Reagan, Vice Chancellor, both of the University of California, Riverside. We are grateful to them for this support.

All members of the organizing committee worked in various ways to make the conference a success. Much of the local work became easier due to the interests of some of the faculty and staff at the University of California, Riverside. Special thanks go to John de Pillis, acting chairman of the Mathematics Department, and Florence Kelly, the administrative assistant of that department, who acted as the conference secretary. Her efficient and enthusiastic work made the meetings a pleasure to attend, and consequently, the subsequent work for the Proceedings became lighter.

We extend thanks and appreciation to the session chairpersons:

L. A. Asimow	G. R. Gordh, Jr.	B. L. McAllister	D. E. Rush
C. E. Burgess	C. L. Hagopian	I. Namioka	F. D. Tall
N. Dinculeanu	R. W. Heath	P. J. Nyikos	H. G. Tucker
J. E. de Pillis	R. P. Hunter	J. W. Petro	E. J. Vought
J. Dugundji	J. L. Kelley	P. Roy	D. C. Wilson
N. E. Gretsky	L. F. McAuley	M. E. Rudin	

All papers were retyped with care and diligence by Joyce Kepler and Patricia Baxter. In addition to the proofreading done by the authors, Dave Holmes proofread all the papers. Also, several graduate students helped with the transportation of guests arriving at and departing from Riverside. We are grateful to all these people for their enthusiastic assistance.

SECTION I

INAUGURAL LECTURE

METRIZATION PROBLEMS*

R. H. Bing

University of Texas, Austin

I. USING EXAMPLES

Burton Jones is the first mathematician with whom I collaborated after receiving my Ph.D. Although R. D. Anderson, E. E. Moise, C. E. Burgess, Mary Ellen Estill (Rudin), Eldon Dyer, Billy Jo Ball were at Texas at this time, they were students and R. L. Moore insisted that his students develop independent work habits. Hence, after receiving my Ph.D., I did not discuss research with these students. However, Jones had received the Ph.D. several years earlier and was just returning from Cambridge where he had done work on underwater sound related to war work. I did not feel restrained in discussing mathematical research with him.

An unsolved problem of considerable interest to me was the question Jones had asked in 1937 [J_1]---is a normal Moore space metrizable? To understand how we worked at unsolved problems, it is necessary to know our modus operandi. Our first approach in attacking a problem was to look for a counterexample. If no one of our vast store of examples worked, we would try modifying known examples to discover a counterexample. It was my gut-reaction (and still is) that there is a real counterexample to the normal Moore space conjecture but it may be more complicated than anything we have examined. I soon learned that Jones had examples in his repetoire that were missing from mine.

*Work on this paper was supported by NSF Grant MCS-790-4709.

ISBN 0-12-481820-X

II. JONES' TIN-CAN-SPACE T

One of these amazing examples is Jones' tin-can-space, which I call T. Jones found the example in the 1930's, showed it to me in the 1940's and wrote up a version of it in the 1960's [J_2]. Points of T are tin cans placed on horizontal shelves. There were ω_1 of these horizontal shelves placed above each other in a best well ordered fashion (each shelf had at most a countable number of other shelves below it). There were only a countable number of tin cans on each level so T had $\aleph_0 \times \aleph_1 = \aleph_1$ points (tin cans).

The bottom level (shelf) L_0 had only one tin can and the can was assigned a size of 0. The next level L_1 had a countable number of disjoint tin cans, and each of these was above the can on the bottom row. These cans were assigned sizes $\frac{1}{2}, \frac{1}{4}, \frac{1}{8}, \frac{1}{16}, \ldots$. There were $\aleph_0$ tin cans on the next level L_2 above each can on L_1 and they were assigned sizes $\frac{1}{2}, \frac{1}{4}, \frac{1}{8}, \ldots$, etc. Inductively cans are also placed on levels $L_2, L_3, \ldots$ and assigned sizes.

Figure 1 shows cans on levels L_0, L_1, L_2 with certain sizes indicated. It was too crowded to show all cans on higher levels, but we note that for levels $L_1, L_2, \ldots, L_\alpha, \ldots$ ($\alpha < \omega_0$), the following properties hold.

1. There are only a countable number of cans on L_i ($0 < i < \omega_0$).
2. If L_α and L_β are levels with L_α below L_β, n is a positive integer, and p is a can on L_α, then there is a stack of cans such that the stack is based on p, has its top in L_β, and the sum of the sizes of the cans in the stack above p (size of p is ignored) is $1/2^n$.

Only a countable number of cans are put on the shelf L_{ω_0} and this causes some difficulty. Stacks of cans lead up to a Cantor set of positions for cans at the L_{ω_0} level, but we ignore most of them and use only a countable number of them---being guided by Properties 1 and 2 of the

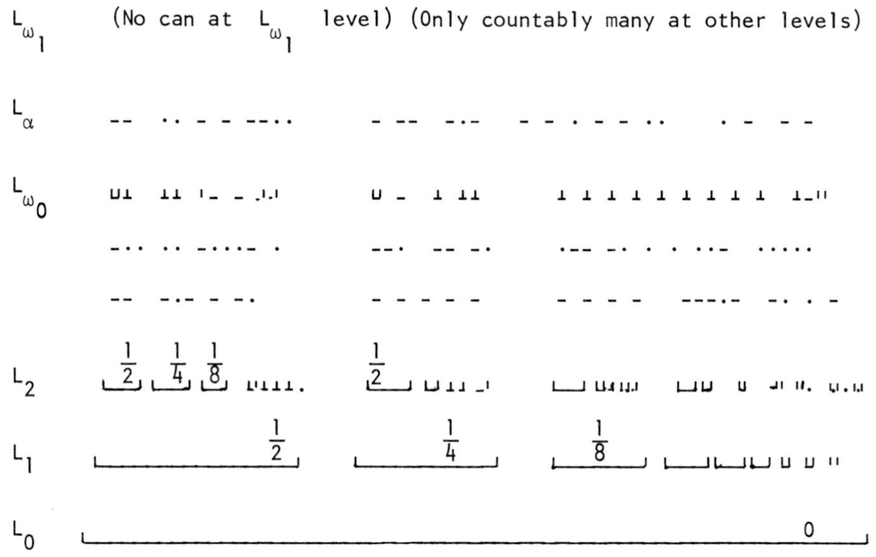

Fig. 1

previous paragraph. This is done as follows. There are only a countable number of cans in the union of all levels below L_{ω_0}. For each such can p on a lower level and each positive integer n, we pick a stack of cans based on p and reaching up to the L_{ω_0} level, so that the sum of the sizes of the cans in the stack above p is $1/2^n$. We then put a can p on the L_{ω_0} level at the top of the stack. We do this for each lower can and each n, but even so---put only a countable number of cans on the L_{ω_0} level. However, conditions 1 and 2 prevail even through level L_{ω_0}. Cans on level L_{ω_0} are assigned a size of 0.

The procedure is continued inductively to get cans on each level L_α with $\alpha < \omega_1$. Pages 116-117 of [J_2] gives details of a variation of the inductive procedure.

For a topology of T, let each can on a nonlimit level be isolated. If p is a can in some limit level, it is the top of some stack. The set of cans in such a stack is a neighborhood.

The space T is locally compact and locally separable. It contains an uncountable collection of mutually disjoint open sets and is not separable.

To show that T is a Moore space, we subdivided T into a countable number of mutually disjoint closed sets $A_{r_1}, A_{r_2}, \ldots$ where r_i is a non-negative rational, and if $p \in A_{r_i}$, the sume of the sizes of cans in the maximal stack topped by p is r_i. To get a development $G_1, G_2, \ldots$, we would let an element $g \in G$ be one of the neighborhoods previously described with the additional restriction that if $p \in A_{r_1} \cup A_{r_2} \cup \cdots \cup A_{r_i}$, and $p \in g \in G_i$, this p is the top can of g.

We show that T is not metrizable by showing that if it were, it would be the union of a monotone increasing sequence of countable sets. With this objective in mind, we suppose T has a bounded metric with diameter less than 1 and for each $p \in T$ let

$$\varepsilon(p) = \mathrm{LUB}\{x \mid x \text{ neighborhood of } p \text{ is countable}\} .$$

For subsets A, B, C of T let

$$f_1(A) = \bigcup_{a \in A} N(a, \varepsilon(a)/2) ,$$

$f_2(B) = \{q$, some element of B lies at same or a higher level than $q\}$, and

$f_3(C) =$ union of $f_2(C)$ and next level.

If X is the one point set which is bottom of T, consider the monotone increasing sequence of countable sets

$$X, f_3f_2f_1(X), (f_3f_2f_1)^2(X), \ldots .$$

Note that f_3 pushes us past isolated levels and f_2f_1 makes short order of limit ones.

If T were normal, it would be an example showing that the normal Moore space conjecture is false. In his 1965 paper, Jones said he had not yet discovered whether T was normal or not. We discuss the normality of T

in the final section of this paper, but mention that some of what we know about T was learned as a result of the study of it as a special Aronszajn tree. See [F and R].

III. EARLY PROOFS

Not all questions are solved with counterexamples. One of the best known metrization results is the one stating that a second axiom regular Hausdorff space is metrizable. Here the proof consists of constructing a distance function. I recall my disappointment when I saw the traditional argument. While it gave a distance function, the distance function constructed had little resemblance to ordinary distance. If a straight line interval were metrized by such a technique, there would be scant hope that the resulting metrized arc could be isometrically embedded in any Euclidean space. It embeds isometrically in a Hilbert cube.

I resolved to try for a more realistic distance function if I had the opportunity. Such an opportunity came when I attacked the following problem.

> Suppose a regular Hausdorff space H has a development $G_1, G_2, \ldots$ such that if two points are covered by a coherent collection of r elements of G_{i+1}, the pair lies in one element of G_i. Describe a metric for H.

I wondered as to the most realistic way to describe a reasonable distance. It would simplify the situation if the elements of G_{i+1} were the open sets with diameters $1/r^i$. Then I would assign a size of 1 to elements of G_1 and a size of $1/r^i$ to elements of G_{i+1}. The size of a chain would be the sum of the sizes of its links and the distance from p to q would be defined as the greatest lower bound of sizes of all chains joining p to q where links of the chains are elements of $G_1 \cup G_2 \cup \cdots$ As shown by Theorem 1 of $[B_1]$, this greatest lower bound is not 0 if

$p \neq q$, so the resulting limit provides a distance---even in the general case.

This method of taking least upper bounds on sizes of chains not only gave a reasonable distance for simple continua, but also was the basis for a method I used later to show that each compact, locally connected metric continuum has a convex metric [B_3].

The following theorem of Bing, Nagata, and Smirnov [B_2N,Sm] has attracted some attention. Bing's version states:

THEOREM. *A regular Hausdorff space* H *is metrizable if and only if there is a sequence* $G_1, G_2, \ldots$ *such that*

each G_i *is a discrete collection of open sets in* H *, and*

$G_1 \cup G_2 \cup \cdots$ *is a basis for* H .

The "only if" part of the proof was facilitated by a result of Stone [St] that a metric space is paracompact, and the "if" part was provided by a construction described below.

Proof. To get a first approximation to a distance, convolute H so that part of it not covered by G_1 lies in a horizontal plane and points in elements of G_1 lie on hills of heights less than or equal to $1/2$. If p and q are two points,

$d_1(p,q)$ = height p if q is on the bottom level,

$d_1(p,q)$ = |height p - height q| if p,q are on the same hill,

$d_1(p,q)$ = (height p + height q) if p and q are on different hills.

Next, turn to G_2 but get d_2 by deciding that the heights of the hills are less than $1/4$. The procedure is continued and the final distance is defined as $d(p,q) = \Sigma d_i(p,q)$. Note that instead of embedding H in a Hilbert cube as was done in the case where H was second countable, it was embedded in the cartesian product of cones.

It is not our purpose here to show that a collectionwise normal Moore space is metrizable [B_2] but we remark that in getting a metric for it, we may use the "hills approach" [B_3].

IV. RISE AND FALL

We once had great expectations that set theory could provide answers to the questions of general topology. Was the continuum hypothesis true? Did topological spaces that are known to have one set of properties necessarily have another set? Is there a Souslin space? Are all normal Moore spaces metrizable?

The axiomatic method is very popular with mathematicians. Most mathematicians have had a pleasant experience in using the axiomatic approach in synthetic geometry. Would it not be wonderful to have a definite set of axioms and procedures and be able to get answers with precision? Perhaps the rise in expectations for the usefulness of the axiomatic set theory approach was based on hope.

Minor problems began to develop early. Russell's paradox showed that the rules of sets applied to some collections did not apply to others. To bypass this problem, these troublesome collections were called classes instead of sets. One might wonder if this is the beginning of a trend to change definitions or concepts. We discuss other variations later. By and large, mathematicians were glad that axiomatic set theory had been rescued from Russell's paradox. If you cannot trust the axioms, whom can you trust?

The concept of the "nonexcluded middle" caused some confusion. In the 1940's and 1950's certain logicians were using certain axioms of set theory to disprove accepted theorems of topology. For example, L. E. J. Brouwer showed on the basis of axioms of set theory used by many logicians of that day that the Jordan curve theorem was false. On another occasion he showed on the same basis that Brouwer's fixed point theorem is false. I am relieved that many of my present day colleagues assure me that these approaches are no longer fashionable. It is not nice to fool Mother Nature!

We still have a sprinkling of people who do not believe in infinite sets. They cite the computer (which knows nothing of continuity, but uses tables rather than graphs) as evidence that this is the wave of the future. They may use axioms to support their points of view. However, this point of view seems at variance with traditional mathematics and the "real world" as most mathematicians envisage it. Presumably, the axioms of set theory suited for studying general topology are not of this sort.

Mathematicians speak of the set of real numbers (or of *the reals*) rather than some set of reals. This suggests the underlying assumption that there is only one set of reals. If one class in some school is studying the reals and another class somewhere else is studying them also, the two classes are studying the same thing. However, we are told that in axiomatic set theory that there are many sets of reals [C]. To distinguish the real reals from such an axiomatic set of reals, I shall call the latter a set of axiomatic reals. I asked the innocent question as to whether there was always a homeomorphism between any two sets of axiomatic reals, but it was suggested that this was not a proper question. Is the cardinality of two different sets of axiomatic reals the same? Is the cardinality of the reals the same as the cardinality of a set of axiomatic reals? Some might duck the questions by saying that they have no meaning.

However, there is a suspicion that a set of axiomatic reals has fewer numbers than the reals, since in a certain sense it is only a countable collection of numbers [C]. However, when I am told that the cardinality of a set is not an intrinsic property of the set but instead is concerned about a shortage of algorithms, I get the uneasy feeling that someone is tinkering with basic traditional concepts.

The terms uncountable, countable, infinite, finite have different meanings to different people. These concepts are used to indicate how many elements a set has. Would one call a drove of hornets infinite or uncountable just because it is impracticable (or impossible) to look inside their nest? I hope not. An uncountable set has more elements than there are integers. To many this would mean that there is no one-to-one correspondence between the uncountable set and the positive integers. There is none whatsoever. It is not enough that there exists none known to man or beasts or even in our list of algorithms in the model. There must be none known to omnipotent powers looking from outer space. There is no such 1-1 correspondence.

Others call an infinite set uncountable if there is no algorithm *in the model* for exhibiting a 1-1 correspondence between the set and the integers. The two concepts differ and there is confusion in using the same term to denote both. For example, on pages 19-20 of [C], Cohen remarks, "This paradox that a countable model can contain an uncountable set is explained by noting that to say that a set in uncountable merely asserts the nonexistence of a one-to-one mapping of the set with the set of integers. The "uncountable" set in M actually has only countably many members in M , but there is no one-to-one correspondence within M of this set with the set of integers." The algorithm approach is the one used in axiomatic set theory and it leads to situations where countable models have uncountable sets [C]. It leads to the statement of results which appear ridiculous if examined from other viewpoints. There is no meeting of minds and we hear, "Why don't you drop your definitions and accept mine?" "I got there first." "Yours makes no sense." "How can sets of measure 0 contain sets of positive measure?" In any case, there is confusion.

What is the beef? One is that if a mathematician uses traditional concepts in asking a question, the axiomatic set theorist may ignore the concepts intended by the questioner, replace them with related concepts, answer the changed questions, and then announce results with verbage that makes it appear that the original question has been answered. For example, if a traditional mathematician asks if the continuum hypothesis is true, he should not get an answer starting with "It is consistent with...," but instead might be told that "Axiomatic set theory is not equipped to cope with such questions since it contains no place for the traditional concepts of the reals, c , uncountable, $\aleph_1$."

Axiomatic set theory is a strong vital subject in its own right, and despite the fact that it tends to mislead general topologists by shifting the meaning of terms, it provides considerable help to general topologists. It warns them what problems to avoid. For example, axiomatic Martin's axiom plus the negation of the axiomatic continuum hypothesis (MA + ¬CH) implies that the Normal Moore Space conjecture is false [F]. To show that the Normal Moore Space conjecture is true in the real world, one would be faced with a formidable task---one with such scant promise of success that most mathematicians would pass it by to try something more promising.

V. PARTIAL SOLUTIONS TO NORMAL MOORE SPACE CONJECTURE

Although considerable effort has been spent in seeking a solution, a real solution eludes us. We list certain partial solutions.

Jones showed [J_1] that if $2^{\aleph_0}<2^{\aleph_1}$, then each separable normal Moore space is metrizable. This tends to case doubt on the existence of a separable counterexample since few would be prepared to argue that $2^{\aleph_0} = 2^{\aleph_1}$.

Traylor showed [T_r] that each metacompact separable, normal Moore space is metrizable.

Reed and Zenor got the very interesting result [RZ] that a normal Moore space is metrizable if it is both locally compact and locally connected.

Bing showed [B_2] that a normal Moore space is metrizable if and only if it is collectionwise normal.

Fleissner showed [R, p. 46] that if each set is constructible (V = L) then each normal locally compact Moore space is metrizable. Although Godel showed that (V = L) is consistent with the usual axioms of set theory, many mathematicians feel that (V = L) does not represent the real situation.

Recall that a Q set is an uncountable subset X of the reals such that if $Y \subset X$, then Y is the intersection of X with a G_δ subset of the reals. There is a question about the existence of Q sets, but Bing has shown [B_2] that if there is one, there is a separable normal Moore space that is not metrizable. Conversely, Heath has shown [H] that if there is a separable, normal, nonmetrizable Moore space, then there is a Q-set.

It is to be hoped that the arguments for the above results are real and not dependent on a variable view of the reals, an algorithmic view of cardinality, the principle of constructibility, nor its big brother forcing. However, axiomatic set theory may make its presence felt unexpectedly.

At first glance it might appear that T was defined without axiomatic restrictions. A closer examination shows that $\aleph_1$ was used. Was this traditional, intrinsic $\aleph_1$ or was it the axiomatic, algorithmic $\aleph_1$? It would be interesting to know which of the metrization results we give hold in real spaces.

Let us consider the separable version of the Normal Moore Space conjecture. Is each normal separable Moore space metrizable? Pick your own poison.

(a) There is a model for set theory in which each separable normal Moore space is metrizable.

(b) There is another model containing a separable normal Moore space which is not metrizable.

First we justify (a). Jones [J_1] has shown that a separable normal Moore space is metrizable if $2^{\aleph_0} < 2^{\aleph_1}$. Perhaps it is not true that

$2^{\aleph_0} < 2^{\aleph_1}$ ---but no matter! Play the axiomatic game and do not be confused by seeking the real truths. Godel has shown that there is a model for set theory where the continuum hypothesis holds. In this model

$$2^{\aleph_0} = c < 2^c = 2^{\aleph_1} .$$

Now for (b). Bing [B_2] has shown how to construct a separable normal, nonmetrizable Moore space if there is a Q set.

Tall reports [T] that Silver discovered that a model described by Soloway in 1967 contained a set of reals with a Q set.

VI. NORMALITY OF T

If one could get a real proof that T is normal, one would have a counterexample to the Normal Moore Space conjecture. However, Fleissner shows in §4 of [F] that T is normal. Why is there no celebration?

Fleissner's definitions and methods are axiomatic. He shows that it is consistent with the axioms, definitions, and methods of axiomatic set theory that T be normal. However, it is also consistent that other results hold ---assumptions like the continuum hypothesis, Martin's axiom, and constructibility that fit into the game of axiomatics, but which are not "self-evident-truths" with regard to the traditional reals. In showing that T was normal, Fleissner used restrictions that Martin axiom holds but the continuum hypothesis does not.

By using other restrictions, Fleissner showed in §3 of [F] that T is not normal.

Both [T] and [R] contain excellent accounts about developments. In summary, known results are partial or "iffy." Where do we go from here?

One might abandon T as a possible counterexample and look in other directions for a counterexample.

Another possibility is to select a reasonable set of axiomatic restrictions and solve the conjecture on the basis of them. It may be difficult for some researchers to select a reasonable set of restrictions since their experience conditions them to accept axiomatic reasoning as a game in which all allowable moves are equally acceptable.

The Normal Moore Space conjecture has been very stimulating. It has been an area in which the findings of those in axiomatic set theory and in general topology have bolstered each other. Recently, most of the exciting results have come from set theorists and we in general topology are envious, interested, and grateful. Keep the good results coming. They add to our understanding of the problem. We need all the help we can get.

REFERENCES

[B_1] Bing, R. H., "Extending a metric," *Duke Math J.* *14*(1947), 511-519.

[B_2] ______, "Metrization of topological spaces," *Canadian J. Math* *3*(1951), 175-186.

[B_3] ______, "Partitioning a set," *Bull. Amer. Math. Soc.* *55*(1949), 1101-1110.

[C] Cohen, Paul J., *Set Theory and the Continuum Hypothesis,* W. A. Benjamin, Inc., Massachusetts, 1966.

[F] Fleissner, "When is Jones' space normal?" *Proc. Amer. Math. Soc.* *50* (1975), 375-378.

[H] Heath, R. W., "Screenability, pointwise paracompactness and metrization of Moore spaces," *Canadian J. Math.* *16*(1964), 764-770.

[J_1] Jones, F. B., "Concerning normal and completely normal spaces," *Bull. Amer. Math. Soc.* *43*(1937), 671-677.

[J_2] ______, "Remarks on the normal Moore space metrization problem," in: *Topology Seminar, Wisconsin, 1965*, R. H. Bing and R. J. Bean, eds. Ann. of Math Studies No. 60, Princeton Univ. Press, New Jersey, 1966, 115-119.

[R] Rudin, Mary Ellen, *Lectures on Set Theoretic Topology, CBMS Regional Conf.*, E. E. Grace, ed. Sem. in Math. No. 23, Amer. Math. Soc., Rhode Island, 1975.

[RZ] Reed, G. M. and P. L. Zenor, "Metrization of Moore spaces and generalized manifolds," *Fund. Math.* *91*(1976), 203-210.

[T] Tall, Franklin D., "The normal Moore space problem," in: *Topological Structures II,* Mathematical Centre Tracts No. 116, 1979, 243-261.

APOSYNDESIS

F. Burton Jones

University of California, Riverside

Aposyndesis arose as a sort of dual to a notion, *local divisibility*, first introduced by G. T. Whyburn in 1931 [40]. Eight years and forty papers later he rediscovered the notion (in a different setting) and called it *semi-local-connectedness* [41]. A connected space is *semi-locally-connected* if each point belongs to an arbitrarily small open set whose complement has a finite number of components. A semi-locally-connected continuum possesses a lot of the properties of a locally connected continuum. This is especially true of structure properties, e.g. cyclic element theory [42]. But the property that first captured my attention was the generalization of the Torhurst Theorem: The boundary of each complementary domain (= component of the complement) of a bounded locally connected continuum in the plane is itself locally connected. Whyburn showed that the boundary of a complementary domain of a bounded semi-locally-connected continuum had to be, not just semi-locally-connected, but locally connected [41]. In other words, he strengthened the Torhurst Theorem by extending it to a much larger class of plane continua; for example, the "book shelf continuum" is not locally connected at any point between 0 and 1 but the boundary of each complementary domain is a simple closed curve.

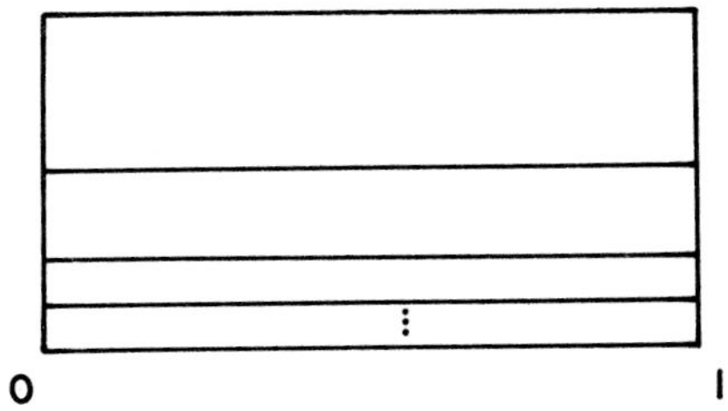

ISBN 0-12-481820-X

I was at the time looking for a pointwise property which would yield local connectedness of the boundary *at a point* regardless of what happened elsewhere. But as a pointwise property semi-local-connectedness at a point is not related to local connectivity. The simple (harmonic) countable fan

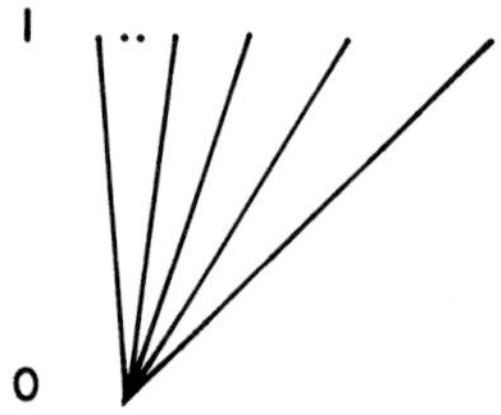

is semi-locally-connected at 1 where it is not locally connected and locally connected at 0 where it is not semi-locally-connected. This example is complicated by the fact that 0 is a separating point. Each connected (relatively) open subset containing 1 must contain 0 . So consider the following modification: (see left).

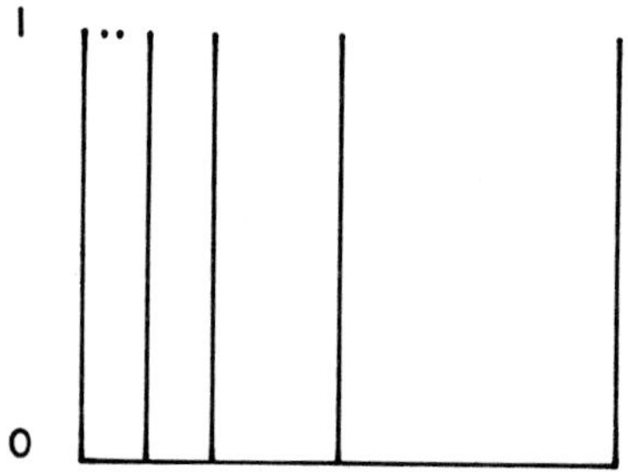

Now 0 is no longer a separating point. Still every connected open subset containing 1 contains 0 in its closure.

Such thoughts as these finally led me to the following definition: A connected space M is *aposyndetic* at a point p of M provided that for every point w of M-{p} there is a closed connected neighborhood of p missing w [21]. This is a generalization of what Moore calls *connected im kleinen:* ...for every closed set W in M-{p} there is a closed connected neighborhood of p missing W . In the above examples, these boundary continua are connected im kleinen (in fact, locally connected) at exactly those points where they are aposyndetic. But I still could not prove the pointwise version of the Torhurst Theorem. And the trouble was that the pointwise version of the Torhurst Theorem itself was not true. Specifically, it is not true that if p is a point of the boundary B of a complementary domain D of a plane continuum M , then B is locally

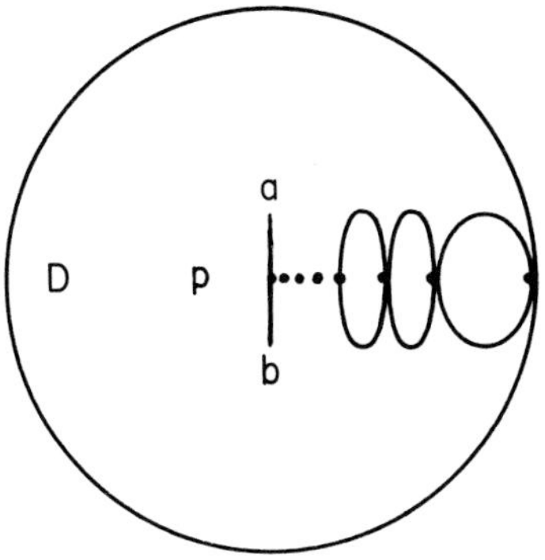

connected at p whenever M is.

Let M be the circle in E^2 together with the tangent *closed* eliptical disks and the vertical interval apb . Clearly M is locally connected at p but B , the boundary of the complementary domain D , is not locally connected or connected im kleinen at p because the interiors of the disks do not belong to B . So while B need not be connected im kleinen at p , it *is* true that E^2-D is connected im kleinen at p .

Specifically, if the continuum M in the plane E^2 is aposyndetic at a point p on the boundary of one of its complementary domains D , then E^2-D is connected im kleinen at p .

This yields Whyburn's stronger Torhurst Theorem for if a continuum M is semi-locally-connected (s-l-c) it is easy to see that M is aposyndetic at each of its points. Suppose that x and y are distinct points of M . Since M is s-l-c at y , there exists an open set U containing y whose closure misses x such that M-U has a finite number of components. The one which contains x is a closed connected neighborhood of x missing y .

On the other hand, if x is a point of M , U is an arbitrarily small (relatively) open set containing x with a compact (rel) boundary F in M and M is aposyndetic at each point of M-{x} , some finite collection H of closed connected neighborhoods of points of F missing x covers F . Hence the complement of U - $\cup H$ has only a finite number of components.

So semi-local-connectivity and aposyndesis are dual notions: A compact connected Hausdorff space X is s-l-c at a point p if it is aposyndetic at every point of X-{p} and aposyndetic at p if it is s-l-c at every point of X-{p} .

Let me return to the opening remark. It is not quite accurate to say that local divisibility and semi-local-connectedness are the same. Local divisibility is a bit stronger. A connected space M is locally divisible at the point p of M, if the complement of the *closure* of an arbitrarily small open set containing p has a finite number of components. (These components would be open.) The dual to this would be: if p is a point and y is a point of $M-\{p\}$ some open connected neighborhood of p has a *closure* missing y. This is slightly stronger than aposyndesis. However, one might expect the difference to disappear if M had the property at all its points (as is the case with local connectivity and connectedness im kleinen). Such is not the case as the following example shows:

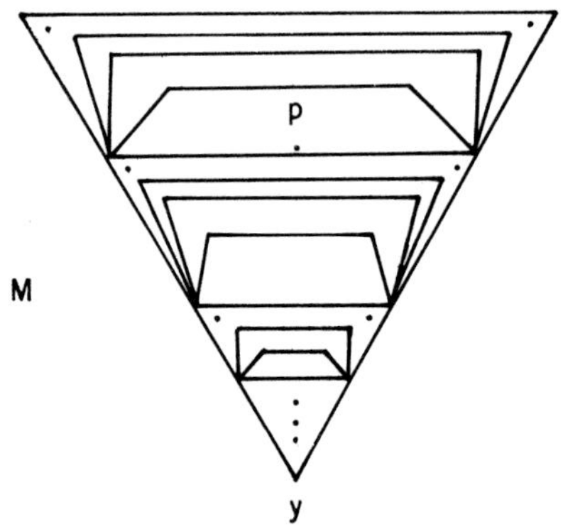

This closed subspace of the plane is aposyndetic at every point. There is a closed connected neighborhood of p which misses y. And there is an open connected neighborhood containing p which misses y. But there is no such neighborhood of p whose *closure* misses y. So there is a difference; but whether it is significant or substantial, I do not know.

How inclusive is the class of aposyndetic continua? All Cartesian products of two or more nontrivial continua are aposyndetic. While the remainder of the Stone-Czech compactification of the nonnegative reals is indecomposable (and hence badly nonaposyndetic), for all other Euclidean spaces it is aposyndetic [1]. This result of Bellamy's is astonishing since none is locally connected. Similarly, the product of two continua is never locally connected unless both factors are. The product of two aposyndetic continua is mutually aposyndetic [12] (each pair of distinct points have disjoint closed connected neighborhoods) and the product of three arbitrary (nondegenerate) continua is mutually aposyndetic [12]. Compact metric continua which contain no cut points (i.e., the complement of each point is continuumwise connected) are almost aposyndetic [22].

Then there is the question of how many aposyndetic continua are also locally connected. We have already seen that an aposyndetic plane continuum which does not separate the plane is locally connected. If the aposyndetic compact metric continuum M is separated by none of its subcontinuum or if only its aposyndetic subcontinua separate, M is locally connected [4,36]. If the compact metric continuum is aposyndetic at each point with respect to subcontinua, then M is locally connected [5]. If M is hereditarily aposyndetic, then M is hereditarily locally connected [38].

Let us turn our attention for the moment to continua which are nonaposyndetic. Indecomposable continua are of this kind for if a connected space is indecomposable the only closed connected neighborhood is the whole thing. While the product of two indecomposable continua is aposyndetic, it will not be mutually aposyndetic if the factors are chainable [12]. Ten years ago, Hagopian questioned the necessity of chainability. Maybe the product of two indecomposable continua never contains two disjoint subcontinua with nonvoid interior.

If a continuum is nonaposyndetic at every point with respect to every other point, it is indecomposable and all its points are cut points (i.e., complements not continuumwise connected). However, if a compact metric continuum is merely nonaposyndetic at each of its points, then it may be rather simple but at least one of its points must be a cut point, although there need not be more than one. In the indicated example, the ray eminating from p and limiting on the circle is replaced by a Cantor pencil of rays doing the same. This cut point theorem does not generalize to

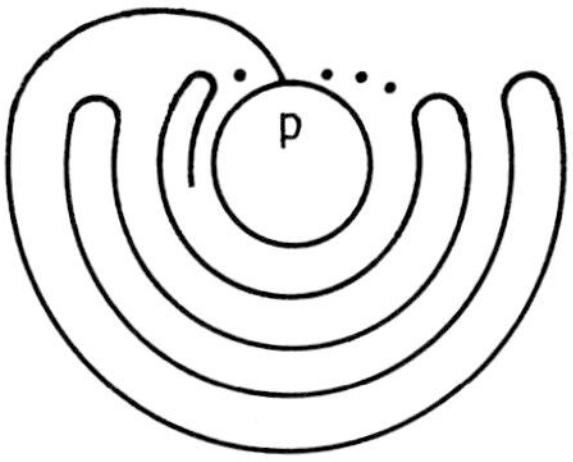

Hausdorff spaces. Grace has shown this and has studied the extent to which it can be generalized [11].

In this connection the contrast with semi-local-connectivity is striking. If a compact metric continuum is semi-locally-connected at no point, its set of cut points is dense in it. Grace has some improvements on this theorem [11] and Hagopian has shown that every point is a condensation point of the set of cut points. It would be nice to know that almost all points were cut points, but this seems to be elusive [31].

Even after all these years I still do not understand this situation. (1) If the continuum M is not aposyndetic at x, there must be a point y in $M-\{x\}$ such that every closed connected neighborhood of x must contain y. (2) If it is non-semi-locally-connected at x, there must exist a point y in $M-\{x\}$ such that every closed connected neighborhood of y must contain x. Why does (2) make M more like an indecomposable continuum than (1) does? Is not there a simple property possessed by indecomposable continua which when coupled with (2) would make the continuum indecomposable? Say, for example, that no subcontinuum separates.

Something approximating this has been done for (1). There is a set function connected with (1) that I called K. If x is a point of a continuum M, K_x is the set of all points y such that y belongs to every closed connected neighborhood of x. If for each x in M, $K_x = M$, M will be indecomposable. Schlais has a very remarkable result for compact metric continua. While $M = K_x$ for some x need not make M indecomposable, it does force M to contain an indecomposable continuum. In fact, if for some x, K_x has nonvoid interior (rel M), M contains an indecomposable continuum [30].

There is a dual set function L. If x is a point of a compact continuum M and L_x is the set of all points y [as in (2) above] such that no closed connected neighborhood of y misses x, then L_x is a continuum. Swingle, some of his students and associates, define a set function T so that $T(x) = L_x$. Or more generally, if A is a subset of M , T(A) is the set of all points y such that every closed connected neighborhood of y intersects A . This has proven to be a fruitful notion and a fruitful notation. They have done so much work that I cannot even do partial justice to it. Recently (at the Greensboro conference in October) Bellamy presented a new result for which $T(A \cup B) = T(A) \cup T(B)$. I shall comment later about another contribution from this group.

Let me return to the consideration of cut points. If the continuum M is semi-locally-connected at x and x is not a separating point of M (i.e., $M-\{x\}$ is connected), then x is not a cut point of M (i.e., $M-\{x\}$ is continuumwise connected). One can easily see this. Let a and b be any two different points of $M-\{x\}$. Each point of $M-\{x\}$ has a closed connected neighborhood which misses x ; so there is a (finite) chain of these neighborhoods from a to b whose union is a continuum missing x . This shows that a compact irreducible continuum is an arc if it is aposyndetic [41]. Many of us remember proving "the arc theorem." With this characterization, the construction would have been somewhat less treacherous.

It may possibly be that this cut point property of semi-local-connectivity led Wilder to call his equivalent notion "almost 0-avoidable." For the cyclic case (= the continuum has no separating point) Wilder apparently got the stronger Torhurst Theorem a little before Whyburn did [43].

From the Torhurst Theorem we see that aposyndetic plane continua must contain arcs since the boundaries of the complementary domains are locally connected. In fact, if it has only a finite number of complementary domains, it is arcwise connected. Hagopian defines a continuum to be semi-aposyndetic if for each pair of points at least one of them has a closed connected neighborhood which misses the other. For semi-aposyndetic continua the same theorem holds, i.e., if a semi-aposyndetic plane continuum has only a finite number of complementary domains, it is arcwise connected even though it may be locally connected almost nowhere. This is the case with the Cantor fan, for example. There are other such theorems [18], but before leaving semi-aposyndesis I wish to raise a question about cut points. If a compact metric continuum is totally non-semi-aposyndetic (i.e., for each point there is another point such that neither has a closed connected neighborhood missing the other), then are not almost all points cut points?

Continua which are neither aposyndetic nor indecomposable are difficult to study. McAuley studied the question of defining the elements of a monotone decomposition which would give an aposyndetic quotient space [24]. His first solution to this problem had a rather surprising twist: the elements of the decomposition which seemed to have the most promise were closed, but not necessarily connected. (The example in his thesis still puzzles me; as simple as it is, I cannot see how anyone would have thought of it.) However, the components of these decomposition elements yield a monotone upper semicontinuous decomposition whose quotient space is aposyndetic. There was one flaw. Some continua which were already aposyndetic were changed by the decomposition. So in a later paper McAuley removed this flaw [25].

Swingle also went after this problem. He defines the elements of the decomposition in an entirely different way. In a rather long paper he and Fitzgerald get several interesting results. To see how (and when) their decompositions coincide with those of McAuley is not easy except in some special cases. Thomas gets an aposyndetic decomposition [33] for certain irreducible continua which seems to be of both types. So does Fugate [10] when there are uncountably many minimal separating subcontinua. Vought has gotten several such theorems for continua whose quotient spaces are rather simple curves [39]. I had one of this kind for certain *homogeneous* continua.

Quite recently Jim Rogers has pursued this last topic. Combining these decompositions with the work of Dyer and Hamstrom and also the work of Wilson on completely regular mappings along with Effros' powerful transformation group theorem, Rogers gets a fairly general theory. Specializing to the plane it follows from this theory that all separating homogeneous subcontinua of the plane are decomposable and these, of course, are really circles. With a little more work he shows if the atriodic, homogeneous, one-dimensional continuum M (not necessarily in the plane) contains an arc, it must be a solenoid. While he was at it, it is too bad he did not show that M had to be circularly chainable.

Of course, some nonaposyndetic continua may more nearly resemble indecomposable continua. For example, in constructing the lakes of Wada example, suppose there was a circular disk of land that was never touched, then the result more nearly resembles an indecomposable continuum than it does a 2-sphere. Except for inverse limit processes there really is no theory for changing nonaposyndetic continua into their nearest indecomposable relatives. Such a theory combined with decomposition theory would complete a sizeable chapter in the structure of continua.

REFERENCES

1. Bellamy, D. P., "Aposyndesis in the remainder of Stone-Čech compatifications," *Bulletin de'l'Academia Polanaise des Sciences 19*(1971), 941-944.

2. Bennett, D. E., "A sufficient condition for countable-set aposyndesis," *Proc. Amer. Math. Soc. 32*(1972), 578-584.

3. ______, "Aposyndetic properties of unicoherent continua," *Pacific J. Math. 37*(1971), 585-589.

4. Bing, R. H., "Some characterization of arcs and simple closed curves," *Amer. J. Math. 70*(1948), 497-506.

5. Davis, H. S., "A note on connectedness im kleinen," *Proc. Amer. Math. Soc.* *19*(1968), 1237-1241.

6. de Groot, J., "Connectedly generated spaces," *Proceedings of the Topological Symposium of Herceg Novi*, Yugoslavia, 1968, 171-175.

7. Dickman, R. F., L. R. Rubin and P. M. Swingle, "Irreducible continua and generalization of hereditarily unicoherent continua by means of membranes," *J. Australian Math. Soc.* *5*(1965), 416-426.

8. Fitzgerald, R. W., "The Cartesian product of nondegenerate compact continua is n-point aposyndetic," *Proc. A.S.U. Topology Conference*, Tempe, 1967, 324-326.

9. Fitzgerald, R. W. and P. M. Swingle, "Core decompositions of continua," *Fund. Math.* *61*(1967), 33-50.

10. Fugate, J. B., "Irreducible continua," *Proc. A.S.U. Topology Conference*, Tempe, 1967, 100-103.

11. Grace, E. E., "Cut points in totally non-semi-locally-connected continua," *Pacific J. Math.* *14*(1964), 1241-1244.

12. Hagopian, C. L., "Mutual aposyndesis," *Proc. Amer. Math. Soc.* *23*(1969), 615-622.

13. ______, "Arcwise connected plane continua," *Proc. Topology Conference*, Emory University, 1970, 41-44.

14. ______, "Concerning arcwise connectedness and the existence of simple closed curves in plane continua," *Trans. Amer. Math. Soc.* *147*(1970), 389-402. [See also *ibid.* *157*(1971), 507-509.]

15. ______, "On generalized forms of aposyndesis," *Pacific J. Math.* *34*(1970), 97-108.

16. ______, "A class of arcwise connected continua," *Proc. Amer. Math. Soc.* *30*(1971), 164-168.

17. ______, "A cut point theorem for plane continua," *Duke Math. J.* *38*(1971), 509-512.

18. ______, "Arcwise connectivity of semi-aposyndetic plane continua," *Trans. Amer. Math. Soc.* *158*(1971), 161-165.

19. ______, "Semi-aposyndetic nonseparating plane continua are arcwise connected," *Bull. Amer. Math. Soc.* *77*(1971), 593-595.

20. Jones, F. B., "Concerning the boundary of a complementary domain of a continuous curve," *Bull. Amer. Math. Soc.* *45*(1939), 428-435.

21. ______, "Aposyndetic continua and certain boundary problems," *Amer. J. Math.* *53*(1941), 545-553.

22. ______, "Concerning nonaposyndetic continua," *ibid.* *70*(1948), 403-413.

23. ______, "Concerning aposyndetic and nonaposyndetic continua," *Bull. Amer. Math. Soc.* *58*(1952), 137-151.

24. McAuley, L. F., "On decomposition of continua into aposyndetic continua," *Trans. Amer. Math. Soc.* *81*(1956), 74-91.

25. ______, "An atomic decomposition of continua into aposyndetic continua," *ibid.* *88*(1958), 1-11.

26. Moore, R. L., Foundations of Point-set Theory, *Amer. Math. Soc. Colloquium Publications 13*, 1932; revised 1962.

27. Rogers, L. E., "Concerning n-mutual aposyndesis in products of continua," *Trans. Amer. Math. Soc.* *162*(1971), 239-251.

28. ______, "Mutually aposyndetic products of chainable continua," *Pacific J. Math.* *37*(1971), 805-812.

29. ______, "Continua in which only semi-aposyndetic subcontinua separate," *Pacific J. Math.* *43*(1972), 493-502.

30. Schlais, H. E., "Nonaposyndesis and nonhereditary decomposability," *Pacific J. Math.* *45*(1973), 643-652.

31. Shirley, E. D., "Semi-local-connectedness and cut points in metric continua," *Proc. Amer. Math. Soc.* *31*(1972), 291-296.

32. Stratton, H. H., "On continua which resemble simple closed curves," *Fund. Math.* *68*(1970), 121-128.

33. Thomas, Jr., E. S., "Monotone decompositions of irreducible continua," *Rozprawy Math.* *50*(1966).

34. Torhurst, M., "Über den Rand der einfach zusammenhängenden gebiete," *Math. Zeit.* *9*(1921), 44-66.

35. Vought, E. J., "n-Aposyndetic continua and cutting theorems," *Trans. Amer. Math. Soc.* *140*(1969), 127-135.

36. ______, "Concerning continua not separated by any nonaposyndetic subcontinuum," *Pacific J. Math.* *31*(1969), 257-262.

37. ______, "A classification scheme and characterization of certain curves," *Colloq. Math.* *20*(1969), 91-98.

38. ______, "Strongly semi-aposyndetic continua are hereditarily locally connected," *Proc. Amer. Math. Soc.* *33*(1972), 619-622.

39. ______, "Monotone decompositions of continua into arcs and simple closed curves," *Fund. Math.* *80*(1973), 213-320.

40. Whyburn, G. T., "The cyclic and higher connectivity of locally connected spaces," *Amer. J. Math.* *53*(1931), 427-442.

41. ______, "Semi-locally-connected sets," *ibid.* *61*(1939), 733-749.

42. ______, Analytic Topology, *Amer. Math. Soc. Colloquium Publications 28*, 1942.

43. Wilder, R. L., "Sets which satisfy certain avoidability conditions," *Casopis pro Pestavani Mathematiky a Pysiky* *67*(1938), 185-198.

SET FUNCTIONS AND CONTINUOUS MAPS

David P. Bellamy[1]

University of Delaware, Newark

Several set functions have been devised for studying aposyndesis and the failure thereof in a delicate and unified way. Several have appeared in the literature, though two, T and K, have been investigated more than the others.

Let X be a continuum (compact, connected Hausdorff space) and suppose $A \subseteq X$. We make the following definitions: A *continuum neighborhood* of a point or set in X is simply a subcontinuum of X containing the point or set in its interior. T(A) is the set of points of X which have no continuum neighborhood missing A. K(A) is the intersection of all continuum neighborhoods of A. aT(A) is the set of points having no finite collection of continuum neighborhoods whose intersection misses A. aK(A) is the intersection of all those closed neighborhoods of A which have only finitely many components. Y(A) is the set of those points which have no connected open neighborhood whose closure misses A.

These functions are formally somewhat similar to the closure function. They may, however, fail to be either idempotent or additive; indeed, the prefix "a" in aT and aK stands for "additive", since these two functions do commute with finite unions. This need not be the case for T, K, and Y. Clearly one can define a number of other set functions in analogous ways; for example, by allowing neighborhoods with countably many

---

[1]Author supported in part by NSF Grant MCS-7908413

ISBN 0-12-481820-X

components, etc. For simplicity we write, e.g. $T(p)$ for $T(\{p\})$.

Some examples may be helpful:

Example 1: Let X denote the harmonic fan, that is, the cone over $\omega + 1$, the first infinite nonlimit ordinal, with v, a, b, c the points indicated:

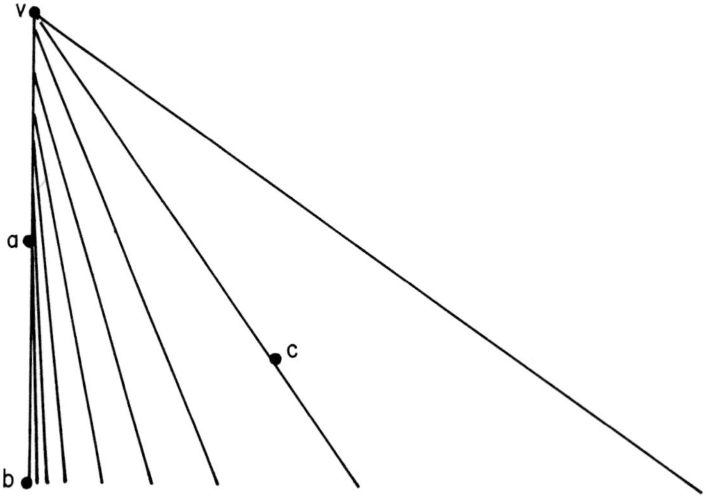

Fig. 1.

Here $T(v)$ is the segment $[v,b]$, $T(a)$ is the segment $[a,b]$, while $T(b) = \{b\}$ and $T(c) = \{c\}$.

Example 2: Let Y denote the closure of the graph of $y = \sin \frac{1}{x}$ for $x \in (0, \frac{1}{\pi}]$.

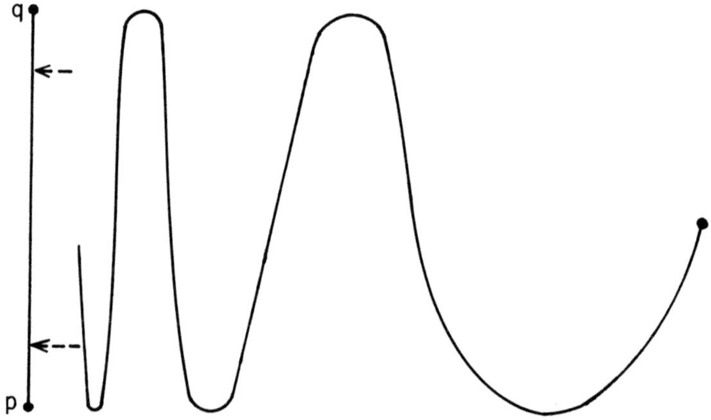

Fig. 2.

Here $T(x) = x$ if x does not lie on the limit segment $[p,q]$ while $T(x) = [p,q]$ if $x \in [p,q]$.

For both these examples, we have $T(A \cup B) = T(A) \cup T(B)$ for any sets A and B. This is not true for the following example.

Example 3: Let Z denote the suspension of the Cantor set, with vertices u and v. Here, $T(x) = \{x\}$ for each $x \in Z$, but $T(\{u,v\})$ is all of Z.

Example 4: Let M be the continuum obtained by identifying p in Y of Example 2 with v in X of Example 1. It is easily seen here that $T(q)$ is not the same as $T^2(q) = T(T(q))$.

T can be used to unify the study of several properties of continua. For instance, we have the following observations, mostly due to F. B. Jones or H. S. Davis. They can be taken as the definitions of the properties considered.

a) S is *connected im kleinen* at $p \in S$ if and only if for each closed $A \subseteq S$, if $p \in T(A)$, then $p \in A$ also.

b) S is a *locally connected* continuum if and only if $T(A) = \bar{A}$ for all $A \subseteq S$, or equivalently if $T(A) = A$ for all closed $A \subseteq S$.

c) S is *semilocally connected* at p if and only if $T(p) = \{p\}$.

d) S is *aposyndetic*, globally, if and only if $T(p) = \{p\}$ for all $p \in S$. (This one has a local version, too.)

e) S is *almost connected im kleinen* at p if and only if for each $A \subseteq S$ such that $p \in \operatorname{Int} T(A)$, it is also true that $p \in \bar{A}$.

f) S is *indecomposable* if and only if $T(p) = S$ for all $p \in S$.

This list could continue somewhat longer, but these suffice to illustrate the point and are the only ones we shall use here. Similar kinds of observations could be made for the other set functions. Part of the importance of the set function T stems from the following theorem, due to H. S. Davis. An analogous theorem holds for Y[18].

THEOREM 1: *Let* S *and* Z *be continua and let* $f\colon S \to Z$ *be a continuous surjection. Then, for all* $A \subseteq S$ *and* $B \subseteq Z$,

a) $T(B) \subseteq f\,T\,f^{-1}(B)$.

b) *If* f *is monotone,* $f\,T(A) \subseteq T\,f(A)$ *and* $T\,f^{-1}(B) \subseteq f^{-1}T(B)$.

c) *If* f *is monotone,* $T(B) = f\,T\,f^{-1}(B)$.

d) *If* f *is open,* $f^{-1}T(B) \subseteq T\,f^{-1}(B)$.

e) *If* f *is both monotone and open,* $f^{-1}T(B) = T\,f^{-1}(B)$.

Proof of c): Suppose $x \in Z$ and $x \notin T(B)$. Let W be a continuum neighborhood of x missing B. Then $f^{-1}(W)$ is a continuum; $f^{-1}(x) \subseteq \mathrm{Int}\, f^{-1}(W)$; and $f^{-1}(W) \cap f^{-1}(B) = f^{-1}(W \cap B) = \emptyset$. Thus, $f^{-1}(x) \cap T\,f^{-1}(B) = \emptyset$. Hence $x \notin f\,T\,f^{-1}(B)$, so that $f\,T\,f^{-1}(B) \subseteq T(B)$. The reverse inclusion follows from part a), a proof of which appears in [1]. The proofs of the other parts are fairly similar.

As an illustration of the use of this result, let X and Y be as in Examples 1 and 2.

THEOREM 2: *There is no monotone mapping of* Y *onto* X .

Proof: Suppose $f\colon Y \to X$ is such a map.

Observe first that for A closed in Y , $T(A) = A$ if $A \cap [p,q] = \emptyset$, while $T(A) = A \cup [p,q]$ if $A \cap [p,q] \neq \emptyset$. In either case, $T(A) \subseteq A \cup [p,q]$.

By c) of Theorem 1, $T(v) = f\,T\,f^{-1}(v) \subseteq f(f^{-1}(v) \cup [p,q]) \subseteq \{v\} \cup f[p,q]$, and, since $T(v) = [v,b]$, $a,b \in f[p,q]$.

$$\begin{aligned} \text{Thus,}\quad \{b\} = T(b) &= f\,T\,f^{-1}(b) \\ &= f(f^{-1}(b) \cup [p,q]) \quad \text{since } f^{-1}(b) \cap [p,q] \neq \emptyset \\ &= \{b\} \cup f[p,q] \end{aligned}$$

so that $f[p,q] = \{b\}$, contradicting the fact that $a \in f[p,q]$. Hence no such map can exist.

Part a) of Theorem 1) is a generalization of the classical result that the continuous image of a locally connected continuum is locally connected. Indeed, Theorem 1.a) can be used to prove that if $f: X \to Y$ is continuous and onto, $y \in Y$ and X is connected im Kleinen at each point of $f^{-1}(y)$, then Y is connected im Kleinen at Y, which is a pointwise version of the better known classical result. I do not know to whom this is due.

Another important result pertaining to the set function T, first proven by F. B. Jones, follows. The proof is not trivial and is given as representative.

THEOREM 3: *If* $T(M) = M$ *for every subcontinuum* M *of* X, *then* M *is locally connected.*

Proof: Let Q be any open set and suppose $p \in Q$. Since $T(p) = \{p\}$ and $X-Q$ is compact, there is a finite collection of continua missing p whose interiors cover $X-Q$. Let $\{W_i\}_{i=1}^{m}$ be such a collection of smallest possible cardinality. Let $U = X - \bigcup_{i=1}^{m} W_i$ and let P be the component of $\bar{U}$ containing p. It will be shown that $p \in \text{Int } P \subsetneq P \subsetneq Q$, and it will follow that the components of Q are open. By the boundary bumping theorem of Janiszewski, each component of $\bar{U}$ meets one of the W_i's. It follows from the minimality of M that no component of $\bar{U}$ except P can meet more than of the W_i's, since taking the union of a component which met more than one W_i with all the W_i's which it met, together with the other W_i's as they were, would yield a smaller covering of $X - Q$. Hence, if for each i, A^i is the union of all those components of $\bar{U}$ which meet W_i, then for $i \neq j$, $A_i \cap A_j \subsetneq P$.

Now, for every i there is a continuum neighborhood K_i of p such that $K_i \cap W_i = \emptyset$, since $p \notin T(W_i)$. Then, $K_i \cap (A_i - P) = \emptyset$ also, for if $x \in K_i \cap (A_i - P)$, let L be the component of $K_i \cap \bar{U}$ containing x. By the boundary bumping theorem again, L meets the frontier relative to K of $K \cap \bar{U}$. This frontier is contained in $\bigcup_{j=i} W_j$, so that

$L \cap W_j \neq \emptyset$ for some $j \neq i$. But $L \subseteq M$ for some component M of $\bar{U}$, $M \neq P$ and $M \subseteq A_i$, so that $M \cap W_j = \emptyset$, a contradiction.

Thus, $\bigcap_{i=1}^{m} K_i \cap \left(\bigcup_{i=1}^{m} (A_i - P) \right) = \emptyset$, and it follows that $\bigcap_{i=1}^{m} K_i \subseteq P$. Since $p \in \mathrm{Int}\left(\bigcap_{i=1}^{m} K_i \right)$, $p \in \mathrm{Int}(P)$, and it follows that P is a connected neighborhood of p containing p and contained in Q. Since $p \in Q$ was arbitrary, each component of Q is open, and the proof is complete.

A local version of this Theorem appears in [8].

Finally, the following beautiful result is due to H. S. Davis [9]. It is stated here only for sequences; however, a similar result is true for arbitrary inverse systems of continua.

THEOREM: *If* X *is expressed as an inverse limit of a sequence of continua* $\{X_n;h_n\}_{n=1}^{\infty}$, *with each* h_n *surjective and if* A *is a closed subset of* X, *and* $p = \langle p_n \rangle_{n=1}^{\infty} \in X$, *then a necessary and sufficient condition that* $p \in T(A)$ *is that for every* n *and every pair of open sets* U_n, $V_n \subseteq X_n$ *with* $p_n \in U_n$, $A_n \subseteq V_n$, *and* $U_n \cap V_n = \emptyset$, *there is a* $k > n$ *such that* U_k *meets more than one component of* $X_k - V_k$. *(Here* U_k *is* $h_{nk}^{-1}(U_n)$ *and similarly for* V_k.*)*

It is my hope and expectation that this theorem will lead to a much deeper understanding of aposyndesis and the set function T in particular.

The discussion so far has concentrated on the properties of the set function T *per se*. It should be noted that this and other set functions have found application on a number of occasions to other problems. See, in particular, [2], [4], [5], [7], [12], [14], and [17]. The literature on the other set functions is less extensive; there is much opportunity for new research. Some basic material can be found in [15], [16], and [18].

The following bibliography is representative but not comprehensive.

REFERENCES

1. Bellamy, D. P., "Continua for which the set function T is continuous," *Trans. Amer. Math. Soc. 151*(1970), 581-588.

2. Bellamy, D. P. and J. J. Charatonik, "The set function T and contractibility of continua," *Bull. Acad. Polon. Sci. Sér. Sci. Math. Astronom. Phys. 25*(1977), 47-49.

3. Bellamy, D. P. and H. S. Davis, "Continuum neighborhoods and filterbases," *Proc. Amer. Math. Soc. 27*(1971), 371-374.

4. Bellamy, D. P. and C. L. Hagopian, "Mapping continua onto their cones," *Colloq. Math. 41*(1979), 53-56.

5. Bellamy, D. P. and L. R. Rubin, "Indecomposable continua in Stone-Čech compactifications," *Proc. Amer. Math. Soc. 39*(1973), 427-432.

6. Bing, R. H. and F. B. Jones, "Another homogeneous plane continuum," *Trans. Amer. Math. Soc. 90*(1959), 171-192.

7. Bennett, D. E., "A characterization of locally connectedness by means of the set function T," *Fund. Math. 86*(1974), 137-141.

8. Davis, H. S., "A note on connectedness im Kleinen," *Proc. Amer. Math. Soc. 19*(1968), 1237-1241.

9. ____________, "Relationships between continuum neighborhoods in inverse limit spaces and separations in inverse limit sequences," *Proc. Amer. Math. Soc. 64*(1977), 149-153.

10. Davis, H. S. and P. H. Doyle, "Invertible continua," *Portugal. Math. 26*(1967), 487-491.

11. Davis, H. S., D. P. Stadtlander, and P. M. Swingle, "Properties of the set functions T^n," *Portugal Math. 21*(1962), 113-133.

12. Davis, H. S., D. P. Stadtlander, and P. M. Swingle, "Semigroups, continua, and the set functions T^n," *Duke Math. J. 29*(1962), 265-280.

13. Davis, H. S., and P. M. Swingle, "Extended topologies and iteration and recursion of set functions," *Portugal Math. 23*(1964), 103-129.

14. Hunter, R. P., "On the semigroup structure of continua," *Trans. Amer. Math. Soc. 93*(1959), 356-368.

15. Moreland, William T., Jr., *Some properties of four set valued set functions*, Masters' Thesis. University of Delaware, Newark, Delaware 19711(1970).

16. Rosasco, J., "A note on Jones' function K," *Proc. Amer. Math. Soc. 49*(1975), 501-504.

17. Schlais, H. E., "Non-aposyndesis and non-hereditary decomposability," *Pacific J. Math. 45*(1973), 643-652.

18. Vandenboss, Eugene L., *Set functions and local connectivity*, Ph.D. Dissertation, Michigan State University (1970), University Microfilms #71-11997.

APOSYNDESIS AND UNICOHERENCE

Donald E. Bennett*

Murray State University, Kentucky

A continuum X is said to be *unicoherent* if for any pair of subcontinua whose union is X , their intersection is also a continuum. Unicoherence for continua was defined by K. Kuratowski in a paper [5] in which the property was used in the characterization of a sphere. The concept itself had been used earlier by L. Vietoris [10] in which he referred to the property as continua "ohne Henkel." All Euclidean spaces, spheres of dimension greater than one, all closed cells, and continua which do not separate the plane are examples of unicoherent spaces. A continuum X is said to be *n-aposyndetic* at a point p if it is aposyndetic at p with respect to each subset of X-{p} that contains n elements. If X is n-aposyndetic for all $p \in X$ and positive integers n , then X is *finitely aposyndetic*. The natural question is what additional properties must be imposed on an n-aposyndetic continuum in order for it to be locally connected.

In [1] Ayres and Whyburn proved that a necessary and sufficient condition for a continuum to be locally connected is that each pair of disjoint closed sets (or continua) can be separated by the sum of a finite number of subcontinua. In [12] E. Vought characterized a 2-aposyndetic continuum as one in which each pair of distinct points can be separated by the sum of a finite number of subcontinua. This characterization together with

*The author was partially supported by a research grant from the Murray State University Committee on Institutional Studies and Research.

ISBN 0-12-481820-X

results of F. B. Jones [4] and R. L. Moore [9] imply the following.

THEOREM 1. *Let* X *be a continuum in the plane. Then* X *is locally connected if and only if* X *is 2-aposyndetic.*

In [2] unicoherent n-aposyndetic continua were shown to be (n+1)-aposyndetic which implies the following proposition.

THEOREM 2. *If* X *is a unicoherent aposyndetic continuum, then* X *is finitely aposyndetic.*

COROLLARY 1: *If* X *is a unicoherent aposyndetic continuum which lies in the plane, then* X *is locally connected.*

A continuum X is said to be *finitely coherent* if for any pair of subcontinua whose union is X, their intersection has at most a finite number of components. It is readily seen that the above theorem and corollary hold for continua which are finitely coherent.

Simple examples show that Corollary 1 does not hold if the continuum does not lie in the plane; for example, consider the product of the unit interval with the cone over the Cantor set. However, the following related question is still open.

Question: Suppose X is a one-dimensional, unicoherent, aposyndetic continuum. Does it follow that X is locally connected?

A dendrite can be characterized as a one-dimensional, unicoherent, locally connected continuum [6]. It follows from Theorem 2 above and Theorem 3 of [11] that a continuum X is a dendrite if and only if X is a unicoherent regular curve (i.e. each point is contained in arbitrary small neighborhoods with finite boundaries).

In [2] a stronger form of unicoherence was defined. A continuum X is said to be *strongly unicoherent* provided for each pair of subcontinua H and K such that $X = H \cup K$, both H and K are unicoherent. T. Mackowiak [7] showed that a continuum X is strongly unicoherent if and only if every non-unicoherent subcontinuum of X has void interior. Thus this class of continua contains the unicoherent continua and is properly contained in the class of hereditarily unicoherent continua. Recall that in order for a continuum X to be hereditarily unicoherent it is necessary and sufficient that if p and q are distinct points of X, there is a unique subcontinuum of X which is irreducible between p and q [8]. In [3] it was shown that strongly unicoherent continua are "hereditarily unicoherent at certain points."

THEOREM 3. *If* X *is a strongly unicoherent continuum,* $p \in X$, *and* X *is both aposyndetic at* p *and semi-locally-connected at* p, *then for each* $q \in X-\{p\}$ *there exists a unique subcontinnum irreducible between* p *and* q.

The relation between aposyndesis and locally connectedness given by the next theorem follows from the fact that a strongly unicoherent continuum is aposyndetic at a point p if and only if it is connected im kleinen at p [2].

THEOREM 4. *A strongly unicoherent continuum is aposyndetic if and only if it is locally connected.*

Together theorems 3 and 4 establish yet another characterization of a dendrite.

COROLLARY 2. *A continuum* X *is a dendrite if and only if* X *is strongly unicoherent and aposyndetic.*

REFERENCES

1. Ayres, W. L., and G. T. Whyburn, "On continuous curves in n-dimensions," *Bull. Amer. Math. Soc. 34*(1928), 349-360.

2. Bennett, D. E., "Aposyndetic properties of unicoherent continua," *Pacific J. Math 37*(1971), 585-589.

3. ____________, "Strongly unicoherent continua," *Pacific J. Math. 60* (1975), 1-5.

4. Jones, F. B., "Aposyndetic continua and certain boundary problems," *Amer. J. Math. 63*(1941), 545-553.

5. Kuratowski, K., "Une characterization topologique de la surface de la sphere," *Fund. Math 13*(1929), 307-318.

6. _____________, *Topology II*, Academic Press, 1968, New York and London.

7. Mackowiak, T., "Some kinds of the unicoherence," *Commentations Math. 20*(1978), 405-408.

8. Miller, H. C., "On unicoherent continua," *Trans. Amer. Math. Soc. 69* (1959), 179-194.

9. Moore, R. L., "A characterization of a continuous curve," *Fund. Math. 7*(1925), 302-307.

10. Vietoris, L., *Proc. K. Akad. Wet. Amsterdam, 29*, 1926.

11. Vought, E. J., "Classification of curves," *Colloq. Math. 20*(1968), 91-98.

12. _____________, "n-Aposyndetic continua and cutting theorems," *Trans. Amer. Math. Soc. 140*(1969), 127-135.

APOSYNDESIS IN HYPERSPACES AND ČECH-STONE REMAINDERS[1]

Eric K. van Douwen
Jack T. Goodykoontz, Jr.

Ohio University, Athens, Ohio
West Virginia University, Morgantown, West Virginia

I. INTRODUCTION

The concept of aposyndesis, introduced by F. B. Jones, [13], has been of considerable interest to continua theorists. In this article we will survey results which deal with aposyndesis and two operations on spaces. We survey the results in the literature which deal with aposyndesis and its variations in the setting of hyperspace theory; here we include some new observations and raise several questions. We also announce some results about aposyndesis in Čech-Stone remainders of n-manifolds; this includes the use of aposyndesis to distinguish Čech-Stone remainders. The second author wishes to thank Professor Sam B. Nadler, Jr. for several conversations concerning the sections of this paper dealing with hyperspaces.

II. APOSYNDESIS IN HYPERSPACES

In this section, by a *continuum* we mean a compact, connected metric space. If X is a continuum, then 2^X $(C(X))$ denotes the hyperspace of closed subsets (subcontinua) of X, each with the Hausdorff metric. An excellent reference on hyperspace theory is the recent text by Nadler [17].

[1] Part of this paper has appeared as [12].

ISBN 0-12-481820-X

Let X be a continuum and (p,q) be a pair of distinct elements of X. Then X is *aposyndetic at* p *with respect to* q provided there exists a continuum M such that $p \in \text{int } M$ and $q \in X - M$. If for each $q \in X - \{p\}$, X is aposyndetic at p with respect to q, then X is *aposyndetic at* p. If X is aposyndetic at each of its points, then X is *aposyndetic*. If for each pair (p,q) of distinct elements of X, X is aposyndetic at p with respect to q or X is aposyndetic at q with respect to p, then X is *semi-aposyndetic*. If for each pair (p,q) of distinct elements of X there exist disjoint continua K and M such that $p \in \text{int } K$ and $q \in \text{int } M$, then X is *mutually aposyndetic*. Let A be a collection of closed subsets of X. If for each $p \in X$ and for each $A \in A$ such that $p \notin A$ there exists a continuum M such that $p \in \text{int } M$ and $M \cap A = \emptyset$, then X is *A-aposyndetic*. If A is the collection of finite (countable closed) (zero-dimensional closed) subsets of X and X is A-aposyndetic, we say that X is finitely (countable closed set) (zero-dimensional closed set) aposyndetic.

A *Whitney map* for $C(X)$ is a continuous function $\mu: C(X) \to [0,\infty)$ such that

(1) for each $x \in X$, $\mu(\{x\}) = 0$, and

(2) if $A, B \in C(X)$, $A \subset B$, and $A \neq B$, then $\mu(A) < \mu(B)$.

If $t \in (0,\mu(X))$, then $\mu^{-1}(t)$ is called a *Whitney level*. It is known [9] that Whitney maps for $C(X)$ are monotone and thus the Whitney levels are subcontinua of $C(X)$.

Let P be a topological property. Then P is a *Whitney property* provided that whenever X is a continuum with property P, then for each Whitney map μ for $C(X)$ and for each $t \in (0,\mu(X))$, $\mu^{-1}(t)$ has property P. The property P is *Whitney-reversible* provided that whenever

X is a continuum such that for each Whitney map μ for $C(X)$ and for each $t \in (0,\mu(X))$, $\mu^{-1}(t)$ has property P, then X has property P. For a detailed discussion of these ideas we refer the reader to Chapter 14 of [17] and to [19].

As a point-wise property, aposyndesis is a generalization of the notion of local connectedness. One of the early results about hyperspaces (actually a combination of results by Vietoris [22] and Wazewski [23]) was that the following statements are equivalent: (1) X is locally connected; (2) 2^X is locally connected; (3) $C(X)$ is locally connected. In contrast, for aposyndesis we have the following basic results:

THEOREM 1. ([10]) *Let* X *be a continuum. Then each of* 2^X *and* $C(X)$ *is aposyndetic.*

THEOREM 2. ([10]) *Let* X *be a semi-aposyndetic continuum. Then each of* 2^X *and* $C(X)$ *is mutually aposyndetic.*

Although the above results were stated in [10] for metric continua, the proofs are also valid for Hausdorff continua. We remark that in the metric case for $C(X)$, shorter proofs than in [10] of the above results can be given using Whitney maps. An example is given in [10] of a non-semi-aposyndetic continuum X for which $C(X)$ fails to be mutually aposyndetic.

In [14] Jones proved that an aposyndetic, planar continuum which fails to separate the plane must be locally connected. Nadler [18] used this result and Theorem 1 to give a short proof that if $C(X)$ is embeddable in the plane, then X must be an arc or a circle.

Bennett [3] has shown that an aposyndetic, unicoherent continuum is finitely aposyndetic. Since it is known (see the discussion in [17]) that 2^X and $C(X)$ are unicoherent, we have the following:

COROLLARY 1. ([10]) *Let* X *be a continuum. Then each of* 2^X *and* C(X) *is finitely aposyndetic.*

The next result generalizes Theorem 1 and Corollary 1 for C(X).

THEOREM 3. ([10]) *Let* X *be a continuum. Then* C(X) *is countable closed set aposyndetic.*

Question 1. Is C(X) zero-dimensional closed set aposyndetic?

Question 2. Is 2^X countable closed set aposyndetic? zero-dimensional closed set aposyndetic?

The above questions seem natural in view of the preceding results. With respect to Question 1, we remark that Krasinkiewicz [15] has shown that no zero-dimensional closed set disconnects C(X) . Nadler has asked ([17], Question 14.63) whether 2^X always admits a monotone Whitney map. An affirmative answer to that question would yield an affirmative answer to the first part (and perhaps the second part) of Question 2.

Petrus [20] has shown that several aposyndesis properties are Whitney properties.

THEOREM 4. ([20]) *Each of the following is a Whitney property: semi-aposyndesis, aposyndesis, mutual aposyndesis, finite aposyndesis.*

Question 3. Is countable closed set aposyndesis a Whitney property? Is zero-simensional closed set aposyndesis a Whitney property?

It is known (see Theorem 14.47 of [17]) that local connectedness is a Whitney-reversible property. The following example will show that aposyndesis is not a Whitney-reversible property.

Example 1. This example is a modification of the harmonic fan in which each arc emanating from the vertex of the fan is replaced by a circle. More precisely, for each positive integer n let A_n denote the circle, in R^3 centered at $(0,1/2,1/2^n)$ of radius $(1/2)[n^2 + 1)/n^2]^{1/2}$

and passing through $v = (0,0,0)$. Let A be the circle in the xy-plane centered at $(0,1/2,0)$ of radius $1/2$ and let $X = A \cup (\bigcup_{n=1}^{\infty} A_n)$. Let μ be any Whitney map for $C(X)$, $t \in (0,\mu(X))$, and $M \in \mu^{-1}(t)$. If $v \in M$ or if for some positive integer n, $M \in C(A_n)$, then it follows from Theorem 2 of [11] that $C(X)$ is connected im kleinen at M. It then follows from Theorem 1.2 of [8] that $\mu^{-1}(t)$ is connected im kleinen at M and hence $\mu^{-1}(t)$ is aposyndetic at M. So we assume that $M \in C(A)$ and that $v \notin M$. Let $K \in \mu^{-1}(t) - \{M\}$. We indicate briefly how to show that $\mu^{-1}(t)$ is aposyndetic at M with respect to K. Let $C(\{v\},X) = \{L \in C(X) \mid v \in L\}$. Let $\mathcal{M}$ be a closed neighborhood in $\mu^{-1}(t)$ of M such that $K \notin \mathcal{M}$ and such that $\mathcal{M} \cap C(\{v\},X) = \emptyset$. Then for each $B \in \mathcal{M}$, $B \in C(A_n)$ or $B \in C(A)$. We observe that for each $B \in \mathcal{M}$, there exists an arc $\mathcal{B}$ in $\mu^{-1}(t) \cap C(A_n)$ (or $\mu^{-1}(t) \cap C(A)$) with endpoints B and some $B' \in C(\{v\},X)$ such that $K \notin \mathcal{B}$. By applying Lemma 14.8.1 of [17] we can show that $C(\{v\},X) \cap (\mu^{-1}(t) - \{K\})$ is arcwise connected. It then follows that a continuum can be constructed in $\mu^{-1}(t)$ which contains M in its interior and misses K. Thus $\mu^{-1}(t)$ is aposyndetic. Since X is clearly not aposyndetic, it follows that aposyndesis is not a Whitney-reversible property.

The argument in the preceding example can be extended to show that the Whitney levels are, in fact, mutually aposyndetic. Thus mutual aposyndesis is not a Whitney-reversible property. If we modify the space X by attaching at $(0,1,0)$ another sequence of circles converging to A from below, then the resulting space fails to be semi-aposyndetic, but it can be shown that its Whitney levels are semi-aposyndetic. Hence semi-aposyndesis also fails to be Whitney-reversible.

Nadler has observed that if Y is a one-dimensional, acyclic, arcwise connected continuum with aposyndetic Whitney levels, then Y must be locally connected. In particular, for the class of dendroids, apo-

$\partial M = \emptyset$ (or if ∂M is compact). The restrictions are essential. Using nothing but the Jordan Curve Theorem for polygons one can prove that $(\mathbb{H}^2)^*$ is not mutually aposyndetic at any point of $(\mathbb{H}^2)^* \cap \mathrm{Cl}\,\partial(\mathbb{H}^2)$. This has two important consequences: It shows that $(\mathbb{R}^2)^*$ and $(\mathbb{H}^2)^*$ are not homeomorphic because $(\mathbb{R}^2)^*$ is mutually aposyndetic but $(\mathbb{H}^2)^*$ is not[4]. It also shows that one can see inside $(\mathbb{H}^2)^*$ which points belong to $\mathrm{Cl}\,\partial(\mathbb{H}^2)$ and which don't, in other words that one can assign a boundary to $(\mathbb{H}^2)^*$.

Unfortunately these proofs only work for $n = 2$, for $(\mathbb{H}^n)^*$ and $(\mathbb{R}^n)^*$ are mutually aposyndetic for $n \geq 3$. This leaves open the problem to find a simple nonalgebraic continuum theoretic property that distinguishes between $(\mathbb{H}^n)^*$ and $(\mathbb{R}^n)^*$. It also leaves open the problem if one can see inside $(\mathbb{H}^n)^*$ which points belong to $\mathrm{Cl}\,\partial(\mathbb{H}^n)$ and which don't.

We conclude this section with some comments on the proofs. If X is locally compact, or, equivalently, if X^* is compact (and perhaps has other nice properties as well), then often X^* has a property $\mathcal{P}$ iff X has "$\mathcal{P}$-modulo compact sets." For example, if X is locally compact, then X^* is connected iff for every two disjoint open sets U and V in X, if $X - (U \cup V)$ is compact then one of U and V has compact closure in X; clearly this latter property becomes connectedness if one replaces "compact" by "empty". A consequence of this is that proofs take place almost entirely inside X, and are directed at doing something modulo compact sets. While parts of the proofs use the fact that we have a manifold under consideration, most of what is being done has nothing to do with manifolds. What we really are looking at is spaces X which are noncompact, σ-compact, locally compact connected and locally connected and

[4] van Douwen and Starbird have subsequently found a different proof that is worth mentioning: $(\mathbb{H}^2)^*$ is unicoherent but $(\mathbb{R}^2)^*$ is not, [7].

which have X^* connected; the temptation to call such spaces *eusyndetic* has proven irresistible. Several of our positive results apply to eusyndetic spaces which don't resemble manifolds at all.

REFERENCES

1. Bellamy, D. P., "A non-metric indecomposable continuum," *Duke Math. J. 38* (1971), 15-20.

2. ______, "Aposyndesis in the remainder of Stone-Čech compactifications," *Bull. Acad. Polon. Sci. Sér. Sci. Math. Astronom. Phys 19* (1971), 941-944.

3. Bennett, D. E., "Aposyndetic properties of unicoherent continua," *Pacific J. Math., 37* (1971), 585-589.

4. Calder, Allan, "The cohomotopy groups of Stone-Čech increments," *Indag. Math. 34* (1972), 37-44.

5. ______, "On the cohomology of βR^n," *Quart. J. Math. Oxford (2), 25* (1974), 385-394.

6. van Douwen, Eric K., "Aposyndesis in Čech-Stone remainders and 2-manifolds," in preparation.

7. ______ and Michael Starbird, "Unicoherence in Čech-Stone remainders," in preparation.

8. Eberhart, Carl, "Continua with locally connected Whitney continua," *Houston J. Math., 4* (1978), 165-173.

9. ______ and Sam B. Nadler, Jr., "The dimension of certain hyperspaces," *Bull. Pol. Acad. Sci., 19* (1971), 1027-1034.

10. Goodykoontz, Jack T., Jr., "Aposyndetic properties of hyperspaces," *Pacific J. Math., 47* (1973), 91-98.

11. ______, "More on connectedness im kleinen and local connectedness in $C(X)$," *Proc. Amer. Math. Soc., 65* (1977), 357-364.

12. ______, "Aposyndesis and hyperspaces," *Proceedings, Conference on metric spaces, Generalized metric spaces, and continua,* Gullford College, 1980.

13. Jones, F. Burton, "Aposyndetic continua and certain boundary problems," *Amer. J. Math., 63* (1941), 545-553.

14. ______, "Concerning aposyndetic and non-aposyndetic continua," *Bull. Amer. Math. Soc., 58* (1952), 137-151.

15. Krasinkiewicz, J., "No 0-dimensional set disconnects the hyperspace of a continuum," *Bull. Pol. Acad. Sci., 19* (1971), 755-758.

16. ______ and P. Minc, "Dendroids and their end points," *Fund.*

Math., *99* (1978), 227-244.

17. Nadler, Sam B., Jr., *Hyperspaces of Sets*, Marcel Dekker, Inc., New York, 1978.

18. ____________, "Some problems concerning hyperspaces," *Topology Conference (V.P.I. and S.U.), Lecture Notes in Math., vol. 375*, Springer-Verlag, New York, 1974, 190-197.

19. ____________, "Whitney-reversible properties," *Fund. Math.*, to appear.

20. Petrus, Ann, "Whitney maps and Whitney properties of C(X)," *Topology Proceedings," 1* (1976), 147-172.

21. Rogers, James T., Jr., "Applications of a Vietoris-Begle theorem of multi-valued maps to the cohomology of hyperspaces," *Mich. J. Math.*, *22* (1976), 315-319.

22. Vietoris, L., "Kontinua zweiter Ordnung," *Monatshefte für Mathematik und Physik*, *33* (1923), 49-62.

23. Wazewski, T., "Sur un continu singulier," *Fund. Math.*, *4* (1923), 214-235.

24. Woods, R. Grant, "Certain properties of $\beta X-X$ for σ-compact locally compact X," *Thesis, McGill University, Montreal, 1968.*

25. ____________, "On the local connectedness of $\beta X-X$," *Canad. Math. Bull.*, *15* (1972), 591-594.

APOSYNDESIS IN HEREDITARILY UNICOHERENT CONTINUA

G. R. Gordh, Jr.

Guilford College, Greensboro, NC and
California State University, Sacramento

Hereditarily unicoherent continua have played an important role in many areas of continua theory. They have been involved in the study of fixed points, homogeneity, contractibility, plane continua, compact semigroups, ordered spaces, mappings, products, hyperspaces, span, inverse limits, and upper semicontinuous decompositions. Among the hereditarily unicoherent continua which have received special attention are arcs, pseudo-arcs, arc-like continua, trees, dendrites, dendroids, λ-dendroids, tree-like continua, nonseparating plane curves, hereditarily equivalent continua, smooth dendroids, smooth continua, solenoids, and hereditarily indecomposable continua.

The purpose of this paper is to discuss the role of aposyndesis and related concepts in the classification and study of these continua. The choice of topics is intended to emphasize the wide applicability of aposyndesis to problems involving hereditary unicoherence; there is no pretense at completeness.

It is assumed that the reader is familiar with aposyndesis (equivalently semi-local-connectivity) and variants such as semiaposyndesis and mutual aposyndesis as well as the aposyndetic set functions T and K. Definitions and other information concerning these notions may be found elsewhere in this volume and in the references.

For the sake of simplicity, we shall restrict our attention to

ISBN 0-12-481820-X

hereditarily unicoherent metric continua. As a consequence, many of the theorems stated here are actually corollaries of much more general results. Most of the theorems have valid analogues in the setting of compact Hausdorff spaces. In many cases the assumption of hereditary unicoherence can be relaxed to that of hereditary unicoherence at a point [12], and in a few cases it is actually superfluous.

By a *continuum* we shall mean a compact connected metric space. A continuum is *hereditarily unicoherent* in case the intersection of any two of its subcontinua is connected.

CONVENTIONS. Throughout the paper, X is assumed to be a hereditarily unicoherent continuum. If x and y are distinct points of X, then xy denotes the unique subcontinuum which is irreducible from x to y (e.g. [32]).

Recall that a *dendrite* is a locally connected continuum which contains no simple closed curve. It follows that X is a dendrite provided X is locally connected.

If X is aposyndetic, then by the cutpoint equivalence theorem stated in [24], every pair of distinct points can be separated by a third; hence X is a dendrite (see page 88 of [45]). C. E. Burgess states this theorem in [2].

THEOREM 1 ([2]). X *is a dendrite if and only if* X *is aposyndetic.*

It is not difficult to show more generally that X is aposyndetic precisely where it is connected im kleinen. In the case of nonseparating plane continua, this follows from a theorem of F. B. Jones [21].

In our setting G. T. Whyburn's characterization of the arc is an immediate corollary.

COROLLARY ([44]). X *is an arc if and only if* X *is aposyndetic and irreducible.*

COROLLARY. X *is a tree if and only if* X *is aposyndetic and irreducible about finitely many points.*

A *dendroid* is an arcwise connected hereditarily unicoherent continuum.

THEOREM 2 ([13]). *If* X *is semiaposyndetic, then* X *is a dendroid.*

For nonseparating plane continua this fact follows from a theorem of C. L. Hagopian [18].

In our setting several results of L. E. Rogers are immediate corollaries.

COROLLARY ([37]). *If* X *is semiaposyndetic, then the following conditions are equivalent: (a)* X *is an arc, (b)* X *is irreducible, and (c)* X *is atriodic.*

COROLLARY ([37]). X *is a tree if and only if* X *is semiaposyndetic and irreducible about finitely many points.*

Simple examples show that semiaposyndesis does not characterize dendroids among hereditarily unicoherent continua. However, an aposyndetic characterization of dendroids follows from Theorem 2. To state it we introduce a definition.

DEFINITION. A continuum is *connected by semiaposyndetic subcontinua* in case each pair of points is contained in a semiaposyndetic subcontinuum.

COROLLARY. X *is a dendroid if and only if* X *is connected by semiaposyndetic subcontinua.*

C. L. Hagopian and L. E. Rogers have a characterization of dendroids whose proof also depends on the semiaposyndetic characterization of arcs given above. It involves the concept of continuum chainability due to I. Rosenholtz.

DEFINITION. A continuum is *continuum chainable* provided each pair of points can be joined by a finite chain of subcontinua having arbitrarily small diameters (see [19] for details).

THEOREM 3 ([19]). X *is a dendroid if and only if* X *is continuum chainable.*

COROLLARY ([19]). *The following conditions are equivalent: (a)* X *is an arc, (b)* X *is continuum chainable and irreducible, and (c)* X *is continuum chainable and atriodic.*

J. J. Charatonik and C. Eberhart [4] define a dendroid X to be *smooth* at the point p if for each convergent sequence of points q_i converging to q, the sequence of arcs pq_i converges to the arc pq. L. Mohler [33] observed that their definition does not need to require arcwise connectivity of X. Such continua are said to be *smooth* [12].

Smoothness at p may be viewed as a restriction on the *weak cutpoint order* $\leq_p$ defined by setting $x \leq_p y$ whenever $px \subseteq py$. It is easy to show that X is smooth at p if and only if $\leq_p$ is a closed [13]. Furthermore, X is a dendroid which is smooth at p if and only if $\leq_p$ is a closed partial order [13].

L. E. Ward, Jr. [43] has characterized smooth dendroids (i.e., *metrizable generalized trees*) in purely order theoretic terms; and L. Lum [27] has obtained an analogous characterization for smooth continua.

The next result follows from Theorem 6 of [4].

THEOREM 4 ([4]). *If* X *is a smooth dendroid, then* X *is semiaposyndetic.*

More generally, it is not difficult to show that every smooth dendroid is n-semiaposyndetic for each natural number n. This fact was observed by L. E. Rogers [37] for the special case when X is a cone.

Semiaposyndesis does not imply smoothness as easy examples show. However a weakened form of aposyndesis called *aposyndesis toward* p [15] precisely characterizes smoothness.

THEOREM 5 ([15]). *X is smooth at p if and only if X is aposyndetic toward p.*

COROLLARY. *X is a smooth dendroid if and only if X is semiaposyndetic and aposyndetic toward some point.*

Theorem 1 shows that mutual aposyndesis, at least in the global sense, is of no interest in hereditarily unicoherent continua. For products of such continua, however, there are a number of notable results.

THEOREM 6 ([37]). *The product of any two smooth dendroids is mutually aposyndetic.*

THEOREM 7 ([37]). *The product of a dendrite and a smooth dendroid is n-mutually aposyndetic for every n.*

"Smooth dendroid" cannot be weakened to "smooth continuum" in the previous two theorems since the product of an arc with a $\sin 1/x$ curve is not mutually aposyndetic as shown by the next theorem.

THEOREM 8 ([38]). *Suppose that X is arc-like. Then X is an arc if and only if* $X \times X$ *is mutually aposyndetic.*

THEOREM 9 ([17]). *Suppose that X is arc-like. Then X is indecomposable if and only if* $X \times X$ *is strictly nonmutually aposyndetic.*

The aposyndetic set function T (introduced as L by F. B. Jones in [22]) and its iterates T^n behave nicely on hereditarily unicoherent continua (e.g. [5],[6],[8]) as the next two theorems illustrate.

THEOREM 10 ([5]). *If A and B are disjoint subcontinua of X, then*

(a) $T^n(A \cup B) = T^n(A) \cup T^n(B)$, *and*

(b) $T^n(A \cap B) = T^n(A) \cap T^n(B)$.

THEOREM 11 ([8]). *If A is a closed subset of X, then* $T(A) = \bigcup \{T(a) : a \in A\}$.

J. J. Charatonik and C. Eberhart [4] used the set function T to characterize smoothness in dendroids. Here we state several generalizations of their results. Other generalizations of this type involving arc-smoothness, near smoothness, and weak smoothness appear in [10],[13],[28],[29].

For each x in X, the *level set* $\{y \in X : x \leq_p y$ and $y \leq_p x\}$ of the quasi-order $\leq_p$ will be denoted by $D_p(x)$.

THEOREM 12 ([13]). X *is smooth at* p *if and only if* $px \cap T(x) \subseteq D_p(x)$ *for each* x.

COROLLARY ([13]). X *is a dendroid which is smooth at* p *if and only if* $px \cap T(x) = \{x\}$ *for each* x.

THEOREM 13 ([13]). X *is a smooth dendroid if and only if given* x *and* y *in* X, *either* $xy \cap T(x) = \{x\}$ *or* $xy \cap T(y) = \{y\}$.

Smooth dendroids are contractible [4], and certain contractible dendroids are smooth (e.g. [9],[10],[33]). D. P. Bellamy and J. J. Charatonik have used the set function T to study the general contractibility properties of dendroids.

THEOREM 14 ([1],[3]). *If* X *is a dendroid which contains points* x *and* y *such that* $T(x) \cap T(y) \neq \emptyset$, $x \notin T(y)$, *and* $y \notin T(x)$, *then* X *is not contractible.*

L. G. Oversteegen has a stronger version of this theorem for fans [35].

J. Krasinkiewicz and P. Minc studied dendroids by means of certain non-negative real valued functions. They used these functions to characterize dendrites and smooth dendroids and to establish the following result.

THEOREM 15 ([26]). *If* X *is a dendroid, then* X *is semi-locally-connected at some end point* e *(i.e.,* $T(e) = \{e\}$ *).*

The proof of the following result of R. P. Hunter makes heavy use of the set function T.

THEOREM 16 ([20]). *If* X *supports the structure of a topological semigroup with unit and zero, then* X *is a dendroid which is smooth at the zero (hence semiaposyndetic).*

On the other hand, if X supports the structure of a topological group (e.g. the solenoids), then X must be indecomposable. This fact follows immediately from the next theorem due to F. B. Jones.

THEOREM 17 ([14],[23]). *If* X *is homogeneous, then* X *is indecomposable (hence strictly nonaposyndetic).*

Theorem 17 can be viewed as a corollary of F. B. Jones's aposyndetic decomposition theorem [25] which states that if Y is a homogeneous continuum, then $\{T(y):y\in Y\}$ is a continuous decomposition of Y whose quotient space is aposyndetic and homogeneous. (Thus in the theorem above, if X were decomposable, then $X/\{T(x):x\in X\}$ would be a nondegenerate homogeneous dendrite.) In the case when Y is one dimensional, J. T. Rogers, Jr. [36] has recently shown that the decomposition elements T(y) must be hereditarily unicoherent (hence indecomposable).

R. W. FitzGerald and P. M. Swingle [7],[8] have developed a very general theory of monotone decompositions which is intimately connected with the set function T . In the setting of hereditarily unicoherent continua, their theory yields the following theorem.

THEOREM 18 ([7],[8]). *For each hereditarily unicoherent continuum* X *, there exists a dendrite* D_X *and a monotone mapping* $f:X\to D_X$ *such that if* $g:X\to E$ *is any monotone mapping onto a dendrite, then there exists a unique monotone mapping* $h:D_X\to E$ *such that* $g = h\circ f$.

The decomposition $\{f^{-1}(y):y\in D_X\}$ is said to be the *core decomposition of* X *with respect to having an aposyndetic quotient space.* In some cases the elements of this decomposition can be described explicitly in terms of the set function T .

THEOREM 19 ([8]). *If for some natural number* n , $T^n(x) = T^{n+1}(x)$ *for all* x *in* X *, then* $\{T^n(x) : x \in X\}$ *is the core decomposition of* X *with respect to having an aposyndetic quotient space.*

E. J. Vought [40] has shown that the hypothesis of Theorem 19 is satisfied in the case when X is irreducible about a finite set and contains no indecomposable subcontinuum with interior. He has also proved a semiaposyndetic version of Theorem 18 in which D_X and E are allowed to be semiaposyndetic dendroids [41]. In case X is smooth at p , the core decomposition of X with respect to having a semiaposyndetic quotient space agrees with the decomposition of X into level sets $D_p(x)$ given in [12].

Another approach to obtaining aposyndetic and semiaposyndetic decompositions uses collections of closed separators and is due to L. F. McAuley (see [8],[30],[31],[40]-[42]).

Dual to the set function T is the set function K , also due to F. B. Jones [22]. In general $K(x)$ need not be connected, however for hereditarily unicoherent continua, the following result is immediate.

THEOREM 20. *For each* x *in* X *, the set* $K(x)$ *is a continuum.*

For nonseparating plane continua, this result follows from a theorem of J. Rosasco [39].

The set function K plays a role in the theory of monotone decompositions (see [42]) and is involved in the study of nearly smooth continua [13].

Many aspects of the theory of hyperspaces involve hereditarily unicoherent continua as the subject index of S. B. Nadler's book [34] illustrates. Furthermore, J. T. Goodykoontz, Jr. [11] has shown that hyperspaces are aposyndetic. As a sample result in this area, let us mention the following recent theorem of J. Grispolakis and E. D. Tymchatyn which does not appear in [34].

THEOREM 21 ([16]). *Let* X *be hereditarily indecomposable and let* D *be an arcwise connected curve. Then* D *is a smooth dendroid if and only if* D *is embeddable in the hyperspace* C(X) .

REFERENCES

1. Bellamy, D. P. and J. J. Charatonik, "The set function T and contractibility of continua," *Bull. Acad. Polon. Sci. Ser. Sci. Math. Astronom. Phys.* *25*(1977), 47-49.

2. Burgess, C. E., "Continua and various types of homogeneity," *Trans. Amer. Math. Soc.* *88*(1958), 366-374.

3. Charatonik, J. J., "The set function T and homotopies," *Colloq. Math.* *39*(1978), 271-274.

4. Charatonik, J. J. and C. Eberhart, "On smooth dendroids," *Fund. Math.* *67*(1970), 297-322.

5. Davis, H. S., D. P. Stadtlander and P. M. Swingle, "Properties of the set functions T^n," *Portugaliae Mathematica* *21*(1962), 113-133.

6. Davis, H. S. and P. M. Swingle, "Extended topologies and iteration and recursion of set-functions," *Portugaliae Math.* *25*(1964), 103-129.

7. FitzGerald, R. W., "Connected sets with a finite disconnection property," *Studies in Topology*, Academic Press, 1975, 139-173.

8. FitzGerald, R. W. and P. M. Swingle, "Core decompositions of continua," *Fund. Math.* *61*(1967), 33-50.

9. Fugate, J. B., G. R. Gordh, Jr. and L. Lum, "On arc-smooth continua," *Topology Proc.* *2*(1977), 645-656.

10. ______, "Arc-smooth continua," *Trans. Amer. Math. Soc.* (to appear).

11. Goodykoontz, J. T., Jr., "Aposyndetic properties of hyperspaces," *Pacific J. Math.* *47*(1973), 91-98.

12. Gordh, G. R., Jr., "On decompositions of smooth continua," *Fund. Math.* *75*(1972), 51-60.

13. ______, "Concerning closed quasi-orders on hereditarily unicoherent continua," *Fund. Math.* *78*(1973), 61-73.

14. ______, "On homogeneous hereditarily unicoherent continua," *Proc. Amer. Math. Soc.* *51*(1975), 198-202.

15. ______, "Aposyndesis and the notion of smoothness in continua," *Conf. on Gen. Topology*, Univ. Calif., Riv., May '80, Academic Press (to appear).

16. Grispolakis, J. and E. D. Tymchatyn, "Embedding smooth dendroids in hyperspaces," *Canadian J. Math.* (to appear).

17. Hagopian, C. L., "Mutual aposyndesis," *Proc. Amer. Math. Soc.* *23*(1969), 615-622.

18. ______, "Arcwise connectivity of semi-aposyndetic plane continua," *Pacific J. Math.* *37*(1971), 683-686.

19. Hagopian, C. L. and L. E. Rogers, "Arcwise connectivity and continuum chainability," *Houston J. Math.* (to appear).

20. Hunter, R. P., "On the semigroup structure of continua," *Trans. Amer. Math. Soc.* *93*(1959), 356-368.

21. Jones, F. B., "Aposyndetic continua and certain boundary problems," *Amer. J. Math.* *63*(1941), 545-553.

22. ______, "Concerning nonaposyndetic continua," *Amer. J. Math.* *70*(1948), 403-413.

23. ______, "Certain homogeneous unicoherent indecomposable continua," *Proc. Amer. Math. Soc.* *2*(1951), 855-859.

24. ______, "Aposyndetic and nonaposyndetic continua," *Bull. Amer. Math. Soc.* *58*(1952), 137-151.

25. ______, "On a certain type of homogeneous plane continuum," *Proc. Amer. Math. Soc.* *6*(1955), 735-740.

26. Krasinkiewicz, J. and P. Minc, *Dendroids and their Endpoints*, Inst. of Math., Polish Acad. of Sci., Preprint #84, 1975.

27. Lum, L., "A quasi order characterization of smooth continua," *Pacific J. Math.* *53*(1974), 495-500.

28. ______, "Weakly smooth continua," *Trans. Amer. Math. Soc.* *214*(1975), 153-167.

29. Mackowiak, T., "Some characterizations of smooth continua," *Fund. Math.* *79*(1973), 173-186.

30. McAuley, L. F., "On decompositions of continua into aposyndetic continua," *Trans. Amer. Math. Soc.* *81*(1956), 74-91.

31. ______, "An atomic decomposition of continua into aposyndetic continua," *Trans. Amer. Math. Soc.* *88*(1958), 1-11.

32. Miller, H. C., "On unicoherent continua," *Trans. Amer. Math. Soc.* *69*(1950), 179-194.

33. Mohler, L., "A characterization of smoothness in dendroids," *Fund. Math.* *67*(1970), 369-376.

34. Nadler, S. B., Jr., *Hyperspaces of Sets*, Marcel Dekker, New York, 1978.

35. Oversteegen, L. G., *Properties of Contractible Fans*, Ph.D. Dissertation, Wayne State Univ., Michigan, 1978.

36. Rogers, J. T., Jr., "Completely regular mappings and homogeneous, aposyndetic continua," *Canadian J. Math.* (to appear).

37. Rogers, L. E., "Concerning n-mutual aposyndesis in products of continua," *Trans. Amer. Math. Soc.* *162*(1971), 239-251.

38. ______, "Mutually aposyndetic products of chainable continua," *Pacific J. Math.* *37*(1971), 805-812.

39. Rosasco, J., "A note on Jones's function K," *Proc. Amer. Math. Soc.* *49*(1975), 501-504.

40. Vought, E. J., "Monotone decompositions into trees of Hausdorff continua irreducible about a finite subset," *Pacific J. Math.* *54*(1974), 253-261.

41. ______, "Monotone decompositions of Hausdorff continua," *Proc. Amer. Math. Soc.* *56*(1976), 371-376.

42. ______, "Monotone decompositions of continua," *Conf. on Gen. Topology*, Univ. Calif., Riverside, May '80, Academic Press (to appear).

43. Ward, L. E., Jr., "Mobs, trees, and fixed points," *Proc. Amer. Math. Soc.* *8*(1957), 798-804.

44. Whyburn, G. T., "Semi-locally-connected sets," *Amer. J. Math.* *61*(1939), 733-749.

45. ______, *Analytic Topology*, Amer. Math. Soc. Colloq. Publ. 28, Providence, Rhode Island, 1942.

APOSYNDESIS AND THE NOTION OF SMOOTHNESS IN CONTINUA

G. R. Gordh, Jr.

Guilford College, Greensboro, NC

I. INTRODUCTION

The purpose of this note is to introduce a form of aposyndesis which lies strictly between aposyndesis at a point and aposyndesis (in the global sense) and to show that it is equivalent to smoothness in the class of dendroids. More generally this concept, called *aposyndesis toward a point*, is equivalent to smoothness in continua which are hereditarily unicoherent at some point. Thus, in a sense, smoothness is shown to be an aposyndetic property.

II. PRELIMINARY DEFINITIONS AND REMARKS

By a *continuum* we mean a compact connected Hausdorff space. Throughout this paper X will denote a continuum.

If the intersection of any two subcontinua of X which contain the point p is connected, then X is said to be *hereditarily unicoherent at the point* p [4]. In such a continuum each point q distinct from p determines a unique subcontinuum, denoted pq, which is irreducible between p and q.

The continuum X is *smooth at the point* p [4] if X is hereditarily unicoherent at p and for each convergent net of points $\{q_n\}$ converging to q, the net of subcontinua $\{pq_n\}$ converges to pq.

This notion of smoothness was studied first for *fans* [1] and then for

ISBN 0-12-481820-X

dendroids (i.e., arcwise connected hereditarily unicoherent metric continua) [3]. The Hausdorff version of smooth dendroids was actually considered earlier in an order-theoretic setting and termed *generalized trees* [16]. It follows from Theorem 2.2 of [5] that arcwise connected smooth metric continua are smooth dendroids. That smoothness makes sense in hereditarily unicoherent continua without any assumption of arcwise connectivity was first observed in [15]. The more general notion of smoothness defined above has been studied in [2],[4]-[11].

III. APOSYNDESIS AND SMOOTHNESS

We shall need the following variants of the well known concept of *aposyndesis* due to F. B. Jones (see this volume for a survey article on this topic).

The continuum X is said to be *aposyndetic at the set* Q *with respect to the set* R provided there exists a subcontinuum H of X such that

$$Q \subseteq \text{Interior } H \subseteq H \subseteq X \setminus R \ .$$

If X is aposyndetic at the closed set Q with respect to each closed set disjoint from Q, we say that X is *connected im kleinen* at Q. If X is aposyndetic at the closed set Q with respect to each point not belonging to Q, we say that X is *aposyndetic at* Q. If X is aposyndetic at each point q, then X is said to be *aposyndetic*.

Theorem 3.1 of [4] implies the following result.

THEOREM 1. *If* X *is smooth at the point* p, *then* X *is connected im kleinen at each subcontinuum containing* p.

Maćkowiak [12] proved the converse for continua hereditarily unicoherent at some point.

THEOREM 2. *If* X *is hereditarily unicoherent at* p *and connected im kleinen at each subcontinuum containing* p, *then* X *is smooth at* p.

Motivated by this fact, Maćkowiak dropped the underlying assumption that X be hereditarily unicoherent at p and defined X to be *smooth at* p if X is connected im kleinen at each subcontinuum containing p (see [12]-[14]). In this paper we shall refer to such continua as *M-smooth at* p .

We define X to be *aposyndetic toward the point* p provided that X is aposyndetic at q with respect to r whenever r does not cut p from q (i.e., whenever p and q can be joined by a subcontinuum missing r).

The next theorem follows immediately from the definitions.

THEOREM 3. *(1) If* X *is aposyndetic toward* p *, then* X *is aposyndetic at* p *. (2)* X *is aposyndetic toward each point if and only if* X *is aposyndetic.*

THEOREM 4. *The continuum* X *is aposyndetic toward* p *if and only if* X *is aposyndetic at each subcontinuum containing* p .

Proof. Suppose X is aposyndetic toward p and let Q be any subcontinuum containing p . Let $r \in X \setminus Q$. If $q \in Q$, then r does not cut p from q , hence there is a subcontinuum H_q containing q in its interior and missing r . By compactness the interiors of finitely many such sets H_q cover Q . The union of these H_q's is a continuum containing Q in its interior and missing r .

If r does not cut p from q , then there is a subcontinuum Q containing p and q and missing r . Since X is aposyndetic at Q , it follows that X is aposyndetic at q with respect to r .

COROLLARY 1. *If* X *is M-smooth at* p *, then* X *is aposyndetic toward* p .

Thus *M-smoothness at* p is related to *aposyndesis toward* p as *connectedness im kleinen at* p is related to *aposyndesis at* p . The next result whose proof is immediate formalizes this observation.

THEOREM 5. *(1)* X *is M-smooth at* p *if and only if for each subcontinuum* Q *containing* p, *the decomposition space* X/Q *is connected im kleinen at the "point"* Q. *(2)* X *is aposyndetic toward* p *if and only if for each subcontinuum* Q *containing* p, *the decomposition space* X/Q *is aposyndetic at the "point"* Q.

THEOREM 6. *Let* X *be hereditarily unicoherent at* p. *Then* X *is smooth at* p *if and only if* X *is aposyndetic toward* p.

Proof. If X is smooth at p, then X is M-smooth at p by Theorem 1. By Corollary 1, X is aposyndetic toward p.

Suppose X is not smooth at p. By Theorem 2.3 of [4] there is a net $\{q_n\}$ converging to q and a net $\{r_n\}$ converging to r such that $r_n \in pq_n$ for each n and $r \in X \setminus pq$. Thus r does not cut p from q, so there is a subcontinuum H containing q in its interior and missing r. Without loss of generality $q_n \in H$ for every n. By hereditary unicoherence at p, it follows that $pq_n \in pq \cup H$. Consequently $r_n \in pq \cup H$ contradicting the fact that $r \notin pq \cup H$.

COROLLARY 2. *The dendroid* X *is smooth at the point* p *if and only if it is aposyndetic toward* p.

Thus for continua hereditarily unicoherent at p, the notions *M-smoothness at* p and *aposyndesis toward* p coincide. This is analogous to the well known and easily proved fact that *connectedness im kleinen at* p is equivalent to *aposyndesis at* p for hereditarily unicoherent continua. In general, of course, *M-smoothness at* p (respectively *connectedness im kleinen at* p) is much stronger than *aposyndesis toward* p (respectively *aposyndesis at* p).

Recall that X is said to be *semi-aposyndetic* if for each pair of distinct points, X is aposyndetic at one of them with respect to the other.

The final theorem generalizes the known result that smooth dendroids are semi-aposyndetic (see ([3], Thm. 6) or ([5], Thm. 3.5)).

THEOREM 7. *If* X *is arcwise connected and aposyndetic toward some point, then* X *is semi-aposyndetic.*

Proof. Let X be aposyndetic toward p , and let q and r be distinct points. Let A be an arc irreducible between p and {q,r} . Then A misses q or r , say r . Consequently r does not cut p from q and X is aposyndetic at q with respect to r .

REFERENCES

1. Charatonik, J. J., *On Fans*, Dissertationes Math. (Rozprawy Mat.) No. 54, Warszawa, 1964.

2. ______, "On irreducible smooth continua," *Proc. Int'l. Symp. in Topology*, Budva, Yugoslavia, 1972, 45-50.

3. Charatonik, J. J. and C. Eberhart, "On smooth dendroids," *Fund. Math.* *67*(1970), 297-322.

4. Gordh, G. R., Jr., "On decompositions of smooth continua," *Fund. Math.* *75*(1972), 51-60.

5. ______, "Concerning closed quasi-orders on hereditarily unicoherent continua," *Fund. Math.* *78*(1973), 61-73.

6. Gordh, G. R., Jr. and L. Lum, "On monotone retracts, accessibility, and smoothness in continua," *Topology Proc.* *1*(1976), 17-28.

7. ______, "Radially convex mappings and smoothness in continua," *Houston J. Math.* *4*(1978), 335-342.

8. Lum, L., "A quasi-order characterization of smooth continua," *Pacific J. Math.* *53*(1974), 495-500.

9. ______, "Weakly smooth continua," *Trans. Amer. Math. Soc.* *214*(1975), 153-167.

10. Maćkowiak, T., "Open mappings and smoothness of continua," *Bull. Pol. Acad. Sci.* *21*(1973), 531-534.

11. ______, "Some characterizations of smooth continua," *Fund. Math.* *79* (1973), 173-186.

12. ______, "On smooth continua," *Fund. Math.* *85*(1974), 79-95.

13. Maćkowiak, T., "Arcwise connected and hereditarily smooth continua," *Fund. Math.* *92*(1976), 149-171.

14. ______, "On decompositions of hereditarily smooth continua," *Fund. Math* *Math.* *94*(1977), 25-33.

15. Mohler, L., "A characterization of smoothness in dendroids," *Fund. Math.* *67*(1970), 369-376.

16. Ward, L. E., Jr., "Mobs, trees, and fixed points," *Proc. Amer. Math. Soc.* *8*(1957), 798-804.

APOSYNDESIS AND WEAK CUTTING

E. E. Grace

Arizona State University, Tempe

I. INTRODUCTION AND DEFINITIONS

Aposyndesis, a generalization of local connectedness (or, more appropriately, of connectedness-im-kleinen), was introduced by F. Burton Jones in [18] as a way of refining G. T. Whyburn's semi-local connectedness [28] in locally-peripherally- compact, Hausdorff continua (in continua generally, it permits a generalization). This is accomplished by shifting the focus of attention away from the point p at which the space is semi-locally-connected to the other points in the space and seeing how they are related to p . This generalization occurred in the context of the study of connectedness properties of the boundaries of complementary domains of continua in the plane.

Sometime after that Jones began to develop a system of classification of the "spectrum" of types of continua between the arcs and simple closed curves at one end and the pseudoarcs at the other end. Whyburn's work with semi-local-connectedness had made it clear that semi-local-connectedness was a natural generalization of local connectedness, since many theorems for locally connected continua generalized to semi-locally-connected continua. Switching to the aposyndesis point of view makes this generalization seem even more natural, since aposyndesis at a point is obviously a generalization of connectedness im kleinen at a point, in a regular space, and connectedness im kleinen is equivalent to local

ISBN 0-12-481820-X

connectedness as a global property. In [6] it is shown that domain aposyndesis, which is related to local connectedness in the way that aposyndesis is related to connectedness im kleinen, is not nearly as interesting as aposyndesis.

Essentially there is only one theorem in the literature that concerns weak cutting in all aposyndetic continua. Before discussing it some definitions are needed.

DEFINITIONS. A *continuum* is a nondegenerate, closed (perhaps not compact), connected subset of a Topological space. A continuum S, in a topological space, is said to be *aposyndetic at* a subset P of S *with respect to* a subset Q of S if there is a subcontinuum H of S and a subset D of S, that is open relative to S, such that $S/Q \supset H \supset D \supset P$. In this terminology, no distinction is made between $\{p\}$ and p, for any point p of S. The continuum S is said to be *nonaposyndetic at* a point p *with respect to* a point q if $p \neq q$ and S is not aposyndetic at p with respect to q. Also S is said to be *totally nonaposyndetic (on* a subset B of S) if it is not aposyndetic at any of its points (in B). The aposyndetic set function of a set variable, T, defined for each subset B of a continuum S by $T(B) = \{x \in S | S$ is not aposyndetic at x with respect to $B\}$, is useful in proving theorems about the existence of weak cut points in nonaposyndetic continua. A continuum S, in a topological space, is *semi-locally-connected at* a point p, if the complement of any neighborhood of p is contained in the union of a finite number of subcontinua contained in $S/\{p\}$ (see Note 2). S is *semi-locally-connected* if S is semi-locally-connected at each point of S. Also, S is said to be *totally non-semi-locally-connected* (on a subset B of S) if it is not semi-locally-connected at any of its points (in B). A point p (a subset P), of a continuum S is a *weak cut point (weak cut set)* of S if there are points x and y in $S\{p\}$ (in $S\setminus\{p\}$) such that each subcontinuum of S that contains $\{x,y\}$ also contains p (intersects P). In this case, the point p (the set P) is said to *cut* S *weakly between* x and y or to *cut* x *weakly from* y *in* S. A point p is a *separating point* of S if $S\setminus\{p\}$ is not connected. (This is sometimes called a cut point in the literature.)

II. WEAK CUTTING IN APOSYNDETIC CONTINUA

The first result concerning weak cut points in aposyndetic continua was proved by Whyburn for semi-locally-connected, metric continua.

THEOREM 1 (Whyburn [28,(6.2),p.737]). *If the point* p *of a semi-locally-connected, metric continuum* M *does not separate* M *between two points* a *and* b *of* M, *then* p *does not cut* M *weakly between* a *and* b.

Whyburn's proof, which shows that the a-component of $M\setminus\{p\}$ is the union of interiors of continua, only uses the semi-local-connectedness at p, and essentially proves the following much more general theorem.

THEOREM 2. *Let* Q *be a subset of a continuum* S *such that, for each point* p *in* $S\setminus Q$, *the continuum* S *is aposyndetic at* p *with respect to* Q. *Then, for every compact subset* P *of the same component of* $S\setminus Q$, *the continuum* S *is aposyndetic at* P *with respect to* Q.

A proof of Theorem 2, for P a pair of points, may be found in [8, Th. 2].

The other results concerning weak cutting in continua near the "nice" end of the spectrum apply explicitly to *non*-locally-connected (or *non*-connected im kleinen) continua. This theory, due originally to Hagopian [13] and in a somewhat more general form to the present author [8], is discussed rather fully in [8]. The reader is referred there for further information. The following theorem is an indication of the nature of these results in their most general setting.

THEOREM 3 (Hagopian-Grace [13;8,Th.5]). *If a continuum* S *is semi-locally-connected at each point of an open set* D *that does not contain any connected set with interior, then the set of weak cut points of* S *does not have a limt point in* D .

Note that any weak cut points in D are in fact separating points, by Theorem 1.

III. WEAK CUTTING IN NONAPOSYNDETIC CONTINUA

There are additional results concerning the relationship between aposyndesis and weak cutting that also relate to Jones' spectrum. They pertain to continua that lie near the "bad" end of the spectrum. Indecomposable continua can be characterized as those continua that are not aposyndetic at any point with respect to any other point [19], Th. 9]. Note also that every point of an indecomposable, compact, metric continuum is a weak cut point, and does a lot of weak cutting, since each point p cuts the continuum weakly between any two points neither of which is in the p-composant and which are not in the same composant. It seems natural, then, to consider both total nonaposyndesis and total non-semi-locally-connectedness as generalizations of indecomposability, and to ask to what extent the weak cutting properties of indecomposable, compact, metric continua are also properties of continua having these more general properties.

It is easily seen that if a continuum S is nonaposyndetic at a point p with respect to a different point q then any small open set containing q will separate S between p and points having p as a limit point. It seems likely then that in a totally nonaposyndetic, compact, metric continuum M there will be points cut weakly from points near them throughout M , since all of the points of M play the role of p in the foregoing. Similarly it seems likely that, if M is totally non-semi-locally-connected instead, then M will have a dense set of weak cut points, since each point of M plays the role of q in the foregoing.

The earliest results of Jones' [19, Th. 6, p. 405] and [19, Th.15, p. 411] can be used to confirm these expectations. It is also shown [19, Ex.1, p. 409] that a totally nonaposyndetic, compact, metric continuum may contain only one weak cut point.

Apparently no systematic study has been made of the way points that are cut weakly from each other lie in a totally nonaposyndetic, compact, metric, continuum M , i.e., of the way $\{(p,q) \mid$ some point of M cuts M weakly between p and $q\}$ lies in $M \times M$. The corresponding question for totally non-semi-locally-connected continua , How do the weak cut points of a totally non-semi-locally-connected, compact, metric, continuum lie in that continuum?, has received intensive attention with limited success. First, Jones showed [19, Th. 15, p.411] that the set of weak cut points is dense. Then the present author got results that showed that in Jones' result it was sufficient to have the continuum be non-semi-locally-connected at each point of a dense G_δ set and also suggested that the set of weak cut points should contain a dense G_δ set[5, Th. 6, p.1244]. Next, Hagopian showed [16, Th.4, p.104] that the set of weak cut points in each open set has cardinal c . Finally, Shirley showed [24, Th. 2.2, p.296] that each set either contains a somewhere dense G_δ set of weak cut points or a subcontinuum of weak cut points. This generalizes the results of the other three authors and is where the problem stands now. Perhaps it should be noted that the existence of a dense G_δ set of weak cut points would follow immediately from the present author's work cited and a positive answer to the following question.

Question 1. Does every compact, metric continuum that is both totally nonaposyndetic, and totally non-semi-locally-connected at each point of an open subset D, contain a G_δ set of weak cut points that is dense in D?

Generalizations of Jones' weak cut point results for totally non-semi-locally-connected continua also occur in [9 and 26]. The theory is discussed from a different point of view in [7 and 8].

The results cited here, concerning totally nonaposyndetic continua, have been generalized in several respects. These have also been discussed from a different point of view in [7 and 8]. In each of the generalizations the heart of the proof is essentially the proof of the following theorem (Jones used L instead of T in [19,Th.6]). (See Note 1).

THEOREM 3. *Suppose* S *is a totally nonaposyndetic continuum that is not the union of* α *or fewer nowhere dense sets and that* S *has an open cover* $\mathcal{C}$ *of cardinal* α *or less such that, for each point* q *of* S *and each subcontinuum* H *of* S *not containing* q, *there is a member of* $\mathcal{C}$ *containing* q *and missing* H. *Then* S *has a weak cut point.*

Proof. Let r and r' be two points of S. For each member C of $\mathcal{C}$, let H(C) be the intersection of T(C) with the union of the r-component and the r'-component of $S\setminus C$ (either of these may be void). Then H(C) is nowhere dense, since no point of T(C) is contained in the interior of a subcontinuum of S contained in $S\setminus C$. But S is not the union of $\mathcal{H} = \{H(C) \mid C \in \mathcal{C}\}$, since the cardinal of $\mathcal{H}$ is α or less than 2 Let $p \in S\setminus(\cup\mathcal{H})$. Then S is nonaposyndetic at p with respect so some other point q, and q cuts p weakly from each point of $\{r,r'\}\setminus\{q\}$.

Generalizations and related results appear in [1,3,4,6,7,11,14,16,20, 21,22,25 and 26]. The principal negative result is an example of a totally nonaposyndetic, compact, Hausdorff continuum with no weak cut point [4]. See [7] for six unanswered research questions related to this section. There have been contributions to Question 4 of [7], reported above, and to Question 3 of [7], reported below.

IV. RESULTS IN THE PLANE AND IN CONTINUA SATISFYING SEPARATION CONDITIONS

THEOREM 4 (Hagopian [16, Th. 1, p.99]). *If a compact, plane continuum* M *is not connected im kleinen at some point* p *but is aposyndetic at* p *with respect to each subcontinuum of* M *in* $M\setminus\{p\}$, *then* p *is a weak cut point of* M.

The following theorem is particularly nice. See Note 1 for a definition of K(p).

THEOREM 5 (Hagopian [17, Th. 2, p.511]). *Suppose* M *is a compact, plane continuum and* p *is a point of* M *such that* K(p) *is not connected. Then each point of* K(p) *that is not in the p-component of* K(p) *cuts* M *weakly.*

DEFINITION (Hagopian [14]). A continuum S is said to be *strictly non-mutually-aposyndetic* if each pair of subcontinua of S, with nonvoid interior (relative to S), intersect.

Howard Cook has asked the following question [University of Houston Problem Book, Problem 19].

Question 2. Does every strictly non-mutually-aposyndetic, compact metric continuum that contains no weak cut point contain uncountably many triods, or, at least, fail to be embeddable in the plane?

Each of Cook and Hagopian [14, Ex.2] has shown that such a continuum can be embedded in R^3. The following theorem gives an affirmative answer to the "at least" part of the question, but the proof gives no insight into the general question.

THEOREM 6 (Grace [10]). *Every strictly non-mutually-aposyndetic, compact continuum in the plane has a weak cut point.*

In [1], Bing showed that a compact, metric continuum M that is not separated by any subcontinuum (i.e., a θ_1-continuum) is locally connected at each point where it is aposyndetic. Then Jones' work in [19] was used to conclude that a θ_1-continuum with no weak cut point is locally connected at two of its points, and, using that, the following characterization of simple closed curves was proved.

THEOREM 7 (Bing [1, Th. 10, p.504]). *A compact metric continuum is a simple closed curve if and only if it is a* θ_1*-continuum with no weak cut point.*

Bing's theorem motivated the following theorem, proved first by Vought [25, Th. 3, p.260] for continua separated by no aposyndetic subcontinuum, and generalized by Rogers.

THEOREM 8 (Vought-Rogers [21, Th. 4, p.498]). *A compact, metric continuum is hereditarily locally connected and cyclically connected if and only if no point cuts* M *weakly and only semiaposyndetic subcontinua separate* M .

DEFINITION (Hagopian [15]). A continuum M is *semiaposyndetic* if for each two points p and q of M , M is either aposyndetic at p with respect to q or aposyndetic at q with respect to p .

V. NOTES

1. In [19, Theorem 3.] Jones used L for the set $T(y)\setminus\{y\}$ where y is a point of a compact metric continuum. Based on that usage it became customary to use L_y or $L(y)$ for $T(y)$. In the mid 1950's, A. D. Wallace found the function useful in the study of topological semi-

groups but changed the notation to T_y or $T(y)$ to avoid confusion with the already established use of L_y for the generators of the principal left ideal generated by y. In [3, p.99] sets $K(B)(=T(B))$ were defined for subsets B of the topological space under consideration. (The K notation resulted from a faulty memory being used in trying to follow Jones' notation. He used K for a different set [19, Theorem 2]). Later Paul Swingle and others [2], apparently unaware of [3] defined $T(B)$ for sets B.

Some authors, intending to follow Jones lead in notation have used $L(B)$ for $T(B)$. A more appropriate use of that notation however would seem to be for $\cup_{y \in B} T(y)$, since Jones used L in defining a set valued function of a set variable in [19, Theorem 6 (proof)] as follows. "Let $L(g)$ denote the set of all points x of $\overline{D}_1$ such that for some point y of g, M is not aposyndetic at x with respect to y." Here M is a compact metric continuum and $\overline{D}_1 \cap g = \phi$. This set is not $\overline{D}_1 \cap T(g)$ as one might first think, but $\overline{D}_1 \cap [\cup_{y \in g} T(y)]$. In view of all of this, the following usage is proposed.

1) Use T as originally defined in [2], i.e., as defined here, on the grounds that it is now the best known notation for the function and that L and K are more appropriately used as follows.

2) Use $L(B) = \cup_{y \in B} T(y)$, which is approximately the way Jones used it in [19, Theorem 6 (proof)]. Note that $L(\{x\}) = T(\{x\})$, by definition.

3) Use $K(B) = \{x \in S \mid S$ is not aposyndetic at B with respect to $x\}$ where S is the continuum under consideration. This is based on Jones' use in [19, Th. 2] and general use in the literature, starting with [23, p.643].

2. Semi-local-connectedness was first used by Whyburn in [27] where

"locally divisible in p " was used for "semi-locally-connected at p ". In [28] the spelling "semi-locally connected" was used in the title and some places in the body of the text. Later Whyburn apparently settled on the preferable spelling with two hyphens.

In [29], Wilder introduced the concept of i-avoidability at a point which, for i = 0 , is equivalent to semi-local-connectedness at the point and the point not separating the continuum, in locally compact, metric continua. In [30] this was generalized to almost-i-avoidability which for i = 0 is equivalent to semi-local-connectedness in locally compact, metric continua.

REFERENCES

1. Bing, R. H., "Some characterizations of arcs and simple closed curves," *Amer. J. Math.* *70*(1948), 497-506.

2. Davis, H. S., D. P. Stadtlander and P. M. Swingle, "Properties of the set function T^n," *Portugal. Math.* *21*(1962), 114-133.

3. Grace, E. E., "Cut sets in totally nonaposyndetic continua," *Proc. Amer. Math. Soc.* *9*(1958), 98-104.

4. ___________, "A totally nonaposyndetic, compact, Hausdorff space with no cut point," *Proc. Amer. Math. Soc.* *15*(1964), 281-283.

5. ___________, "Cut points in totally non-semi-locally-connected continua," *Pacific J. Math.* *14*(1964), 1241-1244.

6. ___________, "Certain questions related to the equivalence of local connectedness and connectedness im kleinen," *Colloq. Math.* *13*(1965), 211-216.

7. ___________, "On the existence of generalized cut points in strongly non-locally connected continua," *Topology Conference, Arizona State University, 1967*, Arizona State University, Tempe, Arizona, 1968.

8. ___________, "Aposyndesis and weak cut points," *Proceedings: Conference on Metric Spaces, Generalized Metric Spaces and Continua* (Honoring F. B. Jones), Guilford College, Charlotte, N.C. 1980, pp. 151-166.

9. ___________, "Cut points in totally non-semi-locally-connected II continua," *Technical Report No. 52*, Arizona State University, 1980.

10. ___________, "A plane cut point theorem," In preparation.

11. Hagopian, C. L., "On non-aposyndesis and the existence of a certain generalized cut point," *Topology Conference, Arizona State University, 1967*, Arizona State University, Tempe, Arizona, 1968, 327-329.

12. _______________, "On non-locally connected continua," *Topology Conference, Arizona State University, 1967*, Arizona State University, Tempe, Arizona, 1968, 330-334.

13. _______________, "Concerning semi-local-connectedness and cutting in non-locally connected continua," *Pacific J. Math. 30*(1969), 657-662.

14. _______________, "Mutual Aposyndesis," *Proc. Amer. Math. Soc. 23*(1969), 615-622.

15. _______________, "Arcwise connected plane continua", *Topology Conference, Emory University, 1970*, Emory University, Atlanta, 1970, pp. 41-44.

16. _______________, "On generalized forms of aposyndesis," *Pacific J. Math. 34*(1970), 97-108.

17. _______________, "A cut point theorem for plane continua," *Duke Math. J. 38*(1971), 509-512.

18. Jones, F. B., "Aposyndetic continua and certain boundary problems," *Amer. J. Math. 63*(1941), 545-553.

19. _____________, "Concerning non-aposyndetic continua," *Amer. J. Math. 70*(1948), 403-413.

20. _____________, "Concerning aposyndetic and non-aposyndetic continua," *Bull. Amer. Math. Soc. 6*(1955), 735-740.

21. Rogers, L. E., "Continua in which only semi-aposyndetic subcontinua separate," *Pacific J. Math. 43*(1972), 493-502.

22. ______________, "Non-n-mutually aposyndetic continua," *Proc. Amer. Math. Soc. 42*(1974), 595-601.

23. Schlais, H. E., "Non-aposyndesis and non-hereditary decomposability," *Pacific J. Math. 45*(1973), 643-652.

24. Shirley, E. D., "Semi-local-connectedness and cut points in metric continua," *Proc. Amer. Math. Soc. 31*(1972), 291-296.

25. Vought, E. J., "Concerning continua not separated by any nonaposyndetic subcontinuum," *Pacific J. Math. 31*(1969), 257-262.

26. ______________, "n-Aposyndetic continua and cutting theorems," *Trans. Amer. Math. Soc. 140*(1969), 127-135.

27. Whyburn, G. T., "The cyclic and higher connectivity of locally connected spaces," *Amer. J. Math. 53*(1931), 427-442.

28. _______________, "Semi-locally connected sets," *Amer, J. Math. 61*(1939), 733-749.

29. Wilder, R. L. "Sets which satisfy certain avoidability conditions," *Časopis Pěst. Mat.* 67(1938), 185-198.

30. ___________, *Topology of Manifolds*, Rev. Ed., Amer. Math. Soc. Colloq. Publ. 32, Amer. Math. Soc., Providence, RI. 1963.

APOSYNDESIS IN THE PLANE

Charles L. Hagopian*

California State University, Sacramento

Sometimes it is difficult to determine whether a given continuum is embeddable in the Euclidean plane E^2. Aposyndesis provides an approach to this problem since many aposyndesis theorems for plane continua do not extend to continua in Euclidean 3-space E^3.

Definitions. A *continuum* is a nondegenerate compact connected metric space. A continuum M is *aposyndetic* at a point p of M with respect to a subset S of M if there exists a subcontinuum H of M that contains p in its interior and misses S. Let n be a positive integer. A continuum M is *n-aposyndetic* if for every point p of M, M is aposyndetic at p with respect to every set of n points in M-{p}.

THEOREM 1. *Every 2-aposyndetic plane continuum is locally connected [4, p. 98][15, p. 140][20, p. 130].*

The product of two pseudo-arcs is 2-aposyndetic (in fact, n-aposyndetic for every n [2, p. 326][18, p. 246]), embeddable in E^3 [1], and nowhere locally connected. Hence, by Theorem 1, the product of two pseudo-arcs is not embeddable in E^2. Since this product contains uncountably many triods, R. L. Moore's theorem [17] also implies that it is not planar.

THEOREM 2. *Suppose* M *is a plane continuum and* F *is a finite set of points in* M *such that for each point* x *in* M-F, M *is not aposyndetic*

*The author was partially supported by NSF Grant MCS79-16811.

ISBN 0-12-481820-X

at x *with respect to* F, *and* M *is aposyndetic at* x *with respect to each point of* $M-(F\cup\{x\})$. *Then* M *is hereditarily arcwise connected* [5, p. 402].

Definition. A point r of a continuum M *cuts* M between two points p and q of $M-\{r\}$ if r belongs to every subcontinuum of M that contains $\{p,q\}$.

THEOREM 3. *Suppose* M *is a plane continuum that satisfies the hypothesis of Theorem 2. If* p *and* q *are distinct points of* M *and no point cuts* p *from* q *in* M, *then there is a simple closed curve in* M *that contains* $\{p,q\}$ [5, p. 405].

To see that Theorems 2 and 3 do not extend to continua in E^3, decompose the product of two pseudo-arcs by identifying two fibers to two points.

Definition. A continuum M is *semi-aposyndetic* at a point x of M if for every point y of $M-\{x\}$, M is either aposyndetic at x with respect to y or aposyndetic at y with respect to x.

THEOREM 4. *If a plane continuum* M *contains a finite set* F *such that for each point* x *of* $M-F$, M *is semi-aposyndetic at* x *and there exists a point* y *of* F *such that* M *is not aposyndetic at* x *with respect to* y, *then* M *is arcwise connected* [6, p. 165].

THEOREM 5. *Suppose* M *is a semi-aposyndetic plane continuum and for each positive number* ε, *there are at most finitely many complementary domains of* M *of diameter greater than* ε. *Then* M *is arcwise connected* [7].

The proof of Theorem 5 involves the following theorem of F. B. Jones.

THEOREM 6. *Suppose* M *is a plane continuum that does not separate the plane. If* p *and* q *are distinct points of* M *and no point cuts* M

between p *and* q , *then* p *and* q *belong to a simple closed curve in* M [*16*].

Theorem 5 is also related to another theorem of Jones.

THEOREM 7. *Suppose a plane continuum* M *with only finitely many complementary domains is aposyndetic at a point* p *(with respect to every point of* M-{p} *). Then* M *is connected* im kleinen *at* p [*13, p. 550*].

THEOREM 8. *Every semi-aposyndetic plane continuum is* λ*-connected* [*10, p. 286*][*11, p. 119*].

For the definition of a λ-connected continuum, see [11] or [12]. In [12] an aposyndetic continuum is defined in E^3 that is not λ-connected.

In [14] Jones defined the functions K and L on a continuum M as follows:

Definitions. For each point x of M , the set K(x) (L(x)) consists of all points y of M such that M is not aposyndetic at x (y) with respect to y (x).

THEOREM 9. *Suppose* M *is a plane continuum. The following three statements are equivalent.*

1. M *is* λ*-connected.*

2. For each point x *of* M , *every continuum in* K(x) *is decomposable.*

3. For each point x *of* M , *every continuum in* L(x) *is decomposable* [*9*].

For every point x of a continuum M , the set L(x) is connected and closed in M [14, p. 405]. The set K(x) is always closed [14, p. 404] but may fail to be connected. Consider the convergent sequence of tangent circles $C_1, C_2, \ldots$ in E^2 pictured below. Let C be the circle that is the limit of $C_1, C_2, \ldots$. There is a point y of C such that for each positive integer n, $C_n \cap C_{n+1} = \{y\}$ and $C_n - \{y\}$ is inside C_{n+1}. Let x be a point of $C-\{y\}$. In the continuum $M_1 = C \cup \bigcup\{C_n : n=1,2,\ldots\}$, the set $K(x)$ is $\{x,y\}$

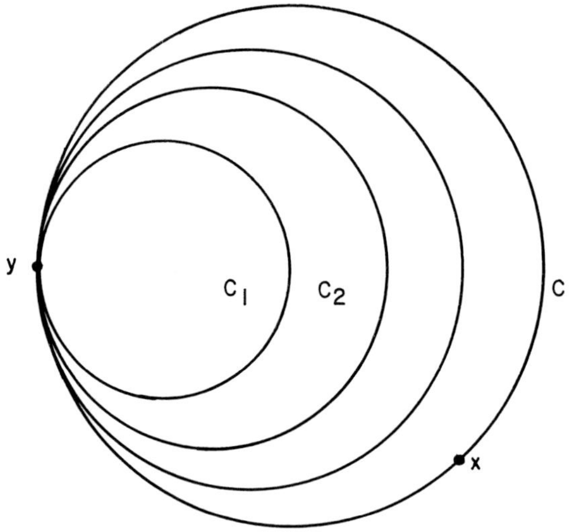

The fact that M_1 has infinitely many large complementary domains in E^2 is consistent with the following theorem of J. Rosasco.

THEOREM 10. *Suppose* M *is a plane continuum and for every positive number* ε, *there are at most finitely many complementary domains of* M *of diameter greater than* ε. *Then for each point* x *of* M, *the set* $K(x)$ *is connected* [*19*].

Note that y cuts M_1 between x and every point of $M_1 - C$.

THEOREM 11. *Suppose* M *is a plane continuum and* x *is a point of* M *such that* $K(x)$ *is not connected. Then each point* y *of* $K(x)$ - (x-component of $K(x)$) *cuts* M *between* x *and a point of* $M-\{x,y\}$ [*8*].

The continuum M_1 defined above can be modified to show that Theorem 11 does not hold for continua in E^3. For each positive integer n, let A_n be an arc segment of diameter $1/n$ in $E^3 - M_1$ that has one endpoint in $C-\{y\}$ and the other in $C_n - \{y\}$. In the continuum $M_2 = M_1 \cup \bigcup\{A_n : n=1,2,\ldots\}$, the set $K(x)$ is $\{x,y\}$. However M_2 does not have a cut point.

Definition. A continuum is *strictly non-mutually aposyndetic* if it does not contain two disjoint continua that have nonvoid interiors.

Recently E. E. Grace announced the following:

THEOREM 12. *Every strictly non-mutually aposyndetic plane continuum has a cut point.*

H. Cook has defined a strictly non-mutually aposyndetic continuum in E^3 that does not have a cut point. The product of two pseudo-arcs also has these properties [3, p. 621].

Question. If M is an indecomposable plane continuum, must the product $M \times M$ be strictly non-mutually aposyndetic?

REFERENCES

1. Bennett, R., "Embedding products of chainable continua," *Proc. Amer. Math. Soc. 16*(1965), 1026-1027.

2. Fitzgerald, R. W., "The cartesian product of nondegenerate compact continua is n-point aposyndetic," *Topology Conference*, Arizona State Univ., 1967, 324-326.

3. Hagopian, C. L., "Mutual aposyndesis," *Proc. Amer. Math. Soc. 23*(1969), 615-622.

4. ______, "On generalized forms of aposyndesis," *Pacific J. Math. 34* (1970), 97-108.

5. ______, "The cyclic connectivity of plane continua," *Michigan Math. J. 18*(1971), 401-407.

6. ______, "Arcwise connectedness of semi-aposyndetic plane continua," *Trans. Amer. Math. Soc. 158*(1971), 161-165.

7. ______, "Arcwise connectivity of semi-aposyndetic plane continua," *Pacific J. Math. 37*(1971), 683-686.

8. ______, "A cut point theorem for plane continua," *Duke Math. J. 38* (1971), 509-512. Erratum, *Duke Math. J. 39*(1972), 823.

9. ______, "Characterizations of λ-connected plane continua," *Pacific J. Math. 49*(1973), 371-375.

10. Hagopian, C. L., "λ-connected plane continua," *Trans. Amer. Math. Soc.* *191*(1974), 277-287.

11. ______, "Mapping theorems for plane continua," *Topology Proc.* *3*(1978), 117-122.

12. ______, "λ-connected products," *ibid.*

13. Jones, F. B., "Aposyndetic continua and certain boundary problems," *Amer. J. Math.* *63*(1941), 545-553.

14. ______, "Concerning non-aposyndetic continua," *Amer. J. Math.* *70*(1948), 403-413.

15. ______, "Concerning aposyndetic and non-aposyndetic continua," *Bull. Amer. Math. Soc.* *58*(1952), 137-151.

16. ______, "The cyclic connectivity of plane continua," *Pacific J. Math* *11*(1961), 1013-1016.

17. Moore, R. L., "Concerning triodic continua in the plane," *Fund. Math.* *13*(1929), 261-263.

18. Rogers, L. E., "Concerning n-mutually aposyndesis in products of continua," *Trans. Amer. Math. Soc.* *162*(1971), 239-251.

19. Rosasco, J., "A note on Jones's function K ," *Proc. Amer. Math. Soc.* *49*(1975), 501-504.

20. Vought, E. J., "n-Aposyndetic continua and cutting theorems," *Trans. Amer. Math. Soc.* *140*(1969), 127-135.

Addendum. In a paper that will appear in the Houston Journal of Mathematics, Lex Oversteegen constructed a continuum-chainable aposyndetic plane continuum that is not arcwise connected.

APOSYNDESIS IN TOPOLOGICAL MONOIDS

R. P. Hunter

Pennsylvania State University, University Park

Let S be a continuum. For any point p let T_p denote the set of all points x such that S is not aposyndetic at x with respect to p. Throughout, we shall assume, in addition, that S is a topological monoid. For many of the results we shall trace, this restriction, that S have an identity element, is not needed. However, usually some algebraic condition is required and for a resumé such as this, the restriction seems appropriate.

It was A. D. Wallace and R. J. Koch who first observed that if T_p meets an ideal it must be contained in that ideal, [11]. This followed from earlier results on the connectedness of ideals and, in particular, on the union of all ideals contained in a given open set. This fact about T_p, which is elementary in nature, is the starting place for the application of aposyndesis.

One of the first applications was the following:

PROPOSITION 1. *Let* S *be irreducible between two points then* S/K *- the hyperspace formed by collapsing the minimal ideal - is an arc,* [4], [11].

Slightly more generally,

PROPOSITION 1 . *Let* S *be irreducible about a finite set. Then* S/K *is a dendrite.*

ISBN 0-12-481820-X

Briefly, one proceeds to obtain an upper semi-continuous decomposition into the sets T_p and then show that all of these (viewed in S/K) are degenerate. (See [5])

The following simple lemma is useful in this connection.

Let X *be a continuum. If* p *weakly cuts the closed subset* A *and* T_p *does not meet* A *then* T_p *separates* X *between some pair of points in* A .

Suppose now that S is a one dimensional compact connected monoid. Furthermore, let us suppose we have already collapsed K , the minimal ideal, so that S has a zero. It follows now from cohomology theory, that S is hereditarily unicoherent. Let M denote the (unique) continuum irreducible from zero to identity. Using the sets T_p and some other lemmas it is shown that M is a sub-monoid and hence an arc. (The structure of M is thus completely determined.) In particular, S is arcwise connected. (The term dendroid, appropriate here, appeared sometime later.) A fairly systematic study of one dimensional monoids was thus made possible. We confine ourselves to some results connected with aposyndesis. See [4] and [8].

1. Let S be as above. If $x \in T_p$ then p is *not* in T_x
2. If p is an endpoint and p is not zero then S is semi-locally connected at p .
3. Define $x \leqq y$ if either x = zero or x weakly cuts zero from y . Then $\leqq$ is an order dense continuous partial order.

Note that these show that

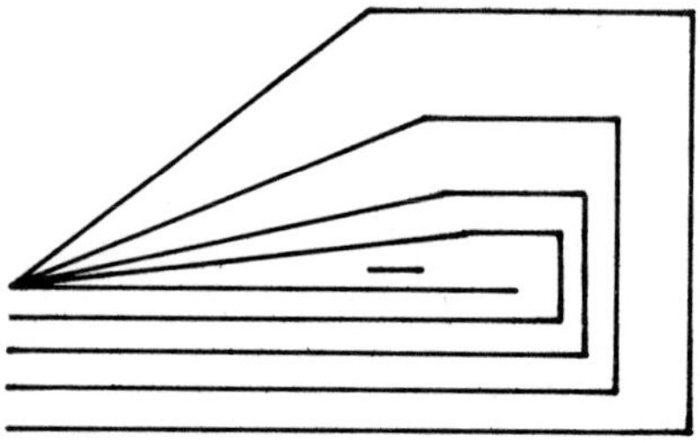

and

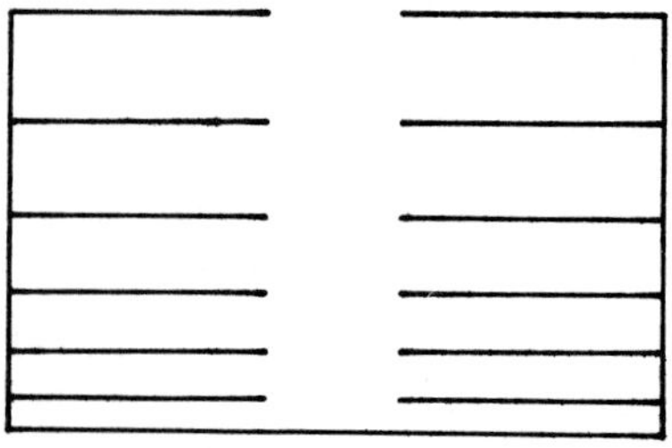

cannot admit the structure of a monoid although

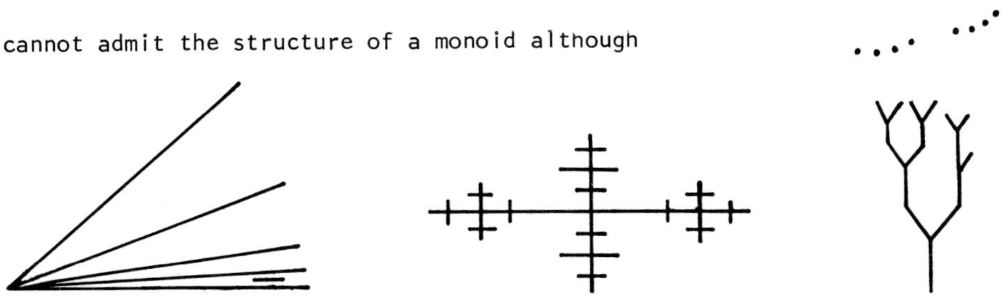

clearly do.

At present, necessary and sufficient conditions that a dendroid be a monoid are not known.

In [12] a class of spaces ruled by arcs is studied. Some of these results are motivated by the non-aposyndetic nature of one dimensional monoids.

As the reader knows, the set T_p has been extended in a variety of ways. ONe useful extension is by Davis, Stadtlander and Swingle [1]. For a set A define $T^0(A) = A$ and $T(T^{n-1}(A)) = T^n(A)$. Using these sets and the notion of n-symmetry the authors succeed in establishing Proposition 1 by totally elementary arguments. (Which is to say, with no recourse to the theory of Lie groups to which the original arguments could be traced).

Other extensions on non-aposyndesis are given in [2]. Here the emphasis is upon idecomposable type properties.

Let us consider now the nature of T_u, where u is the unit element of S. Here the story is as complete as one may desire.

THEOREM. *Let* S *be a compact connected monoid. Then either*

1. $T_u = S$ *in which case* S *is a group and is a solenoid, or*
2. $T_u = \{u\}$. *In this case* S *is, in addition, 1-semi-locally connected at* u.

Note that if Σ is a solenoid then $\Sigma \times \Sigma$ is aposyndetic and

$\Sigma \times \Sigma \to \Sigma$, the projection, is a monotone open homomorphism. Also Σ has sufficiently many light homomorphisms onto the circle group.

We outline the sequence of results needed to establish the theorem above.

LEMMA 1. *Let* G *and* H *be non-degenerate compact connected groups. If* D *is a totally disconnected subgroup of the direct product then* $G \times H/D$ *is aposyndetic.*

LEMMA 2. *Let* A *be an abelian compact connected group, and let* $\alpha \cdot A \to T$ *be a light homomorphism onto a torus which is not the circle group, i.e. of dimension* $\geqq 2$. *Then* A *is aposyndetic.*

Finally, observe that if X has sufficiently many monotone open mappings onto aposyndetic continua X must be aposyndetic.

One then establishes

PROPOSITION 2. *A locally compact connected group which is not aposyndetic must be a solenoid. A compact connected homogeneous space which is not aposyndetic is a solenoid.*

Suppose now that S is a compact connected monoid which is not a group. The group of units G , acts upon S via translation.

One may then appeal to the following result from transformation group theory sometimes referred to as the tube theorem [6].

Let G be a compact group acting on a compact connected space X and let p be a point of X such that $pG \neq X$. If N is a closed normal subgroup such that G/N is a Lie group then there is a subcontinuum such that $F \cap pG \subseteq pN$, $p \in F$ and F meets the complement of pG .

Thus, if the isotropy group G_p is trivial there exist arbitrarily small subcontinua at p which are not contained in the orbit at p .

It now follows that S is aposyndetic at any point of G with respect to $\{u\}$. As we mentioned earlier, if T_u meets an ideal it would be contained in the same. Thus $T_u \subset G$ so we see that $T_u = \{u\}$.

In fact the same type of argument shows that S is 1-semi-locally connected at each point of G.

It might be mentioned that if S is homogeneous, in the usual sense, and is not already a group it must be infinite dimensional and aposyndetic.

There was once considerable interest in algebraic connections between aposyndesis, C-sets, and weak cut points. Recall that D is a C-set if the condition that D meet a continuum implies it contains or is contained in that continuum. Note that if p is in a C-set D then $D \subset T_p$.

The results above show that if S is not a group then the unit is not a weak cut point and can lie in no non-degenerate C-set.

The global problem on C-sets is eliminated in [7]. In fact there is no non-degenerate C-set in S/K and exists as a subset of K under special circumstances.

Finally, we mention a class monoids which provided much of the motivation for the remarks above even in their earliest form: A compact connected monoid is called algebraically irreducible about a closed subset N if no proper compact connected monoid contains N. (See [9])

Thus, for example, S may be the union of a complex unit disc with a half open unit interval winding around.

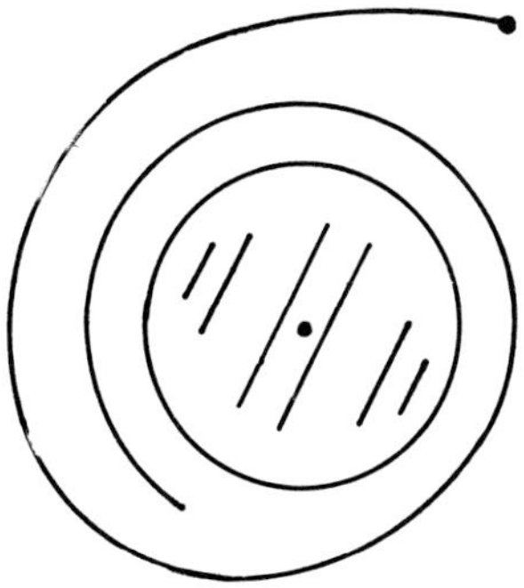

(For some excellent drawings of algebraically irreducible monoids see [3].)

One may then perform a series of constructions obtaining examples such as

Note the second example contains no arc at the unit element. All of these examples are irreducible (algebraically) from zero to unit.

These monoids may be aposyndetic and thus locally connected.

The analysis of these monoids is quite complicated [9,3]. However, it can be shown that a set T_p is either degenerate or is a maximal subgroup and is inaccessible (a C-set) in the submonoid $C\ell\{S\setminus SpS\}$. In any case, one has the following result which includes the algebraically irreducible case and is considerably less general than it could be:

Let S be a compact connected monoid and let H be defined by x and y are H-equivalent if both are in the same orbit of some subgroup of S. Suppose S/H the decompositions space is an arc. Then non-degenerate T_p sets are subgroups and S is quadratically aposyndetic. I.e. $T_p^2 = T_p = T(T_p) = T_p$.

Note that even in the one dimensional case one may have

$$T_p \subsetneq T_p^2 \subsetneq T_p^3 \subsetneq \cdots \subsetneq T_p^n \subsetneq T_p^{n+1} \quad \cdots .$$

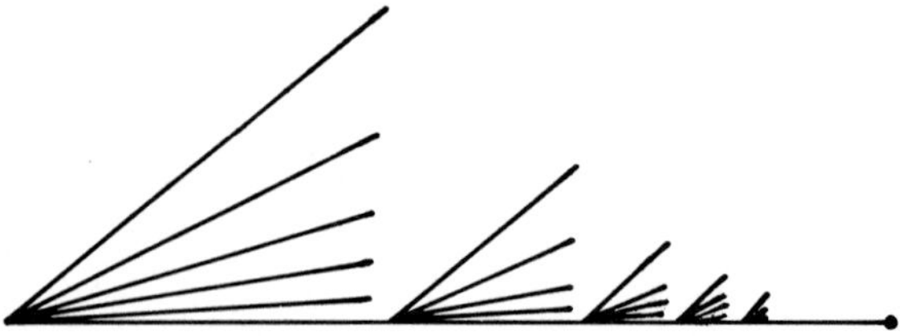

Finally I mention the theory of semigroup actions on continua. An appropriate extension of Proposition 1 is given in [13]. If a compact connected monoid acts appropriately on a continuum X irreducible between two points then X modulo the action of the minimal ideal is an arc.

REFERENCES

1. Davis, H. S., D. P. Stadtlander, and P. M. Swingle, "Semigroups, continua and the set functions T^n," *Duke Math. J.* *29*(1962), 265-280.

2. Dickman, R. F., R. L. Kelley, L. R. Rubin and P. M. Swingle, "Semigroups and clusters of indecomposability," *Fund. Math.* *56*(1964), 21-33.

3. Hofman, K. H. and P. S. Mostert, *Elements of Compact Semigroups,* Chas. E. Merrill, 1966.

4. Hunter, R. P., "On the semigroup structure of continua," *Trans. Amer. Math. Soc.* *93*(1959), 356-368.

5. ______, "Ciertos semigrupos irreducibles sobre un conjunto finito," *Boletin de la Sociedad Mat. de Mexico* *6*(1961), 52-59.

6. ______, "On a conjecture of Koch," *Proc. Amer. Math. Soc.* *12*(1961), 138-139.

7. ______, "Sur la position des C-ensembles dan les demi-groupes," *Bull. Soc. Mat. de Belgique* *14*(1962), 190-195.

8. ______, "On one dimensional semigroups," *Math. Ann.* *146*(1962), 383-396.

9. ______, "On homogroups and their applications to compact connected semigroups," *Fund. Math.* *52*(1962), 69-102.

10. Koch, R. J., "Note on weak cut points in clans," *Duke Math. J.* *24* (1957), 611-616.

11. Koch, R. J. and A. D. Wallace, "Admissibility of semigroup structures on continua," *Trans. Amer. Math. Soc.* *88*(1958), 277-287.

12. Koch, R. J. and L. F. McAuley, "Semigroups on continua ruled by arcs," *Fund. Math.* *56*(1964), 1-8.

13. Stadtlander, D., "Actions with restricted state spaces," *Duke Math. J.* *37*(1970), 199-206.

APOSYNDESIS AND HOMOGENEITY

James T. Rogers, Jr.*

Tulane University, New Orleans

R. H. Bing revived the study of homogeneous continua around 1950 by proving that the pseudo-arc is homogeneous. Previously, some topologists had thought the simple closed curve to be the only homogeneous plane continuum, and Mazurkiewicz had proved this to be true, provided that the continuum is also locally connected.

Perhaps it was natural, then, for Jones to wonder if the same result holds under the weaker assumption that the homogeneous plane continuum M is aposyndetic. Jones showed in [3] that this is the case; in fact, the conclusion is true under the assumption that M contains no cut point.

It follows that a homogeneous plane continuum that does not separate the plane cannot be aposyndetic. In 1951, Jones [1] published a stronger result: a homogeneous, nonseparating plane continuum must be indecomposable. Actually, the result remains true for any homogeneous, hereditarily unicoherent continuum M, planar or not. The proof involves a very clever continuous decomposition of M (where M is assumed decomposable) such that the decomposition space is homogeneous, aposyndetic, and hereditarily unicoherent, an impossible collection of conditions.

Use of the interplay between aposyndesis and cut points is prevalent in the proofs of this and the next theorem.

A little later, Jones [2] honed his decomposition idea even sharper

*The author was partially supported by a research grant from the Tulane University Senate Committee on Research.

ISBN 0-12-481820-X

and allowed aposyndetic decompositions for all homogeneous, decomposable continua. The theorem almost cries out to the reader to construct a circle of pseudo-arcs. Jones' theorem is this.

JONES' APOSYNDETIC DECOMPOSITION THEOREM. *Suppose* M *is a decomposable, homogeneous continuum. Then* $\{L_x : x \in M\}$ *is a nondegenerate collection of homogeneous continua forming a continuous decomposition of* M *such that*

(a) the decomposition space N *is a homogenous, aposyndetic continuum, and,*

(b) if x *is a point of* M, *and if* K *is a subcontinuum of* M *containing a point in* L_x *and a point not in* L_x, *then* K *contains* L_x.

In case M is a plane continuum, it follows from the earlier Jones' results that (1) N is a simple closed curve, and (2) each element of the decomposition is a homogeneous, indecomposable, acyclic continuum.

It clearly is impossible to retain conclusion (1) outside the plane, since, for instance, the product of two homogeneous continua is both homogeneous and aposyndetic. It is possible, however, to retain condition (2), provided M is one-dimensional. This was accomplished in [4] by showing that the quotient mapping π is completely regular.

It is unlikely that aposyndesis will contribute more to the understanding of homogeneous plane continua, but the classification of homogeneous, aposyndetic, nonplanar continua is wide open. Moreover, one is led to the problem in several ways. For instance, it follows from the original Jones' result that an arcwise-connected, homogeneous continuum is aposyndetic, and from [4] that a hereditarily decomposable, homogeneous continuum is aposyndetic. It is not know whether an aposyndetic, homogeneous, one-dimensional continuum must be locally connected.

REFERENCES

1. Jones, F.B., "Certain homogeneous unicoherent indecomposable continua," *Proc. Amer. Math. Soc.* 2 (1951), 855-859.

2. ________, "On a certain type of homogeneous plane continuum," *Proc. Amer. Math. Soc.* 6 (1955), 735-740.

3. ________, "A note on homogeneous plane continua," *Bull. Amer. Math. Soc.* 55 (1949), 113-114.

4. Rogers, J.T., Jr., "Completely regular mappings and homogeneous, aposyndetic continua," *Canad. J. Math.* (to appear).

APOSYNDESIS IN PRODUCT SPACES

Leland E. Rogers
Cook, Washington

I. APOSYNDESIS IN 2-PRODUCTS

The study of aposyndesis in Cartesian products of continua began when G. T. Whyburn showed that in a compact metric space the product of an aposyndetic continuum with an arbitrary continuum is aposyndetic (using the globally equivalent property of semi-local-connectedness) [9]. Later, F. B. Jones showed that the metric, compactness, and aposyndesis of the factors were all unnecessary, by proving that the product of any two Hausdorff continua is aposyndetic [3]. As stronger forms of aposyndesis emerged (n-aposyndesis, mutual aposyndesis, etc.) the quest began for ways of making products even "more" aposyndetic. E. J. Vought showed that the product of an m-aposyndetic continuum with an n-aposyndetic continuum was (m+n+1)-aposyndetic [8]. But this and other attempts were dwarfed by R. W. FitzGerald's proof that for compact Hausdorff continua, products are n-aposyndetic for every n (in fact, countable-set-aposyndetic) [1]. L. E. Rogers was able to remove the compactness requirement (substituting regularity) [4]. Since then no improvement for general two-products has been made.

II. MUTUAL APOSYNDESIS IN 2-PRODUCTS

C. L. Hagopian's introduction of the notion of mutual aposyndesis paved the way for a variety of results dealing with even stronger

ISBN 0-12-481820-X

aposyndesis in products of two continua. First, Hagopian showed that the product of any two aposyndetic continua is mutually aposyndetic [2]. Rogers improved upon this by showing that products of semi-aposyndetic continua are mutually aposyndetic [4]. That result was actually more general in that it stated that for each pair of regular Hausdorff continua that are n-semi-aposyndetic and (n-1)-aposyndetic, the product is (n+1)-mutually aposyndetic. Rogers in [4] and [7] produced various other results in the area of mutually aposyndetic 2-products: the product of an (n-1)-semi-aposyndetic regular Hausdorff continuum with an n-mutually aposyndetic compact continuum is n-mutually aposyndetic; any two regular Hausdorff continua which have no "aposyndetic-terminal-point" will have a mutually aposyndetic product (an aposyndetic-terminal-point is a neighborhood version of a terminal point, namely that for each pair of continua containing the point in their interiors, one contains the other); a certain class of continua (including the simple closed curve) has the property that the product of any member of it with any compact Hausdorff continuum is n-mutually aposyndetic for every n ; for an aposyndetic compact metric continuum M , the product of M with arbitrary continua is always mutually aposyndetic if and only if M is not an arc. Another "anti-arc" theorem is that for a finite-set-aposyndetic compact metric continuum M , the product of M with arbitrary continua is always finite-set-mutually aposyndetic if and only if M is not an arc. Perhaps the most remarkable 2-product result is Rogers' theorem that the product of two compact metric chainable continua is mutually aposyndetic if and only if each factor is an arc [5].

Hagopian began the study of strict non-mutual aposyndesis with the theorem that for chainable continua, 2-products are strictly-non-mutually aposyndetic if and only if each factor is indecomposable [2], and left us with the open question: Is the product of any two indecomposable (whether chainable or not) continua always strictly-non-mutually aposyndetic? He also produced several results regarding quasi-composants in products, including the results that each quasi-composant is dense, and for chainable indecomposable factors, there are c disjoint quasi-composants. Rogers generalized part of Hagopian's theorem by proving that if a 2-product is strictly non-n-mutually aposyndetic for some $n \geqq 2$ then each factor M_i is r_i-indecomposable for some $r_i < n$. For two chainable continua, if one is indecomposable, then the product is strictly-non-3-mutually aposyndetic if and only if the other factor is either indecomposable or 2-indecomposable [5].

III. 3-PRODUCTS

Hagopian provided the first result in the area of Cartesian products of three continua by showing that every 3-product is mutually aposyndetic [2]. Rogers improved the conclusion by proving that the product of any three regular Hausdorff continua is finite-set-mutually aposyndetic [4]. Thus we see that the product of 3 copies of the pseudo-arc is finite-set-mutually aposyndetic, but still far short of being locally connected.

IV. SEPARATION AXIOM ANALOGS

The notions of semi-aposyndesis, aposyndesis, mutual aposyndesis, and connectedness im kleinen are in a sense "continuum" versions of the separation axioms T_0, T_1, T_2, and T_3 as discussed briefly by Rogers in [6]. Perhaps some of the intermediates to the separation axioms will provide new variations of aposyndesis, helping to fill in some gaps in Jones' aposyndesis spectrum. As noted above, 2-products are finite-set-aposyndetic, and 3-products are finite-set-mutually aposyndetic; is there a stronger variation of aposyndesis for which 4-products are special also?

REFERENCES

1. FitzGerald, R. W., "The Cartesian Product of non-degenerate compact continua is n-point aposyndetic," *Topology Conference*, Arizona State University, 1967, Arizona State University, Tempe, Ariz. 1968, 324-329.

2. Hagopian, C. L., "Mutual aposyndesis," *Proc. Amer. Math. Soc. 23*(1969), 615-622.

3. Jones, F. B., "Concerning non-aposyndetic continua," *Amer. J. Math. 70* (1948), 403-413.

4. Rogers, L. E., "Concerning n-mutual aposyndesis in products of continua," *Trans. Amer. Math. Soc. 162*(1971), 239-251.

5. ____________, "Mutually aposyndetic products of chainable continua," *Pacific J. Math 37*(1971), 805-812.

6. ____________, "Aposyndesis and the separation axioms," *Proc. of the University of Oklahoma Topology Conference, 1972*, The University of Oklahoma, Norman, Okla., 1972, 264-273.

7. ____________, "Products with nonlinear finite-set-aposyndetic continua," *Colloq. Math. 32*(1975), 199-206, 309.

8. Vought, E. J., and F. B. Jones, "Stronger forms of aposyndetic continua," *Topology Conference*, Arizona State University, 1967, Arizona State University, Tempe, Ariz., 1968, 170-173.

9. Whyburn, G. T., "Semi-locally connected sets," *Amer. J. Math. 61*(1939), 733-749.

MONOTONE DECOMPOSITIONS OF CONTINUA

Eldon J. Vought

California State University, Chico

The object of this paper is to discuss two aspects of monotone decompositions of compact, metric continua. First of all, for certain types of continua monotone decompositions are obtained so that the elements of the decompositions have void interiors and for which the quotient spaces are locally connected or, in one case, arcwise connected. Secondly, the requirement that the decomposition elements have void interiors is eliminated and techniques are described that yield decompositions in which the quotient spaces have some desired property, e.g. aposyndesis, semi-aposyndesis.

The first type of continuum in which it is possible to characterize decompositions whose elements have void interiors is the θ_n-continuum and, more generally, the θ-continuum. A θ_n-*continuum* is a compact, metric connected space X such that no subcontinuum of X separates it into more than n components. If no subcontinuum of X separates it into an infinite number of components, then X is called a θ-*continuum*. This concept and notation were introduced by R. W. FitzGerald [3, p. 139], the notation being derived from the Greek letter θ, an example of a θ_2-continuum. An indecomposable continuum and a simple closed curve are θ_1-continua, a continuum irreducible between two points is a θ_2-continuum, so is an atriodic continuum. If X is irreducible about n points, then X is a θ_n-continuum. The structure of locally connected θ-continua is very nice as shown by the following two theorems, proved by FitzGerald in his seminal

ISBN 0-12-481820-X

paper [3, pp. 156, 157].

THEOREM 1. *A locally connected* θ*-continuum is a* θ_n*-continuum for some* n .

THEOREM 2. *A locally connected* θ*-continuum is a finite graph.*

A continuum X is *aposyndetic* at a point $p \in X$ if, given $q \in X$, $q \neq p$, there exists a continuum H such that $p \in H^{o} \subset H \subset X \setminus \{q\}$. Note that this notion, which is due to F. B. Jones [8], is a generalization of connectedness *im kleinen* at p where the point q is replaced by a closed set.

In order to study nonlocally connected θ-continua, it is necessary to study those that are nonaposyndetic as the next theorem, due to R. H. Bing [1, p. 499], shows.

THEOREM 3. *If a* θ_1*-continuum is aposyndetic at* p *, then it is locally connected at* p .

Bing's proof extends easily to θ-continua.

There are two notions that extend in different ways the concept of aposyndesis, given above, to subsets of X . These two ideas are useful in what follows and are equivalent for θ-continua. If H is a subset of X , let $\underline{T(H)} = \{x \in X \mid \text{if } Q \text{ is a subcontinuum of } X \text{ such that } x \in Q^{o}, \text{ then } Q \cap H \neq \emptyset\}$ and $\underline{K(H)} = \{x \in X \mid \text{if } Q \text{ is a subcontinuum of } X \text{ such that } H \subset Q^{o}, \text{ then } \{x\} \cap Q \neq \emptyset\}$. It is straightforward to show that for θ-continua, the two set functions T and K are equal.

For nonlocally connected (hence nonaposyndetic) θ-continua, monotone upper semi-continuous decompositions are now obtained so that the quotient spaces are locally connected, hence are finite graphs, and for which the elements of the decompositions have void interiors. The first theorem in this direction is the classical theorem due to Kuratowski [9, p. 216] for irreducible continua.

THEOREM 4. *If the continuum* X *is irreducible between two points and contains no indecomposable subcontinuum with nonvoid interior, then* X *admits a monotone, upper semi-continuous decomposition* $\mathcal{D}$ *such that* $X/\mathcal{D}$ *is an arc and the elements of* $\mathcal{D}$ *have void interiors.*

Furthermore $\mathcal{D} = \{T^2(x) \mid x \in X\}$ *where* $T^2(x) = T(T(x))$ [3, p. 169].

I considered the problem of trying to prove a similar theorem by replacing the irreducible continuum X with a θ_1-continuum, i.e., a continuum X in which no subcontinuum separates X, and concluding that $X/\mathcal{D}$ is a simple closed curve. Although I was able to establish the theorem for plane continua [12, p. 70], surprisingly the requirement that X contains no indecomposable subcontinuum with nonvoid interior turned out to be too weak to obtain the decomposition characterization as the example in [12, pp. 71-73] shows. Requiring the continuum to be hereditarily decomposable guarantees the decomposition but not every such decomposition implies that the continuum is hereditarily decomposable as shown by easy examples. The characterizing condition uses the set function T defined above and is stated in the next theorem [12, p. 74].

THEOREM 5. *If* X *is a* θ_1*-continuum, then* X *admits a monotone upper semi-continuous decomposition* $\mathcal{D}$ *the elements of which have void interiors and such that* $X/\mathcal{D}$ *is a simple closed curve if and only if whenever* H *is a subcontinuum of* X *with void interior, it follows that* $T(H)^{\circ} = \emptyset$. *Furthermore* $\mathcal{D} = \{T^2(x) \mid x \in X\}$.

Recently E. E. Grace and I [6, Thm. 2] extended this result to θ_n-continua.

THEOREM 6. *If* X *is a* θ_n*-continuum, then* X *admits a monotone upper semi-continuous decomposition* $\mathcal{D}$, *the elements of which have void interiors and such that* $X/\mathcal{D}$ *is a finite graph if and only if whenever* H *is a subcontinuum of* X *with void interior, it follows that* $T(H)^{\circ} = \emptyset$.

Furthermore $\mathcal{D} = \{T^{2n}(x) | x \in X\}$.

Grace, in an as yet unpublished paper, has proved this theorem for θ-continua. The elements of the decomposition are $\bigcup_{i=1}^{\infty} T^i(x)$ instead of $T^{2n}(x)$. For each $x \in X$ it is possible to show that there is an integer i_x such that $T^{i_x+1}(x) = T^{i_x}(x)$. It is interesting to note that this theorem yields a condition for which a θ-continuum is a θ_n-continuum for some n . If the θ-continuum X has the property that whenever H is a subcontinuum of X with void interior, then $T(H)^o = \emptyset$, it follows from Grace's extension of Theorem 6 that $X/\mathcal{D}$ is a finite graph, hence is locally connected. By Theorem 1, $X/\mathcal{D}$ is a θ_n-continuum for some n and it is not hard to show that X is then a θ_n-continuum for the same n .

Using the set function K , Schlais has proved the next theorem [11, Thm. 9].

THEOREM 7. *If* X *is a hereditarily decomposable continuum, then if* H *is a subcontinuum such that* $H^o = \emptyset$ *, it follows that* $K(H)^o = \emptyset$.

A continuum X is *δ-connected* if, given two points $x,y \in X$, there is a hereditarily decomposable subcontinuum of X containing x and y . Using this notion, C. L. Hagopian [7, pp. 134, 135] has extended Schlais' theorem.

THEOREM 8. *If* X *is a δ-connected continuum, then if* H *is a subcontinuum such that* $H^o = \emptyset$ *, it follows that* $K(H)^o = \emptyset$.

This theorem has interesting applications to θ-continua. Recall that for θ-continua $K = T$.

COROLLARY. *If* X *is a δ-connected θ-continuum then* X *admits a monotone, upper semi-continuous decomposition* $\mathcal{D}$ *such that the elements of* $\mathcal{D}$ *have void interiors and for which the quotient space* $X/\mathcal{D}$ *is a finite graph.*

Since arcwise connected continua are δ-connected, this result holds for arcwise connected θ-continua.

A continuum X is *λ-connected* if, given two points $x,y \in X$, there exists a continuum $Y \subset X$ irreducible between x and y and such that every indecomposable subcontinuum of Y is a continuum of condensation in Y, i.e., Y is a continuum that in its relative topology is a continuum as in Kuratowski's theorem above. It is an open question as to whether Theorem 8 or the corollary following it can be generalized to a λ-connected θ-continuum.

For the next type of continuum it is always possible to obtain decompositions whose elements have void interiors. A continuum is *hereditarily unicoherent at* p [5] if the intersection of any two continua, each of which contains the point p, is connected. A continuum X is *smooth* [5] in case there is a point p in X such that X is hereditarily unicoherent at p and for each convergent sequence of points $\{a_n\}$ in X, the condition $\lim a_n = a$ implies that $\{pa_n\}$ is convergent and $\lim pa_n = pa$, where px is the unique subcontinuum of X that is irreducible between p and x.

G. R. Gordh, Jr. has proved the following decomposition theorem for continua that are smooth [5].

THEOREM 9. *If* X *is a continuum that is smooth, then there exists a unique minimal monotone, upper semi-continuous decomposition* $\mathcal{D}$ *such that* $X/\mathcal{D}$ *is arcwise connected and for which the elements of* $\mathcal{D}$ *have void interiors. Furthermore* $X/\mathcal{D}$ *is smooth and hereditarily unicoherent.*

Now our second purpose is to consider decompositions in which the quotient spaces exhibit desirable qualities without being concerned whether the elements of the decomposition have void interiors. In fact, the decompositions may be degenerate.

Let K be a collection of closed separators of X [10, p. 2] with the following property:

If $k \in K$ and $X \setminus k = A \cup B$, a separation, then if $a \in A$, $b \in B$, $c \in k$, there exists $k' \in K$, an open set Q, and a continuum W such that $a \in Q \subset W$ and k' separates $\{c\} \cup W$ from b. Let $\tilde{K}$ be the unique maximal such collection and let $S_x = \{y \in X \mid \nexists\ k \in \tilde{K}$ that separates x from $y\}$. McAuley [10, p. 3] has utilized these ideas to obtain decompositions with aposyndetic quotient spaces.

THEOREM 10. *If* X *is a continuum, then* $\mathcal{D} = \{S_x \mid x \in X\}$ *is the unique minimal decomposition of* X *such that* $\mathcal{D}$ *is monotone, upper semi-continuous and* $X/\mathcal{D}$ *is aposyndetic.*

A totally different approach due to FitzGerald and Swingle [4] yields the same decomposition as Theorem 10. If P is a function defined on the subsets of X, such that if $A \subset X$, then $P(A) \subset X$, P is *expansive* [4, p. 34] if (i) $A \subset P(A)$, (ii) $P(A) \subset P(B)$ whenever $A \subset B$. A subset A of X is P-*closed* if $P(A) = A$.

THEOREM 11. *If* X *is a Hausdorff continuum, then* X *has a unique minimal decomposition* $\mathcal{D}$ *with respect to being upper semi-continuous and having* P-*closed elements.*

The next decomposition theorem, due to FitzGerald and Swingle [4, p. 37] gives exactly the same decomposition as Theorem 10.

THEOREM 12. *If* $P = T$ *in Theorem 11, then* X *admits a unique, minimal, monotone upper semi-continuous decomposition* $\mathcal{D}$ *having* T-*closed elements and such that* $X/\mathcal{D}$ *is aposyndetic.*

A decomposition $\mathcal{D}$ of a continuum is *admissible* [2, p. 115] if whenever L is a layer of an irreducible subcontinuum of X and $D \in \mathcal{D}$ such that $L \cap D \neq \emptyset$ then $L \subset D$. Charatonik [2, pp. 116, 117] obtains hereditarily arcwise connected quotient spaces by means of the next theorem.

THEOREM 13. *If* X *is a continuum then there exists a unique, minimal, admissible, monotone upper semi-continuous decomposition* $\mathcal{D}$. *Furthermore* $X/\mathcal{D}$ *is hereditarily arcwise connected.*

For certain continua, e.g. hereditarily unicoherent, atriodic, those irreducible about n points; monotone upper semi-continuous decompositions with hereditarily arcwise connected quotient spaces are admissible so, by Theorem 13, for these continua there exist unique minimal upper semi-continuous monotone decompositions with hereditarily arcwise connected quotient spaces.

Question: Can those continua that admit this type of decomposition be classified?

Using the notion of closed separators [14, p. 74], an equivalent description of Charatonik's decomposition can be obtained for certain continua such as hereditarily unicoherent continua. Let K be a collection of closed separators of X with the following property:

If $k \in K$ and $X \backslash k = A_1 \cup A_2$, a separation, and if $a_1 \in A_1$, $a_2 \in A_2$, $c \in k$, then for either $i = 1$, $j = 2$ or $i = 2$, $j = 1$ there exists $k' \in K$ and a continuum Q such that $\{a_i, c\} \subset Q$ and k' separates Q from a_j .

Let $\tilde{K}$ be the unique maximal such collection and let $S_x = \{y \in X \mid \nexists\, k \in \tilde{K}$ that separates y from $x\}$.

THEOREM 14. *For certain continua, e.g. hereditarily unicoherent, those irreducible about* n *points;* $\mathcal{D} = \{S_x \mid x \in X\}$ *coincides with Charatonik's decomposition and is therefore the unique minimal monotone upper semi-continuous decomposition such that* $X/\mathcal{D}$ *is hereditarily arcwise connected.*

Question: Is it possible to classify those continua for which $\mathcal{D} = \{S_x \mid x \in X\}$ coincides with Charatonik's decomposition?

An affirmative answer to this question would probably yield an affirmative answer to the previous question.

The continuum X is *semi-aposyndetic* if, given two distinct points $x,y \in X$, there exist open sets U_x, U_y and continua H_x, H_y such that $x \in U_x \subset H_x \subset X\setminus\{y\}$ or $y \in U_y \subset H_y \subset X\setminus\{x\}$.

Now the notion of closed separator used earlier by McAuley to obtain aposyndetic quotient spaces will be varied slightly in order to obtain semi-aposyndetic quotient spaces for hereditarily unicoherent continua [13, pp. 374,375]. Let K be a collection of closed separators of X with the following property: If $k \in K$ and $X\setminus k = A_1 \cup A_2$, a separation, then if $a_1 \in A_1$, $a_2 \in A_2$, $b \in k$, there exist $k' \in K$ and for either $i = 1, j = 2$ or $j = 1, i = 2$, an open set Q and continuum W such that $a_i \in Q \subset W$ and k' separates a_j from $W \cup \{b\}$.

Let $\tilde{K}$ be the unique maximal such collection and let $S_x = \{y \in X \mid \nexists\, k \in \tilde{K}$ that separates y from $x\}$.

THEOREM 15. *If* X *is a hereditarily unicoherent continuum, then* $\mathcal{D} = \{S_x \mid x \in X\}$ *is the unique minimal monotone upper semi-continuous decomposition such that* $X/\mathcal{D}$ *is semi-aposyndetic.*

For hereditarily unicoherent continua, let $\mathcal{D}_A, \mathcal{D}_S, \mathcal{D}_{ARC}$ be the unique minimal monotone upper semi-continuous decompositions such that $X/\mathcal{D}_A$, $X/\mathcal{D}_S$, $X/\mathcal{D}_{ARC}$ are aposyndetic, semi-aposyndetic and arcwise connected, respectively. Since aposyndetic, hereditarily unicoherent continua are locally connected, $X/\mathcal{D}_A$ can alternately be described as the unique minimal monotone upper semi-continuous decomposition such that $X/\mathcal{D}_A$ is locally connected. Also Gordh has proved that a semi-aposyndetic, hereditarily unicoherent continuum is arcwise connected, so it follows that $\mathcal{D}_{ARC} \leq \mathcal{D}_S$ where $\leq$ means refinement. Clearly $\mathcal{D}_S \leq \mathcal{D}_A$ so we have the sequence of refinements $\mathcal{D}_{ARC} \leq \mathcal{D}_S \leq \mathcal{D}_A$.

It is easily seen with simple examples that these refinements cannot be reversed. Also by forming a hereditarily unicoherent continuum that includes a $\sin(1/x)$ curve and its limit bar, a simple fan, the union of two simple fans in "opposite" directions with a common limit interval, it is instructive to see how the successive application of these three decompositions to this continuum "improves" the quotient space.

REFERENCES

1. Bing, R. H., "Some characterizations of arcs and simple closed curves," *Amer. J. Math. 70*(1948), 497-506.

2. Charatonik, J. J., "On decompositions of continua," *Fund. Math. 79* (1973), 113-130.

3. FitzGerald, R. W., "Connected sets with a finite disconnection property," *Studies In Topology*, Academic Press, New York, 1975, 139-173.

4. FitzGerald, R. W. and P. M. Swingle, "Core decompositions of continua," *Fund. Math. 61*(1967), 33-50.

5. Gordh, Jr., G. R., "On decompositions of smooth continua," *Fund. Math. 75*(1972), 51-60.

6. Grace, E. E. and E. J. Vought, "Monotone decompositions of θ_n-continua," to appear in *Trans. Amer. Math. Soc.*

7. Hagopian, C. L., "λ-connectivity and mappings onto a chainable indecomposable continuum," *Proc. Amer. Math. Soc. 45*(1974), 132-136.

8. Jones, F. B., "Aposyndetic continua and certain boundary problems," *Amer. J. Math. 63*(1941), 545-553.

9. Kuratowski, C., *Topology, Vol. 2*, 3d ed, Monografie Mat. No. 21, PWN, Warsaw, 1961; English transl.: Academic Press, New York, 1968.

10. McAuley, L. F., "An atomic decomposition of continua into aposyndetic continua," *Trans. Amer. Math. Soc. 88*(1958), 1-11.

11. Schlais, H. E., "Nonaposyndesis and nonhereditary decomposability," *Pacific J. Math. 45*(1973), 643-652.

12. Vought, E. J., "Monotone decompositions of continua not separated by any subcontinua," *Trans. Amer. Math. Soc. 192*(1974), 67-78.

13. ______, "Monotone decompositions of Hausdorff continua," *Proc. Amer. Math. Soc. 56*(1976), 371-376.

14. ______, "On decompositions of hereditarily unicoherent continua," *Fund. Math. 102*(1979), 73-79.

SECTION III

ALGEBRAIC, DIFFERENTIAL, AND GENERAL TOPOLOGY

MONOTONE MAPPINGS - SOME MILESTONES

Louis F. McAuley

State University of New York, Binghamton

Dedicated to F. Burton Jones on the occassion of his retirement from the University of California at Riverside.

I. INTRODUCTION

The subject of Monotone Mappings has a relatively long history with a large literature containing numerous important theorems. It continues to attract good minds and new exciting work is being produced. The impossibility of covering the field in an hour talk should be clear. Therefore, I shall review the history by choosing among the many milestones, survey some recent contributions, and mention problems that now actively engage researchers. Choices must be made. The work mentioned here does not imply that (even) I impart less importance to some results not mentioned.

The thesis of J. R. Walker on "Monotone Mappings," Syracuse University, 1966 [96] is an excellent historical source. My own article "Some Fundamental Theorems and Problems Related to Monotone Mappings" prepared for the *Proceedings of the (First) Conference on Monotone and Open Mappings*, SUNY-Binghamton, 1970 [69], contains a reasonable historical survey up to that date. Free copies of [70] are available. Other articles which provide additional historical surveys include McAuley [67;68], Lacher [58], Armentrout [10;11], and Whyburn [117].

The origin of the *formal* concept of monotone mapping was probably due

ISBN 0-12-481820-X

to Charles B. Morrey, Jr. [79] in 1935. Clearly, the concept of monotone increasing (decreasing) real valued functions had something to do with the naming of the concept.

In the early part of this century, R. L. Moore began an axiomatic approach to the study of E^2 and plane-like spaces. He constructed new spaces from old ones by using "chunks" of the old space for "new points". These "new points" were "point-like" with respect to the old space. Thus, Moore began the study of "monotone upper semicontinuous decompositions of continua." His fundamental paper [74] appeared in 1925.

There are monotone (indeed, one-to-one) mappings which are not closed. An easy example is a 1-1 mapping f of $(0,1)$ onto $(-1,0) \cup \{(x,y) \mid x^2+y^2 = 1\}$ which can be obtained in an obvious way. Thus, we restrict our attention to closed (or quasi-compact) mappings.

The usual definitions by Moore [74] and Whyburn [117] of upper semicontinuous decompositions of a metric space (X,d) yield closed mappings (the quotient maps) from X to the decomposition space.

Ideally, I would have liked to have chosen *theorems and examples* which have had a major impact on the history of monotone mappings either by the *nature of the result or by the methods and ideas used in the proofs and constructions*. It would have been appropriate to discuss these various ideas which have so influenced the development of the subject. Unfortunately, I have not had the time to complete this task.

II. UPPER SEMICONTINUOUS DECOMPOSITIONS - A GENERAL PROBLEM

Let G denote a collection of pairwise disjoint compact subsets of X which covers (fills up) X where (X,d) is a metric space. We say that G is an upper semicontinuous (usc) decomposition of X if and only if for each open set U in X, $G(U) = \{g \mid g \in G \text{ and } U \supset g\}$ has the property that $G(U)^*$, the union of the elements of $G(U)$, is open in X. An usc decomposition G of X is *monotone* iff for each $g \in G$, g is connected. The "points" of the decomposition space X/G are the elements of G and the topology is the collection of all $G(U)$, U open in X.

If $f : X \to Y$ is a closed (or quasi-compact [117;122]) mapping where each of (X,d) and (Y,ρ) is a metric space and f is point compact ($f^{-1}f(x)$ is compact for each $x \in X$), then $G = \{f^{-1}f(x) | x \in X\}$ is an usc decomposition of X and X/G is homeomorphic to Y.

A continum C in X is *point-like* iff for some point $x \in X$, $X - C$ is homeomorphic to $X-\{x\}$.

Problem. Characterize all (point-like) monotone usc decompositions of a given metric space (X,d). In particular, characterize all closed (point-like) monotone mappings of nice spaces (manifolds, polyhedra, ANR's, Peano continua, etc.).

Special Case: Require the range space (image) to be homemorphic to the original space (X,d).

Now, we shall mention some of the major results concerning this general problem and the special case.

1925 R. L. Moore [74]

Suppose that f is a monotone mapping of $S^2(E^2)$ onto a nondegenerate Hausdorff space (Y,T) such that $f^{-1}f(x)$ is point-like (non separating) for each $x \in X$. Then Y is homeomorphic to $S^2(E^2)$.

1929 R. L. Moore [75]

Suppose that f is a monotone mapping of S^2 onto a Hausdorf space (Y,T). Then Y is a cactoid.

A Peano continuum is a *cactoid* iff each true cyclic element is homeomorphic to S^2 [117].

1968 Sosinskiĭ [90]

A necessary and sufficient condition that a Hausdorf space (Y,T) be the image of a monotone *open* mapping of S^2 is that Y be a cactoid.

1938 Roberts and Steenrod [85]

Suppose that P is a generalized cactoid (Peano continuum for which each true cyclic element is a 2-manifold [at most a finite number are different from S^2] with a finite number of identifications). Then there is a monotone mapping f of some 2-manifold M^2 onto P. Each monotone image of a 2-manifold M^2 is such a continuum P.

Pioneering work on monotone point-like decompositions of E^3 was begun by R. H. Bing in the early 1950's. Perhaps, the most important of his early work is the following:

1952 R. H. Bing [16]. Dogbone Space

Here, Bing constructed an example of an usc decomposition G of E^3 such that the collection H of all nondegenerate elements consisted of polyhedral arcs (indeed, each one made up of three straight line intervals [39]) such that E^3/G is not homeomorphic to E^3 and $P(H)$ is a Cantor set.

It was shown [18] by Bing that $(E^3/G) \times E^1$ is homeomorphic to E^4. In 1962, he constructed [19] an example of an usc decomposition G of E^3 such that the collection H of all non degenerate elements consisted of *straight* line intervals such that $P(H)$ is a Cantor set and E^3/G is not homeomorphic to E^3. (This was proved by Eaton [31] in 1975.)

III. SPECIAL CONSTRUCTIONS OF MONOTONE MAPPINGS

In the early 1930's, R. L. Moore developed a method for constructing closed monotone mappings which was particularly useful in mapping "thin" spaces onto acyclic Peano continua.

1932 R. L. Moore [78]

Suppose that (M,d) is a metric continuum (compact). For each point $p \in M$, let $M(p) = \{x|$there does not exist an uncountable point set each point of which separates X from p in M} . The collection G of all such sets M(p) is a monotone usc decomposition of M and M/G is an acyclic Peano continuum.

1954; 1955 L. F. McAuley [62;63;64]

In my thesis (written under the supervision of F. B. Jones), I obtained various generalizations of Moore's sets M(p) . In fact, I described methods for constructing usc decompositions of connected metric spaces (M,d) (often separable but not necessarily compact) such that the decomposition spaces M/G had selected properties such as local connectivity, semilocal connectivity, aposyndesis (a concept defined and developed by F. B. Jones), etc. In 1955, I gave a method for obtaining an "atomic" decomposition of spaces (M,d) into aposyndetic metric spaces M/G which left aposyndetic spaces (M,d) invariant, i.e., each element of G is a point.

1967 R. W. Fitzerald and P. M. Swingle [37]

These authors found another (and simpler) method for describing atomic usc decompositions G of certain connected metric spaces (M,d) into aposyndetic metric spaces M/G .

IV. MONOTONE MAPPINGS - ALL POINT INVERSES NONDENGERATE

A major problem which has intrigued researchers since the early twenties concerns mappings with all point inverses non degenerate. Indeed, there is an impressive history of important results concerning this problem. We list some of these.

Problem. When do there exist closed monotone mappings f (usc decompositions G) of a given metric space (X,d) onto a metric space (Y,p) with all point inverses nondegenerate (elements of G are nondegenerate)? In particular, when do there exist closed monotone mappings $f : X \to X$ with $f^{-1}f(x)$ nondegenerate for *each* $x \in X$? When do there exist closed monotone mappings $f : X \to Y$ with $f^{-1}f(x)$ homeomorphic to a given nondegenerate continuum C ?

1929 J. H. Roberts [82]

There is a closed monotone mapping f of E^2 onto E^2 such that for each $x \in E^2$, $f^{-1}f(x)$ is a nondegenerate point-like continuum (non separating).

1935 J. H. Roberts [83]

It was *announced* by Roberts that in the above result, f could be constructed so that each $f^{-1}f(x)$ is homeomorphic to $[0,1]$.

1936 J. H. Roberts [84]

Contradicting the announcement mentioned above, Roberts proved: There does not exist a closed monotone mapping $f : E^2 \to E^2$ such that for each $x \in E^2$, $f^{-1}f(x)$ is homeomorphic to $[0,1]$.

1950 R. D. Anderson [3]

It was *announced* by Anderson that there is a closed monotone *open* mapping $f : E^2 \to E^2$ such that for each $x \in E^2$, $f^{-1}f(x)$ is a pseudo arc. No proofs were given and no manuscript has ever been available.

1949 Rozanskaya [87]

It was claimed in [87] that there is no closed monotone open mapping $f : E^2 \to E^2$ such that each point inverse is nondegenerate. (Compare this claim with the work of Lewis and Walsh listed below.)

1955 Eldon Dyer [26]

There does not exist a closed monotone open mapping $f : E^n \to Y$ such that for each $x \in E^n$, $f^{-1}f(x)$ is a $\begin{cases} \text{1-dimensional AR} \\ \text{1-dimensional decomposable cell-like set} \end{cases}$.

1977 Wayne Lewis and John J. Walsh [59]

These researchers constructed a closed monotone open mapping $f : E^2 \to E^2$ such that for each $x \in E^2$, $f^{-1}f(x)$ is a pseudo arc. This verified the earlier claim by Anderson.

1968 S. L. Jones [48]

There is no closed monotone mapping $f : E^n \to E^n$ such that for each $x \in E^n$, $f^{-1}f(x)$ is homeomorphic to $[0,1]$.

1970 S. L. Jones [49]

There is no closed monotone open mapping $f : E^n \to E^n$ such that for each $x \in E^n$, $f^{-1}f(x)$ is homeomorphic to a k-cell, $k \geqq 1$.

V. A SPECIAL CASE OF THE GENERAL PROBLEM

In 1935, Knaster solved a special case of the problem stated in §4.

1935 Knaster [51]

There is a (compact) irreducible (between two fixed points) metric continuum (X,d) and a monotone open mapping $f : X \to [0,1]$ such that for each $x \in X$, $f^{-1}f(x)$ is nondegenerate.

He raised the following question:

Question (Knaster). Is there an irreducible continuum (X,d) and a monotone open mapping $f : X \to [0,1]$ such that for each $x \in X$, $f^{-1}f(x)$ is homeomorphic to a fixed nondegenerate continuum K?

1949 E. E. Moise [73]

If (X,d) is a continuum (compact metric) and $f : X \to [0,1]$ is a monotone open mapping and $f^{-1}f(x)$ is homeomorphic to $[0,1]$ for each $x \in X$, then X is *not* irreducible.

1951 M. E. Hamstrom [41]

It was shown by Hamstrom that if $f^{-1}f(x)$ is a nondegenerate Peano continuum (rather than $[0,1]$ in Moise's Theorem), then X is not irreducible.

1953 Eldon Dyer [25]

The theorem of Moise was extended further by requiring that for each $x \in X$, $f^{-1}f(x)$ is a decomposable continuum.

Additional work on monotone decompositions of irreducible continua has been done by Thomas [92] and Vought. Generalizations to irreducible Hausdorff continua have been made by Gordh [40].

VI. ANOTHER SPECIAL CASE OF THE GENERAL PROBLEM

Suppose that each of (X,d) and (Y,p) is a metric continuum and f is a monotone open mapping of X onto Y. How are the dimensions of X and Y related? In particular, can a monotone open mapping on a finite dimensional continuum raise dimension? It was proved by Alexandroff that a countable-to-one open mapping does not raise dimension. For a discussion and generalization of Alexandroff's theorem, see the work [47] by Hodel.

1952 R. D. Anderson [4]

There is a monotone open mapping f of the universal (Menger) 1-dimensional Peano continuum U onto the Hilbert cube I^{∞}

In this paper, Anderson introduced new and extremely important techniques using "partitions" or a "kind of triangulation" which have been developed by David C. Wilson and John J. Walsh. In fact, Anderson's ideas have led to some fundamental theorems concerning the existence of various kinds of mappings between continua.

1956 Eldon Dyer [27]

Suppose that each of (X,d) and (Y,p) is a compact metric space and that $f : X \to Y$ is a monotone mapping such that for each $x \in X$, $f^{-1}f(x)$ is a nondegenerate Peano continuum. Furthermore, f maps no "small" simple closed curve to a point. Then $\dim X = \dim Y + 1$.

1956 R. D. Anderson [9]

It was *announced* by Anderson that there is a monotone open mapping f of the (Menger) universal 1-dimensional Peano continuum U onto any given Peano continuum P such that for each $x \in U$, $f^{-1}f(x)$ is homeomorphic to U .

It was also *announced* [9] that there is a monotone open mapping of a compact connected m-manifold M^m onto an n-cell I^n where $m \geqq 3$ and $n \geqq 2$.

No proofs were given and no manuscript has ever been available.

1957 Keldys [50] .

There is a monotone mapping $f : I^3 \to I^k$, $k > 3$.

Contrast this result with the fact that monotone mappings on 2-manifolds do not raise dimension.

1958 R. D. Anderson

Conjecture: There is a light open mapping of U (universal 1-dimensional Peano continuum) onto I^n . (A mapping f is *light* [117] iff each $f^{-1}f(x)$ is totally disconnected.)

1969 D. C. Wilson (Ph.D. Thesis) [126]

(1) For each Peano continuum P , there is a monotone open mapping f of U onto P such that for each $x \in U$, $f^{-1}f(x)$ is homeomorphic to U .

(2) For each nondegenerate Peano continuum P , there is a light open mapping f of U onto P such that $f^{-1}f(x)$ is homeomorphic to the Cantor set for each $x \in U$.

Wilson's work marked the beginning of new activity in this area. The earlier ideas of Anderson were thoroughly understood and developed further by Wilson. His work is deep and difficult.

Anderson's failure to publish his ideas appears to have delayed work in this area by some fifteen to twenty years.

VII. MORE HISTORY AND PROBLEMS

An ambitious paper by Samuel Eilenberg entitled "On the Problems in Topology" was published in the *Annals* in 1949 [35]. Problems numbered

41 and 42 concerned light open mappings and are listed below:

Problem 41. Suppose that each of M^n and M^k are manifolds of dimension n and k, respectively. Does there exist a monotone open mapping of M^n onto M^k where $k > n$?

Problem 42. Does there exist a light open mapping f of an n-manifold M^n onto a metric space (Y,p) such that $f^{-1}f(x)$ is homeomorphic to the Cantor set for *some* $x \in M^n$?

A somewhat related conjecture is the following:

1950 G. T. Whyburn [119]

Conjecture. Suppose that f is a light open mapping of an n-cell I^n onto I^n such that $f(\mathrm{Bd} I^n) = \mathrm{Bd} I^n$, $f(\mathrm{Int}\ I^n) = \mathrm{Int}\ I^n$, and $f|\mathrm{Bd}\ I^n$ is one-to-one. Then f is a homeomorphism. (Here, Bd = boundary and Int = interior.)

Whyburn proved the conjecture for $n = 2$ (the case $n = 1$ is obvious).

1960's L. F. McAuley [66]

Various partial results were obtained by me and others. Perhaps, the best of these is the following [130].

Let M^m be a compact connected m-manifold with boundary. Let f be a discrete $(f^{-1}f(x)$ has no limit point for each $x \in M^n)$ open mapping of M^m onto M^m such that $f(\mathrm{Bd}\ M^m) = \mathrm{Bd}\ M^m$ and $f(\mathrm{Int}\ M^m) = \mathrm{Int}\ M^m$. If $f|\mathrm{Bd}\ M^m$ is a homeomorphism then f is a homeomorphism.

The proof follows easily from the work in [66] and [95].

The paper by Eilenberg [35] was updated some six years later by W. S. Massey in a paper entitled "Some Problems in Algebraic Topology and the Theory of Fibre Bundles" which appeared in the *Annals* in 1955 [61].

1955 No progress reported on Problems numbered 41 and 42.

1971 A. V. Bondar [22]

It was claimed that if f is a light open mapping of M^n onto M^k (manifolds without boundary) then $f^{-1}f(x)$ is discrete for each $x \in M^n$.

This seemed to confirm a claim by Stoilow [91] in which a mistake was found. The claim would also answer Problem 42.

1971 D. C. Wilson [128]

It was proved that if M^3 is any compact connected triangulated 3-manifold and $m \geqq 3$, then there is a monotone open mapping $f : M^3 \to I^m$.

Thus, Wilson answered the question in Problem 41 originally listed by Eilenberg in 1949 (although raised earlier).

1971 D. C. Wilson [130]

If M^3 is any compact connected triangulated 3-manifold and $m \geqq 3$, then there is a light open mapping $f : M^3 \to I^m$ such that for each $x \in M^3$, $f^{-1}f(x)$ is homeomorphic to the Cantor set.

At last, Wilson gave a *correct* answer to Problem 42 which contradicts the earlier claims by Bondar and Stoilow!

1971 D. C. Wilson [130]

In addition to the results above, Wilson constructed a counterexample to the Whyburn conjecture for $n \geqq 3$.

I would like to take this opportunity to make a few remarks. At one time, Problems 41 and 42 were considered to be interesting, important, and *clearly extremely difficult*. Various outstanding mathematicians including members of the National Academy of Sciences attacked these problems and *gave up in failure*. Now after being ignored for more than a decade, an unknown new Ph.D. student comes along and solves these problems. What happens? Almost no one recognized this remarkable achievement! What would have happened if he had been born earlier and had solved these problems in 1950? He can not get an NSF grant today!! This is an incredible injustice. It is a fickleness that one would not expect from mathematicians.

VIII. CONSTRUCTION AND EXISTANCE OF CERTAIN MAPPINGS

In the late sixties and early seventies, I raised the following question with students. Among these students was John J. Walsh.

Question. When does there exist a $\left\{\begin{array}{l}\text{monotone}\\ \text{monotone open}\\ \text{light open}\end{array}\right\}$ mapping $f : X \to Y$

where each of (X,d) and (Y,p) is a metric space. In particular, consider the nice compact metric spaces: manifolds, Peano continua, etc.

There should be some way to decide whether such mappings exist. In particular, when do such mappings exist with $f^{-1}f(x)$ homeomorphic to a continuum K for each $x \in X$ where K is specified. There should be a set of instructions for producing the desired mappings (when they exist). In the case of continuous mappings, there are some nice theorems and characterizations.

1973 John J. Walsh [100]

Walsh began a systematic study of the existence of monotone, open, monotone open, and light open mappings from manifolds onto polyhedra. He wanted to know: When is a mapping "homotopic to" or "can be approximated by" one of these *four types* of mappings?

THEOREM (Walsh). *A mapping* f *from a compact connected* PL *manifold* $M^m (m \geqq 3)$ *to a connected polyhedron* Q *is homotopic to a monotone mapping* $g : M \to Q$ *if and only if* $f_* : \pi_1(M^m) \to \pi_1(Q)$ *is surjective.*

The "only if" part was proved by S. Smale in 1957.

Walsh began to use tools from algebraic and geometric topology to obtain new and interesting results.

1975 John J. Walsh [101]

Walsh obtained general criteria for the existence of *open* mappings and used these to show that *monotone* mappings can be *approximated* by *monotone open* mappings.

1976 John J. Walsh [103]

The result in Walsh's thesis mentioned above was extended by replacing the polyhedron Q by a connected ANR. In addition, M^m can be replaced by any compact connected Hilbert Cube manifold.

1976 John J. Walsh [102]

A mapping from a compact connected PL manifold $M^m (m \geqq 3)$ to a compact connected polyhdron Q is homotopic to an *open* mapping if and only if the index of $f_*(\pi_1(M^m))$ in $\pi_1(Q)$ is finite.

Later, Walsh replaced Q by a connected ANR.

IX. THE GENERAL PROBLEM AGAIN

Let us return to the general problem discussed in §4.

When do there exist monotone mappings (Y,p) with all point inverses nondegenerate?

1979 John J. Walsh [105]

For $n \geqq 3$, a mapping $f : M^n \to Y$ can be approximated by an open mapping $f : M^n \to Y$ iff f is quasi-monotone. Furthermore, g exists with $g^{-1}g(x)$ nondegenerate for *each* $x \in M^n$ such that shape $g^{-1}g(x) =$ shape $f^{-1}f(x)$. In fact, for a closed subset A in Y, shape $g^{-1}(A) =$ shape $f^{-1}(A)$.

It follows that a homeomorphism (or identity map) on M^n can be approximated by an open monotone mapping g on M^n with $g^{-1}g(x)$ *nondegenerate and point-like* for each $x \in M^n$. Compare this remarkable result with others in §4!

X. MONOTONE MAPPINGS AND ORBIT MAPPINGS

I raised the following question with students in the early and mid-seventies. In particular, I raised it with Ronald Fintushel.

Question What open mappings $f : X \to Y$ are equivalent to the orbit mapping ϕ of the action of some group of homeomorphisms on X ?

1975 Ronald Fintushel Ph.D. Thesis [36]

Suppose that each of M^3 and N^2 are PL manifolds of the indicated dimensions, $f : M^3 \to N^2$ is a PL open mapping, and for each $x \in M^3$, $f^{-1}f(x)$ is either a point or homeomorphic to S^1. Then f is equivalent to the orbit mapping of a local S^1 action on M^3.

1976 Fintushel and Edmonds [33]

The result above from Fintushel's thesis was extended to separable metric spaces (X,d) where f is a completely regular mapping on the complement of $F = \{x | f^{-1}f(x) = x$ for $x \in X\}$

XI. REGULAR MAPPINGS

The concept of regular convergence was introduced by G. T. Whyburn in 1936 [114]. This idea is of fundamental importance in the theory of closed monotone mappings. In my opinion, the idea has its roots in the work of R. L. Moore [76;77] and H. Whitney [113]. However, it was Whyburn who identified and formalized the concept into a most useful tool. There is a voluminous literature which is crucial to the theory. Various milestones are due to Arnold [12], Begle [15], and White [108;109;110; 111;112].

A paper of outstanding significance is the one by Dyer [29] entitled "Regular Mappings and Dimension," which appeared in the *Annals* in 1958. In my opinion, this paper has not been fully understood by almost all of us. We have yet to realize its full significance and impact on the theory.

A number of important concepts have developed from Whyburn's idea of regular convergence. These include (in addition to the regular mappings and n-regular mappings in the references above) completely regular mappings (Dyer and Hamstrom [30]), strongly regular mappings (Addis [1;2], and pseudo-regular mappings (Ungar [93]).

Applications have been made to fiber spaces by me, P. Tulley (McAuley), and my students, in particular.

Robinson and I have continued the development of "convergence" concepts related to regular convergence in a paper" On Inverse Convergence of Sets, Inverse Limits, and Homotopy Regularity."

XII. EXTENDING MONOTONE DECOMPOSITIONS OF MANIFOLDS

The problem of prescribing certain "elements" of a possible decomposition of a metric space (X,d) and considering whether these elements are, indeed, members of *some* usc decomposition G of X was first raised by R. H. Bing [19].

Problem. Suppose that C_1 and C_2 are two disjoint linking circles in $E^3(S^3)$. Is there a monotone usc decomposition G of $E^3(S^3)$ such that $C_i \in G$, $i = 1, 2$, and E^3/G is homeomorphic to E^3? That is, does there exist a closed monotone mapping $f : E^3(S^3) \to E^3(S^3)$ such that C_i is a point inverse under f for $i = 1, 2$?

Bing answered this question in the affirmative with an example [19]. Consider the following theorem which has evolved as indicated below.

THEOREM. *Suppose that* X *is a proper closed subset of a compact connected p.l. manifold* M^m, $m > 2$, *such that for each open connected subset* $U \subset M^m$ *either* $U - (X \cap U)$ *is connected or* $X \cap \mathrm{Bd}U = \emptyset$. *Then there is a monotone open mapping* f *of* M^m *onto the m-sphere* S^m *with each component of* X *being a point inverse under* f.

1970 R. J. Bean [14]

Bean proved this theorem for $M^m = S^3$,

1970 R. H. Bing [21]

Bing generalized Bean's result to M^m.

1974 D. Coram [24]

The result was proved for $m \geqq 4$.

1979 John J. Walsh [106]

Walsh generalized the theorem above in two ways as follows:

With M^m and X as in the hypothesis of the theorem above, let P be a simply connected and connected polyhedron with $\dim P \geqq 3$. Then there exists a closed monotone mapping f from M^m onto P with each component of X being a point inverse of f . In case M^m is oriented and P is an m-sphere S^m , there exists such a monotone mapping of *each* degree.

For related results, see also the paper [104] by Walsh.

The maps by Bing, Bean, and Coram are "almost" PL. and have degree ± 1 .

Recall that closed monotone mappings between ANR's induce surjections between fundamental groups [101;102]. In fact, from Walsh's work, for mappings f from M^m . $m \geqq 3$, onto ANR's this is the only obstruction to being homotopic to monotone mappings.

The main tool used here is in the paper [104] by Walsh. Roughly stated, his theorem is that if a mapping from the boundary of an m-manifold, $m \geqq 3$, to a simply connected ANR has an extension to the entire manifold, then it has an extension which is monotone.

Earlier results of a related nature are due to Floyd and Fort.

1953 Floyd and Fort [38]

Suppose that a 2-sphere S^2 is the boundary of a 3-cell B^3 . A mapping f of S^2 onto S^2 is monotone iff f has a continuous extension to the interior of B^3 which maps Int B^3 homeomorphically onto itself.

The space of all homeomorphisms of S^2 onto S^2 is dense in the space of all monotone mappings of S^2 onto S^2 . Each monotone mapping of S^2 onto S^2 is a cellular mapping.

1970 W. E. Haver [45;46]

A mapping f of $S^n = \mathrm{Bd}\, B^{n+1}$, $n \neq 4$, onto itself is cellular iff f has a continuous extension which maps $\mathrm{Int}\, B^{n+1}$ homeomorphically onto itself.

This is a generalization of the Floyd-Fort result given above. The converse problem of extending homeomorphisms on the $\mathrm{Int}\, B^n$, $n = 2$, 3, to monotone and nonalternating mappings on S^1 and S^2 was studied by Floyd.

One of my students, John Kavanagh, has generalized the results of Floyd to higher dimensions. He is studying mappings and extensions related to a general problem that can be formulated.

XIII. REMARKS

I shall not have time to discuss the extremely important work concerning spaces of mappings (homeomorphisms, monotone, cellular, cell-like, PL. monotone, etc.). A number of researchers have made outstanding contributions. These include Hamstrom, Haver, Siebenmann, and others.

Also, the problems concerning whether a closed monotone mapping on manifolds (and other spaces) is compact (proper) must be passed over. Such a question was originally raised by Whyburn [120] in 1959. Contributions have been made by Bing, K. Whyburn, Glaser, Martin, and Wilson [128].

XIV. n-MONOTONE MAPPINGS

I would like to mention the important work of Wilder which has not (in my opinion) received the attention that it deserves. A mapping $f : A \to B$ is called *n-monotone* iff $H^r(f^{-1}(b)) = 0$ for each $b \in B$

and $r \leqq n$.

1957; 1958 R. L. Wilder [124;125]

Let S be an orientable n-gcm and $f : S \to S'$ be an (n-1)-monotone (over the integers) mapping where S' is a finite dimensional nondegenerate Hausdorff space. Then S' is an orientable n-gcm of the same homology type as S .

XV. CONCLUDING REMARKS

I regret that I have not been able to cover some of the new work in the area of monotone usc decompositions of E^3 and generalizations to E^n . The seventies have brought a tremendous revival of interest and activity in this area, particularly with the *application of usc decomposition theory* in the proof of the double suspension theorems by R. Edwards, J. W. Cannon, and Giffen.

In this talk, I have tried to restrict the discussion to the more general aspects of monotone mappings rather than to confine it to Euclidean spaces.

It is an exciting period for the theory of (closed) monotone mappings. I have not mentioned the important new work of John Walsh, Kozlowski, and others [52;53;54;55;56] because we shall have an opportunity to hear John tell us about it in his lecture.

Again, I repeat that we are not able to cover either here or in this conference some of the most interesting new work in monotone usc decompositions (closed mappings) of E^n , $n \geqq 3$. The list of researchers who are making great strides is long and includes Armentrout, Bing, J. W. Cannon, Daverman, Eaton, Edwards, Everett, Glaser, Lacher, McMillan, Miller, Row, Siebenmann, Starbird, Woodruff, and many others (who are no less important).

REFERENCES

1. Addis, David, *Generalizations of completely regular mappings and lifting spaces of maps through a light map*, PhD Dissertation, Rutgers University, 1970.

2. ________, "A strong regularity condition of mappings", *General Topology and App.* *2*(1972), 199-213.

3. Anderson, R. D., "On collections of pseudo-arcs", Abstract 337t, *Bull. Amer. Math. Soc.* *56*(1950), 350.

4. ________, "Monotone interior dimension raising mappings", *Duke Math Jour.* *19*(1952), 359-366.

5. ________,"On monotone interior mappings in the plane", *Trans. Amer. Math. Soc.* *73*(1952), 211-222.

6. ________, "Continuous collections of continuous curves in the plane", *Proce. Amer. Math. Soc.* *3*(1952), 647-657.

7. Anderson, R. D. and M. E. Hamstrom, "On spaces filled up by continuous collections of atriodic continuous curves", *Proc . Amer. Math. Soc.* *6* (1955), 766-769.

8. Anderson, R. D., "On continuous curves admitting monotone open maps onto all locally connected metric continua" (abstract), *Bull. Amer. Math. Soc* *62*(1956), 264-265.

9. ________, "Open mappings of compact continua", *Proc. Nat. Acad. Sci, USA* *42*(1956), 347-349.

10. Armentrout, Steve, "Monotone decompositions of E^3", *Annals of Math Studies No.* *60*(1965), 1-25.

11. ________, "Cellular decompositions of 3-manifolds that yield 3-manifolds", *Memoirs Amer. Math. Soc. No.* *107*(1971).

12. Arnold, H. A., "On regular convergence of sets", *Bull. Amer. Math. Soc. Abstract*, 47-7-326.

13. Bean, Ralph J., "Decomposition of E^3 which yields E^3", *Pac. Jour. Math.* *20*(1967), 411-413.

14. ________, "Repairing embeddings and decompositions in S^3", *Duke Math Jour.* *36*(1970), 379-385.

15. Begle, E. G., "Regular convergence", *Duke Math. Jour.* *11*(1944), 441-450.

16. Bing, R. H., "A decomposition of E^3 into points and tame arcs such that the decomposition space is topologically different from E^3 ", *Annals of Math.* *65*(1957), 484-500.

17. ________, "Upper semi-continuous decomposition of E^3", *Annals of Math* *65*(1957), 363-374.

18. ________, "The cartesian product of a certain non-manifold and a

line is E^4", *Annals of Math.* *70*(1959), 399-412.

19. Bing, R. H.,"Decompositions of E^3", *Topology of 3-Manifolds and Related Topics*, edited by Fort, Prentice-Hall, 1962, 5-21.

20. ___________, "Point-like decompositions of E^3", *Fund. Math.* *50*(1962), 437-563.

21. ___________, "Extending monotone decompositions of 3-manifolds", *Trans. Amer. Math. Soc.* *149*(1970), 351-369.

22. Bondar, A. V., "A solution to a problem of Stoilow in Metric Problems in the Theory of Functions and Mappings",III, Kiev, *Ukranian Acad. Sci.*, (1971), 12-30 (Russian).

23. Cannon, J. W., "$\Sigma^2 H^3 = S^5/G$", *Rocky Mt. Jour. Math* *8*(1978), 527-532.

24. Coram, D., "Semi-cellularity, decompositons and mappings on manifolds", *Trans. Amer. Math. Soc.* *191*(1974), 227-244.

25. Dyer, Eldon, "Continuous collections of continua", *Duke Math. Jour.* *20* (1953), 589-592.

26. ___________, "Continuous collections of decomposable continua on a spherical surface", *Proc. Amer. Math. Soc.* *6*(1955), 351-360.

27. ___________, "Certain transformations which lower dimension", *Annals of Math.* *63*(1956), 15-19.

28. ___________, "Irreducibility of the sum of the elements of a continuous collection of continua", *Duke Jour. Math.* *20*(1953), 589-592.

29. ___________, "Regular mappings and dimension", *Annals of Math.* *67*(1958), 119-149.

30. Dyer, E. and M. E. Hamstrom, "Completely regular mappings", *Fund. Math.* *45*(1957), 103-118.

31. Eaton, W. T., "Application of a mismatch theorem to decomposition spaces", *Fund. Math.* *89*(1975), 199-224.

32. Edmonds, A. L. and R. Fintushel, "Singular circle fiberings", *Math. Zeitschrift* *151*(1976), 86-99.

33. ______________________________, "Topological singular circle fiberings", *Duke Math. Jour.* *45*(1978), 619-635.

34. Edwards, R. D. and Leslie C. Glaser, "A method for shrinking decompositions of certain manifolds", *Trans. Amer. Math. Soc.* *165*(1972), 45-56.

35. Eilenberg, Samuel, "On the problems of topology", *Annals of Math. Vol.* *50*(1949), 247-260.

36. Fintushel, Ronald, *Orbit Maps Of Local S^1-actions On Manifolds Of Dimension Less Than Five*, PhD Thesis, SUNY Binghamton, May, 1975.

37. Fitzgerald, R. W. and P. M. Swingle, "Core decompositions of continua", *Fund. Math. 61*(1967), 33-50.

38. Floyd, E. E. and M. K. Fort, "A characterization theorem for monotone mappings", *Proc. Amer. Math. Soc. 4*(1953), 828-830.

39. Fort, M. K., Jr., "A note concerning a decomposition space defined by Bing", *Annals of Math. 65*(1957), 501-504.

40. Gordh, G. R., Jr., "Monotone decompositions of irreducible Hausdorff continua", *Pac. Jour. Math. 36*(1971), 647-658.

41. Hamstrom, M. E., "Concerning continuous collections of continuous curves", *Bull. Amer. Math. Soc. 58*(1952), 204.

42. ____________, "Regular mappings whose inverses are 3-cells", *Amer. Jour. Math. 82*(1960), 393-429.

43. ____________, "Regular mappings and the space of homeomorphisms on a 3-manifold", *Memoirs of the Amer. Math. Soc. 40*(1961).

44. ____________, "Completely regular mappings whose inverses have LC^0 homeomorphism group: a correction", *Proceedings Conference on Monotone Mappings and Open Mappings*, 1970, SUNY at Binghamton, 255-260.

45. Haver, W. E., *Cellular Mappings On Manifolds*, PhD Dissertation, State University of New York at Binghamton, 1970.

46. ____________, "A characterization theorem for cellular maps", *Bull. Amer. Math. Soc. 76*(1970), 1277-1280.

47. Hodel, R. E., "Open functions and dimension", *Duke Jour. Math. 30* (1963), 461-467.

48. Jones, S. L., "The impossibility of filling E^n with arcs", *Bull. Amer. Math. Soc. 74*(1968), 155-159.

49. ____________, "Continuous collections of compact manifolds", *Duke Math. Jour. 37*(1970), 579-587.

50. Keldys, L. V., "Transformation of a monotone irreducible mapping into a monotone-interior mapping of the cube onto the cube of higher dimension", (Russian), *Dokl. Akad. Nauk SSSR (N.S.),114*(1957), 472-475.

51. Knaster, B., "Un continu irreductible a decomposition continue en tranches", *Fund. Math. 25*(1935), 568-577.

52. Kozlowski, G. and J. J. Walsh, "Cell-like mappings on 3-manifolds", to appear.

53. __________________, "The cell-like mapping problem", to appear in *Bull. Amer. Math. Soc.*

54. __________________, "The finite dimensionality of cell-like images of 3-manifolds", to appear.

55. Kozlowski, G., W. H. Row and J. J. Walsh, "Cell-like mappings with 1-dimensional fibers on 3-dimensional polyhedra", to appear.

56. Kozlowski, G., J. Van Mill, and J. J. Walsh, "AR-maps obtained from cell-like maps", to appear.

57. Lacher, R. C., "Cell-like mappings I", *Pac. Jour. Math.* *30*(1969), 717-731.

58. ____________, "Cell-like mappings and their generalizations", *Bull. Amer. Math. Soc.* *83*(1977), 495-552.

59. Lewis, Wayne, and J. J. Walsh, "A continuous decomposition of the plane into pseudo-arcs", *Houston Jour. Math.* *4*(1978), 209-222.

60. Martin, V., "Monotone transformations of non-compact two-dimensional manifolds", *Duke Math. Jour.* *8*(1941), 136-153.

61. Massey, W. S., "Some problems in algebraic topology and the theory of Fibre Bundles", *Annals of Math. Vol.* *62*(1955), 327-359.

62. McAuley, L. F., *Decompositions Of Continua Into Aposyndetic Continua,* PhD Thesis, University of North Carolina, 1954.

63. ______________, "On decomposition of continua into aposyndetic continua", *Trans. Amer. Math. Soc.* *81*(1956), 74-91.

64. ______________, "An atomic decomposition of continua into aposyndetic continua", *Trans. Amer. Math. Soc.* *88*(1958), 1-11.

65. ______________, "Some upper semi-continuous decompositions of E^3 into E^3", *Annals of Math.* *73*(1961), 437-457.

66. ______________, "Condition under which light open mappings are homeomorphisms", *Duke Math. Jour.* *33*(1966), 445-452.

67. ______________, "Open mappings and open problems", *Proceedings of the Conference on Point Set Topology,* Arizona State University, March, 1967.

68. ______________, "More about open mappings and open problems", *Topology Conference,* Auburn University, 1969, 57-70.

69. ______________, "Some fundamental theorems and problems related to monotone mappings", *Proc. Conf. on Monotone Mappings and Open Mappings,* SUNY-Binghamton, 1970, 1-36.

70. ______________, Editor, *Proc. Conf. On Monotone Mappings and Open Mappings,* SUNY-Binghamton, 1970, 444 pages.

71. McAuley, L. F. and E. E. Robinson, "On inverse convergence of sets, inverse limits, and homotopy regularity", submitted.

72. Moise, E. E., "A monotonic mapping theorem for simply connected 3-manifolds", *Ill. Jour. of Math.* *12*(1968), 451-474.

73. ____________, "A theorem on monotone interior transformations", *Bull. Amer. Math. Soc.* *55*(1949), 810-811.

74. Moore, R. L., "Concerning upper semi-continuous collections of continua", *Trans. Amer. Math. Soc.* *27*(1925), 416-428.

75. Moore, R. L., "Concerning upper semi-continuous collections", *Monatshefte für Math. und Physik 36*(1929), 81-88.

76. __________, "Concerning certain equicontinuous systems of curves", *Trans. Amer. Math. Soc. 22*(1921), 41-55.

77. __________, "On the generation of a simple surface by means of a set of equicontinuous curves", *Fund. Math. 4*(1923), 106-117.

78. __________, "Foundations of Point Set Theory", *Amer. Math. Soc. Colloq. Pub. XIII,* 1932.

79. Morrey, C. B., Jr., "The topology of (path) surfaces", *Amer. Jour. Math 57*(1935), 17-50.

80. Puckett, W. T., Jr., "Regular transformations", *Duke Math. Jour. 6* (1940), 80-88.

81. __________, "On 0-regular surface transformations", *Trans. Amer. Math. Soc. 47*(1940), 95-113.

82. Roberts, J. H., "On a problem of C. Kuratowski concerning upper semi-continuous collections", *Fund. Math. 14*(1929), 96-102.

83. __________, Abstract No. 196, *Bull. Amer. Math. Soc. 41*(1935), 330.

84. __________, "Collections filling a plane", *Duke Math. Jour. 2* (1935), 10-19.

85. Roberts, J. H. and N. E. Steenrod, "Monotone transformations of 2-dimensional manifolds", *Annals of Math. (2)*39(1938), 851-862.

86. Robinson, Eric, *Characterizations And Properties Of Some Light-open Mappings,* PhD Theses, SUNY Binghamton, May, 1975.

87. Rozanskaya, Yu A., "Open mappings and dimension", (Russian) *Uspehi Matematiceskih Nauk N.S. 4*(1949), 178-179.

88. Seidman, S. B., *Completely Regular Mappings,* PhD Dissertation, University of Michigan, 1969.

89. __________, "Completely regular mappings with locally compact fibers", *Trans. Amer. Math. Soc. 147*(1970), 1-11.

90. Sosinskii, A. B., "Monotonically open mappings of a sphere", *Amer. Math. Soc. Translations (Series 2) Vol. 78*(1968), 67-101.

91. Stoïlow, S., "Sur les transformations continues et la topologies des fonctions analytiques", *Ann. Sci. Ecole Norm Sup., 45*(1928), 347-382.

92. Thomas, E. S., "Monotone decompositions of irreducible continua", *Rozprawy Matematyczne 50,* Warszawa, 1966.

93. Ungar, Gerald S., "Pseudoregular mappings", *Colloquium Mathematicum XIX*(1968), 225-229.

94. Ungar, Gerald S., "Completely regular maps, fiber maps, and local n-connectivity", *Proc. Amer. Math. Soc. 21*(1969), 549-553.

95. Vaisala, J., "Discrete open mappings on manifolds", *Ann. Acad. Sci. Fenn. (A), I, Math 392*(1966), 3-9.

96. Walker, J. R., *Monotone Mappings*, PhD Thesis, Syracuse University, 1966.

97. Wallace, A. D., "Monotone transformations", *Duke Math Jour. 8*(1942), 487-506.

98. ______________, "On 0-regular transformations", *Amer. Jour. Math. 62* (1940), 277-284.

99. Walsh, John J., "Fiber preserving cellular decompositions", *Bull. Amer. Math. Soc. 78*(1972), 746-748.

100. ______________, *Monotone, Monotone Open, And Light Open Mappings On Manifolds*, PhD Dissertation, SUNY Binghamton, 1973.

101. ______________, "Monotone and open mappings on manifolds I", *Trans. Amer. Math. Soc. 209*(1975), 419-432.

102. ______________, "Light open and open mappings on manifolds II", *Trans. Amer. Math. Soc. 217*(1976), 271-284.

103. ______________, "Monotone and open mappings onto ANR's", *Proc. Amer. Math. Soc. 60*(1976), 286-289.

104. ______________, "Extending mappings to monotone mappings", *Houston Jour. Math. 3*(1977), 579-592.

105. ______________, "Isotoping mappings to open mappings", *Trans. Amer. Math. Soc. 250*(1979), 121-145.

106. ______________, "Extending monotone decompositions of manifolds", *Proc. Amer. Math. Soc. 74*(1979), 197-201.

107. ______________, "A general method for constructing UV^k - mappings on manifolds with applications to spheres", to appear.

108. White, P. A., "On r-regular convergence", Abstract *Bull. Amer. Math. Soc. 49*(1943), 699.

109. ______________, "r-regular convergence of spaces", *Am. Jour. Math. 66* (1944), 69-96.

110. ______________, "Regular convergence in terms of Cech cycles", *Annals of Math. 55*(1952), 420-432.

111. ______________, "Regular convergence of manifolds with boundary", *Proc. Amer. Math. Soc. 4*(1953), 482-485.

112. ______________, "Regular convergence", *Bull. Amer. Math. Soc. 60*(1954), 431-443.

113. Whitney, H., "Regular families of curves I", *Proc. Nat'l. Acad. of Sci, USA 18*(1932), 275-278.

114. Whyburn, G. T., "On sequences and limiting sets", *Fund. Math. 25*(1935), 408-426.

115. ______________, "Regular convergence and monotone transformations", *Am. Jour. Math. 57*(1935), 902-906.

116. ______________, "On sequences and limiting sets", *Fund. Math. 25*(1935), 408-426.

117. ______________, "Analytic Topology", *Amer. Math. Soc. Colloquium Publications Vol. 28*(1942).

118. ______________, "Boundary alternations of monotone mappings", *Duke Math. Jour. 12*(1945), 663-667.

119. ______________,"An open mapping approach to Hurwitz's Theorem", *Trans. Amer. Math. Soc. 71*(1951), 113-119.

120. ______________, "Compactness of certain mappings", *Amer. Jour. Math. 81*(1959), 306-314.

121. ______________, "Monotoneity of limit mappings", *Duke Math. Jour. 29* (1962), 465-470.

122. ______________, "Generic and related mappings", *Bull. Amer. Math. Soc. 69*(1963), 757-761.

123. ______________, "Dynamic Topology", *Math. Monthly 77*(1970), 556-570.

124. Wilder, R. L., "Monotone mappings of manifolds", *Pacific Jour. of Math. 7*(1957), 1519-1527.

125. ______________, "Monotone mappings of manifolds II", *Mich. Jour. of Math. 5*(1958), 19-23.

126. Wilson, David C., *Monotone Open And Light Open Dimension Raising Mappings*, PhD Thesis, Rutgers University, 1969.

127. ______________, "Completely regular mappings and dimension", *Bull. Amer. Math. Soc. 76*(1970), 1057-1061.

128. Wilson, David C., "Monotone mappings of manifolds onto cells", *Proceedings Conference on Monotone vappings and Open Mappings*, SUNY Binghamton, 1970, 37-54.

129. ______________, "Open mappings of the universal curve onto continuous curves", *Trans. Amer. Math. Soc. Vol. 168*(1972), 497-515.

130. ______________, "Open mappings on manifolds and a counterexample to the Whyburn Conjecture", *Duke Math. Jour. Vol. 40*(1973), 705-716.

131. Woodruff. E. P., "Decomposition spaces having arbitrarily small neighborhoods with 2-sphere boundaries", *Trans. Amer. Math. Soc. 232* (1977), 195-204.

CYCLIC CONNECTEDNESS THEOREMS

David P. Bellamy[1]
University of Delaware, Newark

Lewis Lum[2]
Salem College, Winston-Salem, NC

In his Ph.D. dissertation R. L. Wilson [8] proved that no uniquely arcwise connected Hausdorff continuum supports the structure of a topological group. Since he used mainly algebraic techniques, his proof revealed little about the topological nature of such continua. In private conversation with the authors, Wilson asked whether uniquely arcwise connected Hausdorff continua can be homogeneous. In this note we state two theorems; the latter of which provides a negative answer to Wilson's question for metric continua. A complete version of this paper will appear in the Transactions of the American Mathematical Society.

A continuum X is *cyclicly connected* [6] provided each pair of points in X lie together on some simple closed curve in X. In [6] G. T. Whyburn proved that a locally connected plane continuum is cyclicly connected if and only if it contains no separating points. This theorem was fundamental in his original treatment of cyclic element theory. Since then numerous authors have obtained extensions of Whyburn's theorem. W. L. Ayres

---

[1]The first author was supported during the latter stages of this research by NSF Grant MCS79-08413.

[2]This research was begun while the second author was a visitor at the University of Delaware in 1978-79. He wishes to thank the members of the Mathematical Sciences Dept there for their hospitality during that year.

ISBN 0-12-481820-X

[1] (see also [7] and [2]) proved the same theorem without the planar assumption. The absence of cut points was seen by F. B. Jones [4] to be sufficient for cyclic connectedness in nonseparating plane continua. And C. L. Hagopian [3] obtained a sufficient condition in aposyndetic plane continua.

Our first theorem characterizes cyclic connectedness in the class of all Hausdorff continua. In a Hausdorff continuum, an *arc* is a subcontinuum with exactly two nonseparating points and a simple closed curve is the union of two arcs meeting only at their nonseparating points.

Let x be a point in a Hausdorff continuum X. Then x is an *end point* if it is a nonseparating point of each arc in X containing x. And x is an *arc-cut point* if $X-\{x\}$ is not arcwise connected. Note that in locally connected metric continua separating points and arc-cut points are identical.

THEOREM. *A Hausdorff continuum* X *is cyclicly connected if and only if* X *is arcwise connected and no point in* X *is an arc-cut point or an end point.*

THEOREM. *Arcwise connected homogeneous metric continua contain no arc-cut points or end points. And hence, all such continua are cyclicly connected.*

We conclude with some related questions. Wilson's original question is still of interest.

(a) [Wilson] Does there exist a uniquely arcwise connected homogeneous Hausdorff continuum?

(b) Suppose X is an arcwise connected homogeneous metric continuum which is itself not a simple closed curve. Does X contain simple closed curves of arbitrarily small diameter?

In conversation with the authors, G. R. Gordh, Jr. raised the following question:

(c) [Gordh] Suppose X is an arcwise connected homogeneous metric continuum and *some* pair of its points separates X . Is X necessarily a simple closed curve?

Finally, we include a question which appears in [5].

(d) [K. Kuperberg] Are arcwise connected homogeneous continua locally connected?

REFERENCES

1. Ayres, W. L., "Concerning continuous curves in metric spaces," *Amer. J. Math 51*(1929), 577-594.

2. ______, "A new proof of the cyclic connectivity theorem," *Bull. Amer. Math. Soc. 48*(1942), 627-630.

3. Hagopian, C. L., "Concerning arcwise connectedness and the existence of simple closed curves in plane continua," *Trans. Amer. Math. Soc. 147* (1970), 389-402. [See *ibid. 157*(1971), 507-509.]

4. Jones, F. B., "The cyclic connectivity of plane continua," *Pacific J. Math 2*(1961), 1013-1016.

5. Kuperberg, K., "A locally connected micro-homogeneous nonhomogeneous continuum," *Bull. Acad. Polon. Sci.*, to appear.

6. Whyburn, G. T., "Cyclicly connected continuous curves," *Proc. Nat'l. Acad. Sci. USA 13*(1927), 31-38.

7. ______, "On the cyclic connectivity theorem," *Bull. Amer. Math. Soc. 37* (1931), 429-433.

8. Wilson, R. L., *Intrinsic Topologies On Partially Ordered Sets and Results On Compact Semigroups*, Ph.D. Dissertation, Univ. of Tennessee, 1978.

SOME REMARKS ON INTRINSIC GEOMETRY

Karol Borsuk

Mathematics Institute, Warsaw, Poland

A metric space X with a metric ρ is said to be geometrically acceptable (notation: $X \in GA$) provided for every two points $x,x' \in X$ there exists in X an arc with a finite length, joining x with x', and if for every $\varepsilon > 0$ there is a neighborhood U of x in X such that every point $x' \in U$ can be joined in X with x by an arc with length $< \varepsilon$. Setting

$$\rho_X(x,x') = \text{The lower bound of lengths of arcs joining in } X \text{ the points } x,x',$$

one gets another metric ρ_X for $X \in GA$, called the intrinsic metric in X. It is clear that the metric ρ_X induces in X the same topology as the given metric ρ. Evidently

$$\rho(x,x') \leqq \rho_X(x,x') \quad \text{for every two points } x,x' \in X .$$

A map f of the space $X \in GA$ onto another space $Y \in GA$ is said to be an intrinsic isometry, if

$$\rho_X(x,x') = \rho_Y(f(x),f(x')) \quad \text{for every } x,x' \in X .$$

By an intrinsic geometry one understands the theory of all properties which are preserved by all intrinsic isometries.

In the classical geometry of Riemann spaces, one considers the maps which preserve lengths of arcs, but the limitation to maps of Riemann spaces onto Riemann spaces implies that the class of all invariants is

ISBN 0-12-481820-X

different from the class of the invariants in the sense of the intrinsic geometry given here. For instance, a Riemannian surface lying in the euclidean 3-space E^3 isometric to the sphere S^2 is intrinsically isometric to S^2, but there exist in E^3 surfaces not isometric to S^2, but intrinsically isometric to S^2. In order to see this, assume that S^2 is the sphere with radius 1 and a is a point of S^2. Let A denote the subset of S^2 consisting of all points $x \in S^2$ whose distance from a is less than a given positive number $\alpha < 1$. Setting

$f(x)$ = identity for $x \notin A$

$f(x)$ = the point which is symmetric to x with respect to the plane containing the boundary of A for $x \in A$,

one gets an intrinsic isometry f mapping S^2 onto a subset of E^3 which is not isometric to S^2.

In the intrinsic geometry in the sense just given, one meets many new problems and also several theorems with a clear intuitive sense. In particular, one proves the following

THEOREM. *For every* $\varepsilon > 0$ *there exists an intrinsic isometry mapping the n-dimensional euclidean space* E^n *onto a subset* $f(E^n)$ *of the 2n-dimensional euclidean space* E^{2n} *such that the diameter of* $f(E^n)$ *(by the usual metric in* E^{2n}*) is less than* ε.

Let us add, that one can construct the isometry f so that it is of the class C^∞ (i.e., f has all partial derivatives of all orders).

The question whether the number $2n$ appearing in this theorem can be replaced by a smaller one, for instance by $n+1$, is an open question.

Among many open problems of the intrinsic geometry, let us mention the following one:

Problem. Is it true that for every n-dimensional GA-space X

there exists an intrinsic isometry f mapping X onto a subset of the space E^{2n+1} ?

A positive answer to this problem would constitute an analogue to the well known theorem of Menger and Nöbeling.

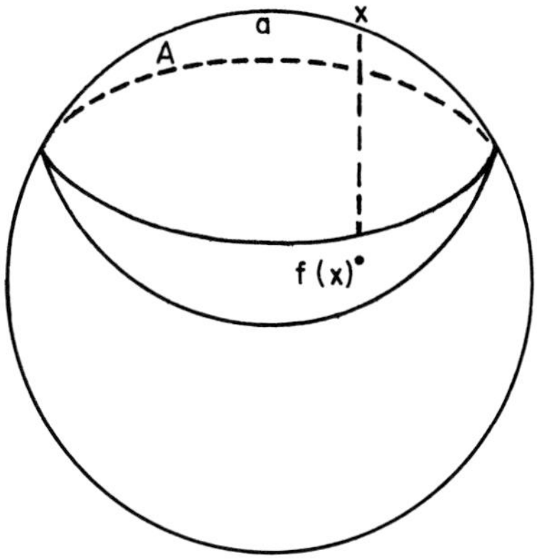

THE PRIME END STRUCTURE OF INDECOMPOSABLE CONTINUA AND THE FIXED POINT PROPERTY

Beverly Brechner
John C. Mayer

University of Florida, Gainesville

I. INTRODUCTION

The purpose of this paper is to present a new approach for studying fixed point problems in the plane; in particular, the well known *Scottish Book* problem #107, due to Sternbach, is the following: Do nonseparating plane continua have the fixed point property for continuous maps? for homeomorphisms? The answer is "yes" for extendable homeomorphisms by the Cartwright-Littlewood-Bell theorem [6 and 2]. We are concerned here with the more general case.

Now, it follows from another important theorem, due independently to Bell [1] and Sieklucki [14], that a nonseparating continuum which admits a fixed-point-free map must have a *(s-d-c) simple dense canal* (that is, a Lake-of-Wada type of channel) in its boundary *for every embedding into the plane*.

Thus we are basically interested in embedding problems, and we will distinguish among certain embeddings by the differences in their prime end structures.

In Section 2, we make some definitions, note some consequences of the Bell-Sieklucki theorem and its proofs, and relate this to prime ends. Included are some important theorems of prime end theory. In particular, we show that the existence of a s-d-c is equivalent to the existence of a

ISBN 0-12-481820-X

prime end E such that I(E) = P(E) = BdX . See Theorem 2.9.

In Section 3, we obtain embeddings of the "Three Point Continuum" (a continuum chainable alternately between the endpoints a and b , b and c , and c and a ; see p. 142 of [7]), both *with* and *without* simple dense canals. It follows that the pseudo arc may be embedded in the plane with a simple dense canal.

In Section 4, we show that Ingram's example of a nonchainable, tree-like, atriodic continuum [8] has the fixed point property.

Finally in Section 5, we partially answer some questions from [5]. Additional questions are raised in Sections 3, 4, and 5.

II. PRIME ENDS AND THE BELL-SIEKLUCKI THEOREM

We refer the reader to Section 2 of [4], for a summary of the basic definitions and theorems of prime end theory, as well as illustrative examples. Many references are also included there.

The important theorems of this section are the Embedding Corollary 2.5, the equivalence between the existence of a s-d-c and the existence of a prime end E such that I(E) = P(E) = BdX (Theorem 2.9), and Theorem 2.11, which says that equivalently embedded continua have identical prime end structures.

Theorem 2.11 has already been implicitly assumed in the literature (for example, in [4,9]), but is not explicitly stated anywhere (as far as we know).

NOTATION. A double arrow $\twoheadrightarrow$ means an onto map. The *impression* of the prime end E is denoted by I(E) , while the set of *principal points* is denoted by P(E) .

2.1 *DEFINITION* (Sieklucki [14]). Let X be a nonseparating continuum and let D be a set homeomorphic to [0,1) in S^2-X , where

$\alpha: D \twoheadrightarrow [0,1)$ is a given homeomorphism. Then D will be called a *simple canal in* X iff the following three conditions are satisfied:

(1) $\bar{D} - D \subseteq \mathrm{Bd}X$.

(2) For each $p \in D$, there is a "bridge" to X ; that is, a crosscut to X , which (crosscut) is transverse to D , and intersects D at exactly one point.

(3) If $p_i \to \infty$ (i.e., $\alpha(p_i) \to 1$), then there is a sequence of bridges $\{Q_{p_i}\}$ such that $Q_{p_i} \cap D = \{p_i\}$ and diam $Q_{p_i} \to 0$.

If, in addition, condition (4) holds, where (4) is:

(4) $\bar{D} - D = \mathrm{Bd}X$,

we call D a *simple dense canal (s-d-c)* .

2.2 LEMMA (Sielucki, Lemma 4.4 of [14]). *If* X *is a nonseparating plane continuum, and* D *is a s-d-c in* X , *then* BdX *is indecomposable.*

2.3 THEOREM (Bell [1] and Sieklucki [14, p. 270]). *Let* $X \subseteq E^2$ *be a nonseparating plane continuum, and let* $f: X \to E^2$ *be a map such that* $f(\mathrm{Bd}X) \subseteq X$ *and* f *is fixed-point-free. Then there exists a simple canal* D *in* X *such that* $f(\lim D) \subseteq \lim D$.

2.4 COROLLARY (Bell and Sieklucki). *Let* $f: X \to X$ *be a fixed point free map of* X *into itself, where* X *is a nonseparating continuum. Then (by Zorn's lemma) there exists a minimal invariant continuum* $M \subseteq \mathrm{Bd}X$. *It follows from Theorems 2.3 and 2.2 that* M *is indecomposable.*

We note that if X is one-dimensional and nonseparating, then we may assume that f is onto, since the intersection of a maximal tower of invariant subcontinua would have to be minimal invariant and nonseparating, so that we could always restrict our attention to that subcontinuum.

2.5 EMBEDDING COROLLARY. *Let* X *be a nonseparating plane continuum, and* $f: X \to X$ *a fixed-point-free map such that* $f(\mathrm{Bd}X) = \mathrm{Bd}X$ *and* BdX *is minimal invariant in* BdX . *Then there exists a s-d-c in* X *for every embedding of* X *into* E^2 .

Our embedding corollary leads us to make the following definition.

2.6 *DEFINITION.* Let X be an (abstract) nonseparating plane continuum, and let $\alpha : X \to E^2$ be an embedding. We will say that α is a *principal embedding* of X into E^2 and that $\alpha(X)$ is *principally embedded* iff there is a s-d-c in $\alpha(X)$. (If $X \subseteq E^2$ then X is *principally embedded* iff the identity is a principal embedding.) The continuum X is a *principal continuum* iff *every* embedding of X is principal. X will be called an *n-principal continuum* iff for every embedding α of X into E^2 , $\alpha(X)$ contains at least n s-d-c's, and at least one embedding has exactly n s-d-c's.

2.7 LEMMA. *A simple canal* D *is a ray (half open arc) defining a prime end* E , *such that each point of* lim D *is a principal point of* I(E) . *If* D *is a s-d-c, then* I(E) = P(E) = BdX .

Proof. Let $x \in \lim D$. Then there exists $\{x_i\}$ in D such that $x_i \to \infty$, $x_{i+1} > x_i$, and $\lim x_i = x$. Let L_i be a *bridge* from x_i to X such that diam $L_i \to 0$. That is, L_i is a crosscut of S^2-X , transverse to D , and intersecting D at exactly one point.

We show that $\{L_i\}$ is a chain of crosscuts defining a prime end E . $L_1 \cup X$ defines two complementary domains U_1 and V_1 in S^2 . Let U_1 be that complementary domain that contains $\{x_2, x_3, \ldots\}$. Let $\varepsilon_1 = d(x_1, \{x_2, x_3, \ldots\})$, and let $x_{n_1} = x'_2$ be the first point of the sequence such that there exists a bridge L_2 from x'_2 to X with diam $L_2 < \varepsilon_1$.

Let U_2 be that complementary domain of $L_2 \cup X$ which contains $\{x_{n_1+1}, x_{n_1+2}, \ldots\}$, and continue the above process inductively, obtaining a subsequence $\{x'_i\}$ of $\{x_i\}$ such that $x'_i \to \infty$ on D , $\lim x'_i = x$, and also obtaining a sequence $\{L_i\}$ such that L_i is a crosscut of S^2-X at

x_i' with $\{L_i\}$ a *chain* of crosscuts of S^2-X . Then $\{L_i\}$ defines some prime end E . Further, x was an arbitrary point of lim D and x ∈ P(E) .

If D is a s-d-c, then lim D = BdX . By the first part of the proof, lim D ⊆ P(E) . But (see [4, bottom p. 632]) P(E) ⊆ lim D for any ray D converging to E . Thus, lim D = P(E) ⊆ I(E) ⊆ BdX . The theorem follows.

2.8 LEMMA. *Let* X *be a nonseparating plane continuum, and let* E *be a prime end of* S^2-X *such that* I(E) = BdX = P(E) . *Then there exists a s-d-c* D *in* S^2-X .

Proof. Let $h: S^2-X \twoheadrightarrow \text{Ext } B$ be a C-map. We may assume that h is an extension of the inverse of the homeomorphism $h: S^1 \times [0,\infty) \twoheadrightarrow Q_0 - X$ obtained by Sieklucki in the third paragraph on page 270 of [14].

Let e ∈ BdB be the point corresponding to the prime end E . Let $h(D^*)$ be any half open arc with endpoint in $h(S^1 \times \{0\})$ and satisfying the following additional conditions:

(1) $\overline{h(D^*)}$ is an arc in the annulus $\overline{h(S^1 \times [0,\infty))}$ whose other endpoint is e , and

(2) $h(D^*)$ misses the straight line segments of the crosscut structure used in constructing h , and illustrated as the straight line segments in $\overline{h(S^1 \times [0,\infty))}$ of S^2-Int B in Figure 2 below.

Then D^* is a ray in S^2-X converging to the prime end E .

Now the retraction of Lemma 5.5(iv) of [14] is a deformation retraction, and takes some connected open set containing D^* onto an infinite tree which must contain a simple canal D . (This is the same argument used by Sieklucki in the proof of his Auxiliary Theorem. See [14, pp. 273-274] for details.) Since the retraction is defined on an open set which misses the straight line segments illustrated in Figure 1 below, it follows that D converges to the same prime end E that D^* converges to.

Thus D is also a ray in S^2-X converging to the prime end E , and satisfying conditions (1) and (2) of paragraph two above. Further, $P(E) \subseteq \lim D \subseteq I(E) \subseteq BdX$. But $P(E) = I(E) = BdX$. Thus $\bar{D}-D = BdX$ and the simple canal D is a s-d-c.

REMARK. Lemma 2.8 could be strengthened to show that there exists a simple canal D converging to any prime end E , and necessarily $P(E) = \lim D \subseteq I(E)$. The same argument works.

In Figures 1 and 2 below, $h:C(X)$ in Figure 1 $\rightarrow C(B)$ in Figure 2.

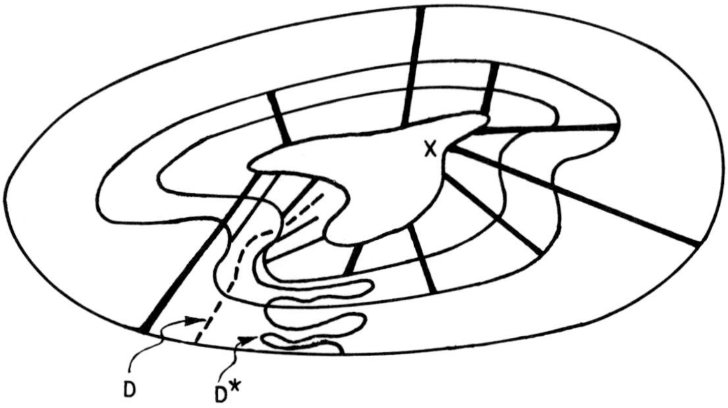

Fig. 1

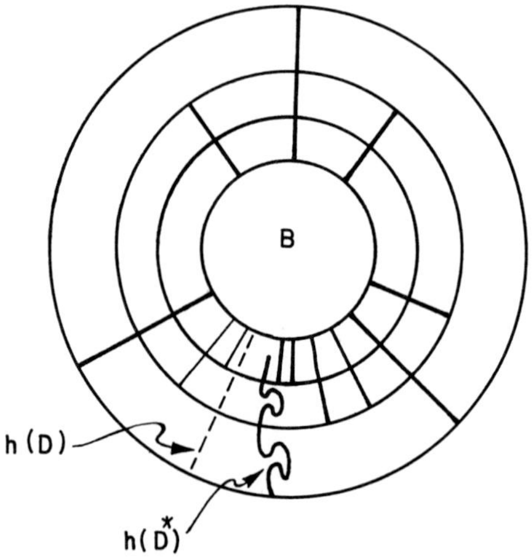

Fig. 2

2.9 THEOREM. *For any nonseparating continuum* $X \subseteq S^2$ *, the existence of a s-d-c in* X *(roughly, a Lake-of-Wada) is equivalent to the existence of a prime end* E *such that* $I(E) = \text{BdX} = P(E)$. *That is, the impression of* E *is all of the boundary, and each point of the impression is principal. (Hence the term principal continuum, above.) Further,* BdX *is indecomposable.*

Proof. This follows immediately from Lemmas 2.7, 2.8, and 2.2.

2.10 *DEFINITION.* Let $X, Y \subseteq S^2$ be nonseparating continua, and let $\phi: S^2 - X \twoheadrightarrow \text{Ext B}$ and $\psi: S^2 - Y \twoheadrightarrow \text{Ext B}$ be C-maps. We will say that the prime end structures of $S^2 - X$ and $S^2 - Y$ are *traditionally identical* iff there exists a homeomorphism $g: \text{BdB} \twoheadrightarrow \text{BdB}$ such that, if e corresponds to a prime end E of $S^2 - X$, and E is of the ith kind, then g(e) corresponds to a prime end E′ of $S^2 - Y$ and E′ is also of the ith kind, i = 1,2,3,4 . (See Collingwood and Lohwater, *The Theory of Cluster Sets*, Cambridge Univ. Press, 1966, for classification of prime ends.)

2.11 THEOREM. *Let* X, Y, ϕ , *and* ψ *be as in Definition 2.10. Let* $h: S^2 \twoheadrightarrow S^2$ *be a homeomorphism such that* $h(X) = Y$. *Then the prime end structures of* $S^2 - X$ *and* $S^2 - Y$ *are identical.*

Proof. We define a homeomorphism $g: \text{Ext B} \twoheadrightarrow \text{Ext B}$ by: $g(x) = \psi h \phi^{-1}(x)$ for $x \in \text{Ext B}$.

We show that g extends to a homeomorphism $\bar{g}: \overline{\text{Ext B}} \twoheadrightarrow \overline{\text{Ext B}}$. Let $e \in \text{BdB}$, and let $\{Q_i\}$ be a chain of crosscuts defining the prime end E of $S^2 - X$, corresponding to e . Now $\{h(Q_i)\}$ will be a chain of crosscuts of $S^2 - Y$, and thus defines a prime end E′ of $S^2 - Y$. But $\{\psi(h(Q_i))\}$ is then a chain of crosscuts of Ext B , converging to the point e′ corresponding to E′ . Let $\bar{g}(e) = e'$. Clearly $\bar{g}: \overline{\text{Ext B}} \twoheadrightarrow \overline{\text{Ext B}}$ is a homeomorphism, and takes points corresponding to prime ends of the ith kind to (other) points corresponding to prime ends of the same kind.

Thus $\bar{g}|BdB$ is the homeomorphism required for our theorem.

2.12 *REMARK*. The converse of this theorem is false. It is shown by Mayer in [11], that there exist uncountably many inequivalent embeddings of the $\sin \frac{1}{x}$ curve, M, into the plane, such that for exactly one prime end E, $I(E)$ = limit segment, $P(E)$ is a singleton (a "bad" endpoint), and $I(F)$ is a singleton for every other prime end F of S^2-M.

III. EMBEDDINGS AND s-d-c's

In view of the results of Section 2, the following questions are of interest: (1) Do there exist principal continua in the plane? (If not, then the fixed point property follows immediately.) Or equivalently, can every nonseparating plane continuum be (re-)embedded without a s-d-c?

(2) Can a chainable continuum be principal? (If not, then the f.p.p. for chainable continua follows as a corollary---giving us a difficult proof of a simple theorem, but perhaps with an important insight.)

(3) Do there exist nonseparating, atriodic, nonchainable, *principal* continua? (If a counter-example to f.p.p. exists, such a continuum is a likely candidate.)

In this section we will construct embeddings of two chainable continua ---the "3 point continuum" and the pseudo arc---both with s-d-c's and without s-d-c's. Our *approach to re-embedding* will be to change a prime end of the third kind (s-d-c) into a prime end of the second kind (exactly one principal point in $I(E)$), or vice versa. We will include diagrams of the continuum as it appears in the plane, as well as a schematic "inverse limit-with-directions-for-embedding" system, which simplifies such descriptions in general.

We note that substantial partial solutions to the above questions have been obtained by Mayer. In particular, in [11] he proves that every chainable continuum M with at least one endpoint can be embedded in the plane without a s-d-c. Thus no such continuum is principal. (This is still unknown in case M has no endpoints.) Further, for every integer n, $1 \leq n \leq \infty$, there exists a chainable continuum with at least one endpoint having an n-principal embedding. These are constructed like the U-continuum, but with different numbers of loops and in different places. He also observes that n-principal embeddings may also be obtained by a slight modification of Lewis' examples [10], so that they admit period n homeomorphisms interchanging the channels.

In [12], Mayer constructs an example of an atriodic, tree-like, nonchainable continuum $M \subseteq S^2$ such that every proper subcontinuum of M is an arc, M has positive surjective span, and M is principally embedded. (It *appears* that M is a principal continuum, but this question is still open.)

3.1 THEOREM. *There exist embeddings of the three-point continuum into the plane, both with a s-d-c and without a s-d-c.*

Proof. The diagrams below are self-explanatory.

3.2 THEOREM. *There exist embeddings of the pseudo arc into the plane, both with and without a s-d-c.*

Proof. The standard embedding (the Moise embedding [13]) has no s-d-c. We can obtain an embedding with a s-d-c by alternating crooked chains with the pattern of Figure 3.

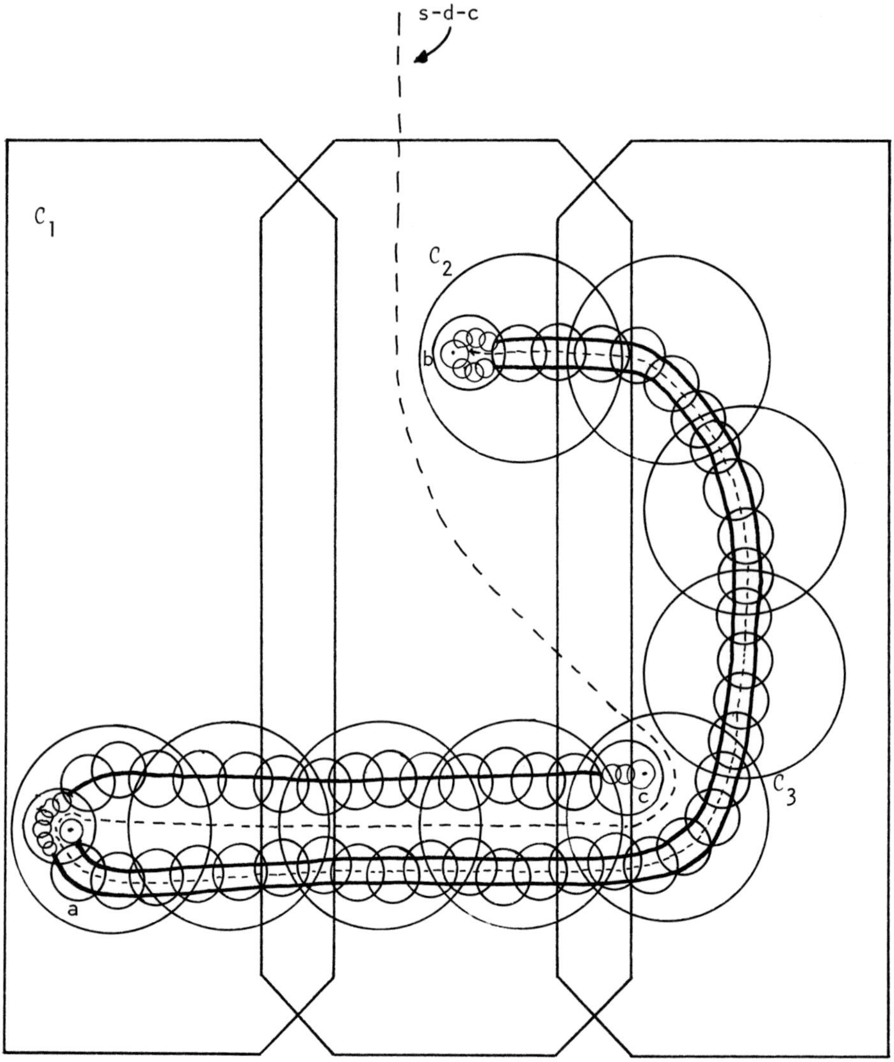

Fig. 3. 3-Point Continuum with s-d-c.

C_1: a-b-c

C_2: b-c-a

C_3: c-a-b

C_4: a-b-c

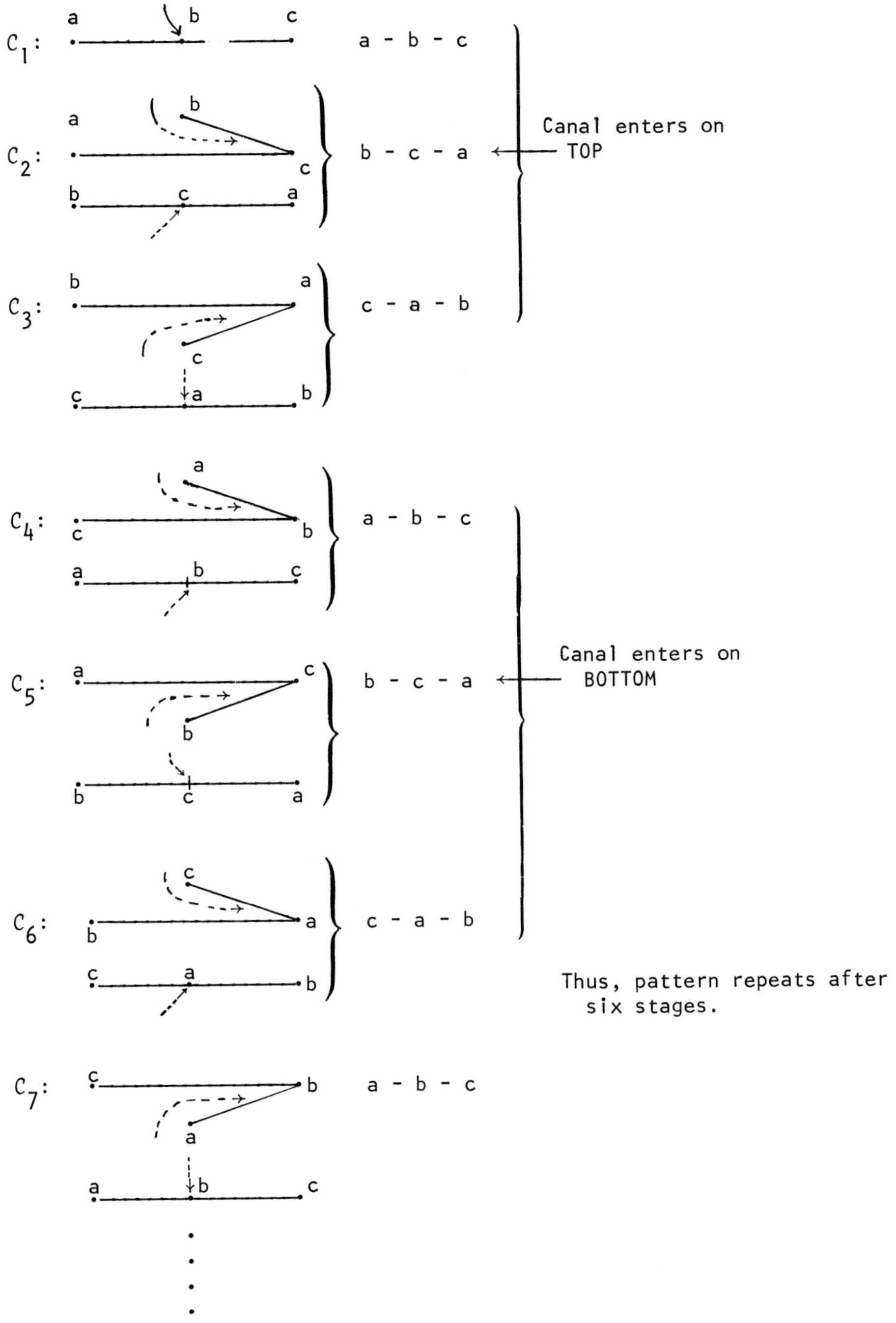

Fig. 4. SCHEMATIC DIAGRAM: 3-Point Continuum with s-d-c.

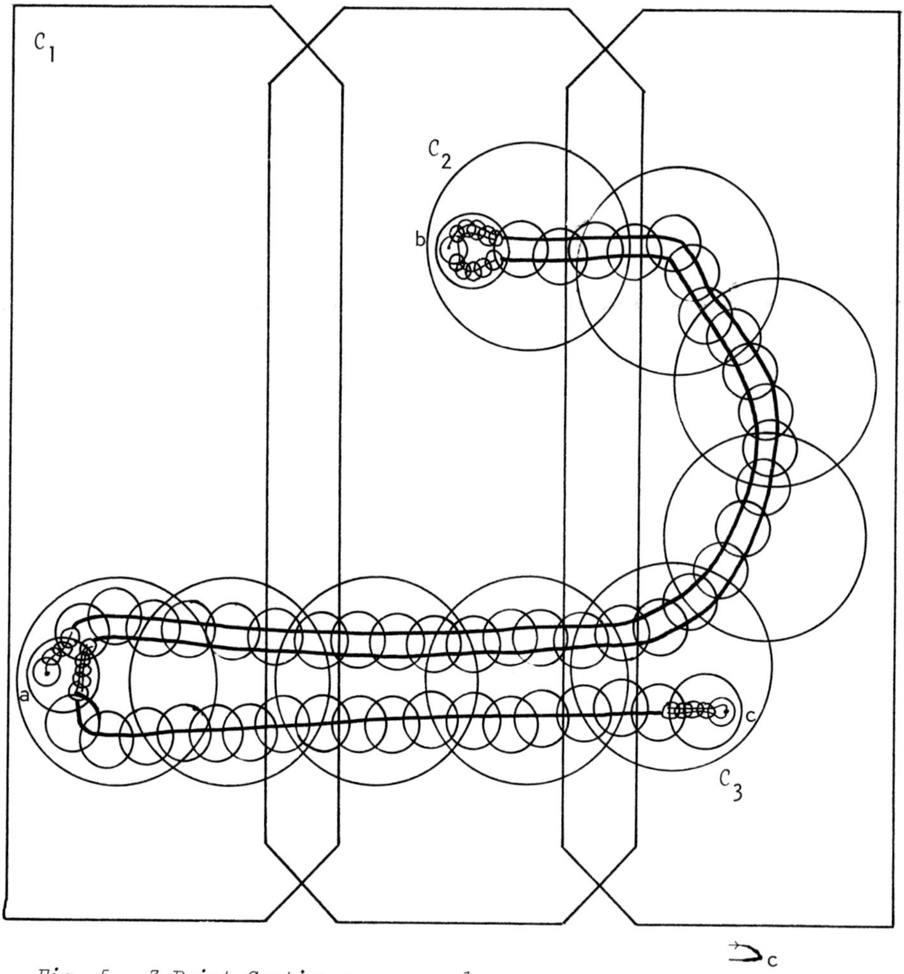

Fig. 5. 3-Point Continuum - no s-d-c.

C_1: a-b-c

C_2: b-c-a

C_3: c-a-b

C_4: a-b-c

C_5: b-c-a

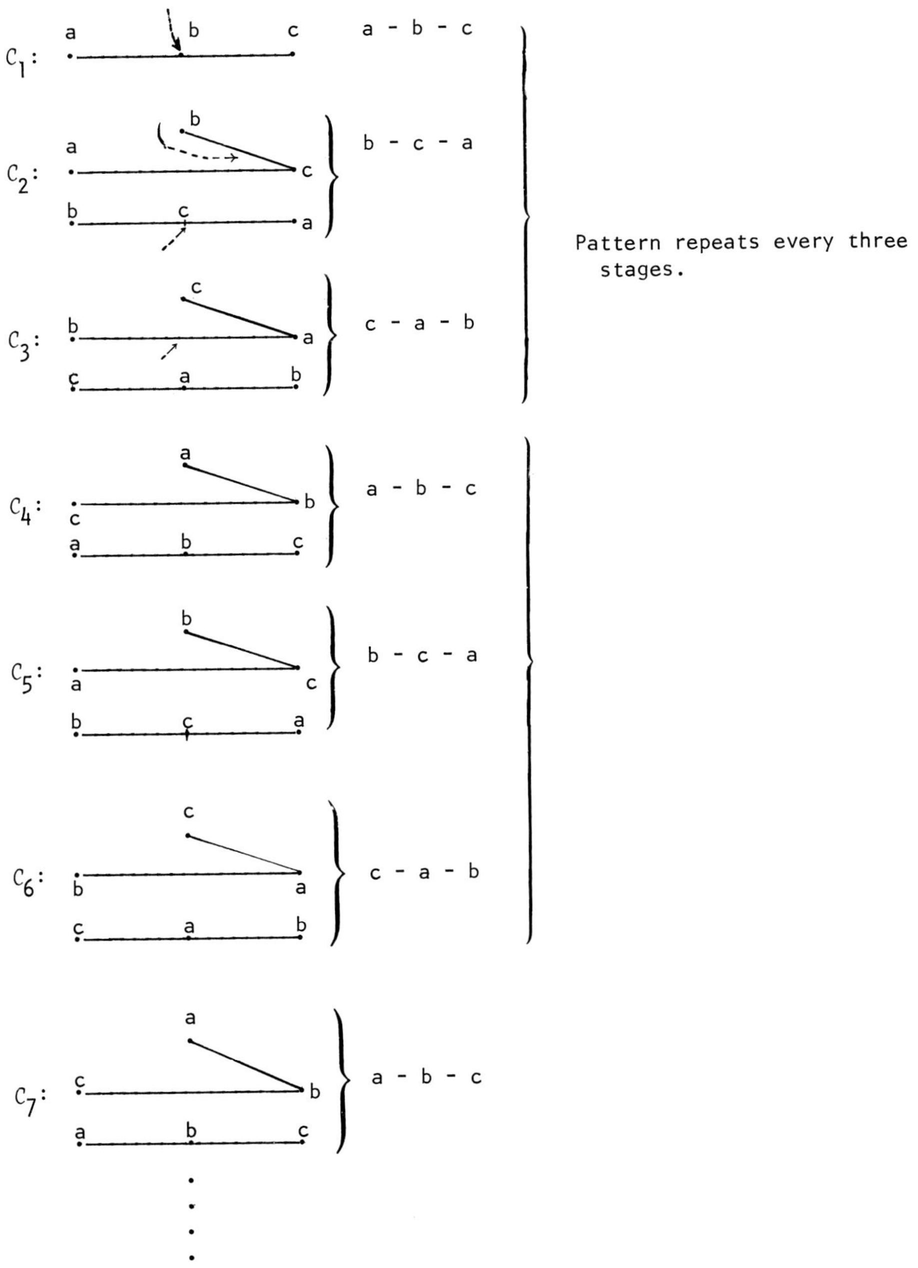

Fig. 6. SCHEMATIC INVERSE LIMIT-WITH-EMBEDDING DIAGRAM: 3-Point Continuum without s-d-c.

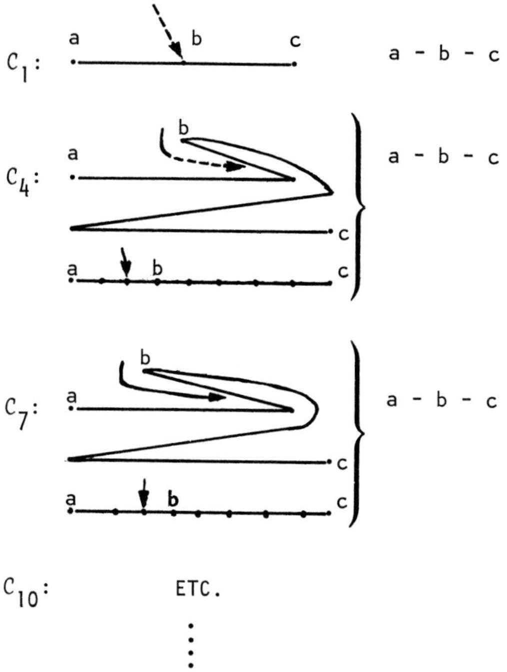

Fig. 7. ANOTHER SCHEMATIC DIAGRAM: 3-Point Continuum without s-d-c.

IV. ATRIODIC CONTINUA AND THE FIXED POINT PROPERTY

In [8], Ingram constructed an example of a nonchainable, tree-like, atriodic continuum with positive span, and such that every proper subcontinuum is an arc. This example is illustrated both geometrically and with an inverse limit---with embedding diagram in Figures 8, 9, and 10.

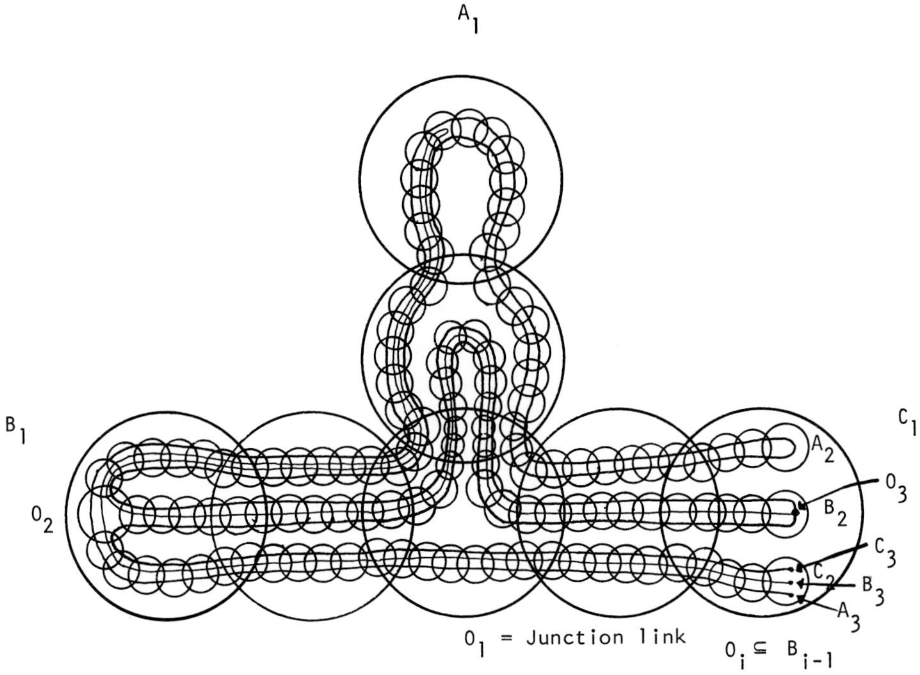

Fig. 8. INGRAM'S EXAMPLE OF AN ATRIODIC, NONCHAINABLE, TREE-LIKE CONTINUUM *(with positive span) such that every proper subcontinuum is an arc.*

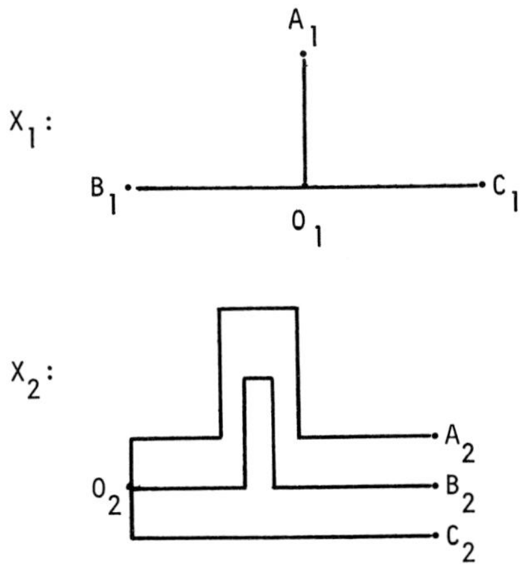

Fig. 9. SCHEMATIC DIAGRAM OF INVERSE LIMIT MAP.

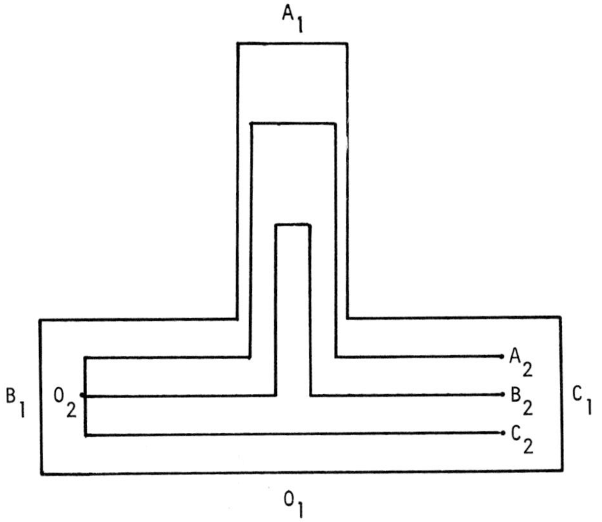

Fig. 10. INGRAM'S EXAMPLE: Description of inverse limit.

4.1 THEOREM. *Let* X *be Ingram's example described above. Then* X *has the fixed point property.*

Proof. Suppose that $f:X \to X$ is fixed point free. Then no proper subcontinuum A of X can be invariant, since A is an arc and has the fixed point property. Thus X is minimal invariant under f.

Now, since $\mathrm{Bd}X = X$ is minimal invariant, there is a s-d-c in X, by our Embedding Corollary 2.5. But it is clear that the embedding described in Figures 8-10 has no s-d-c. This is a contradiction.

4.2 *REMARK AND QUESTION.* We note that Mayer's continuum M [12] has all of the properties of Ingram's example, and it also has a s-d-c. Is M a principal continuum? Does it have the fixed point property?

V. SOME REMARKS

We obtain partial answers to questions 5, 6, and 7 of [5].

Question 5, due to Bing, asks whether a single Lake-of-Wada continuum has the f.p.p., and whether it admits a homeomorphism with exactly one fixed point. We have shown that the pseudo arc may be embedded as a Lake-of-Wada continuum (i.e., with a s-d-c). Thus, such a continuum *may* have the f.p.p. and *may* admit a homeomorphism with exactly one fixed point (a period two rotation about the center point, in the *standard* embedding).

Question 6 asks whether every homeomorphism of a single Lake-of-Wada continuum may be extended to the plane. By [3], the pseudo arc is homogeneous. Thus, if h is a homeomorphism taking an accessible point to a nonaccessible point, then h cannot be extended.

Question 7 asks whether a single Lake-of-Wada continuum can be embedded so that the "channel" or "Lake" disappears. The answer is "at least sometimes," as the 3 point continuum and the pseudo arc may both be re-embedded without a Lake.

5.1 *QUESTION.* Let X be a nonseparating continuum with exactly one s-d-c, which represents the prime end E. If E is the *only* prime end for which $I(E) = \mathrm{Bd}X$, then is there only one accessible composant?

5.2 *QUESTION.* Can a chainable continuum without endpoints always be embedded without a s-d-c? See [11].

5.3 *QUESTION.* Does Mayer's example [12] have the f.p.p.? Is it a principal continuum?

5.4 *REMARK.* James Rogers has asked us in conversation whether the Knaster U-continuum (i.e., bucket handle) can be embedded with a s-d-c. We observe that the following inverse limit with embedding directions accomplishes this:

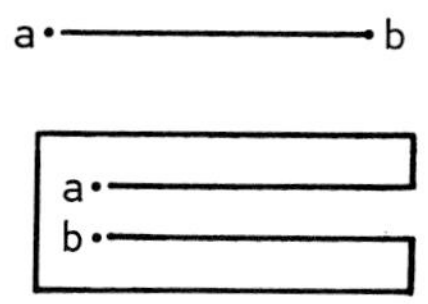

Fig. 11.

The bonding map is the square of the standard two-to-one map.

REFERENCES

1. Bell, H., "On fixed point properties of plane continua," *Trans. Amer. Math. Soc. 128*(1967), 539-548.

2. ______, "A fixed point theorem for plane homeomorphisms," *Fund. Math. 100*(1978), 119-128. See also *Bull. Amer. Math. Soc. 82*(1976), 778-780.

3. Bing, R. H., "A homogeneous indecomposable plane continuum," *Duke Math. J. 15*(1948), 729-742.

4. Brechner, B., "On stable homeomorphisms and imbeddings of the pseudo arc," *Ill. J. Math. 22*(1978), 630-661.

5. ______, "Prime ends, indecomposable continua, and the fixed point property," *Topology Proceedings*, Ohio Univ. Conference, March, 1979.

6. Cartwright, M. L. and J. E. Littlewood, "Some fixed point theorems," *Ann. Math. 54*(1951), 1-37.

7. Hocking, J. G. and G. S. Young, *Topology*, Addison-Wesley, Mass., 1961.

8. Ingram, W. T., "An atriodic tree-like continuum with positive span," *Fund. Math. 77*(1972), 99-107.

9. Lewis, I. W., "Embeddings of the pseudo arc in E^2," preprint.

10. ______, "Periodic homeomorphisms of chainable continua," preprint.

11. Mayer, J. C., "Embeddings and prime end structure of chainable continua," submitted for publication.

12. ______, "Principal embeddings of atriodic plane continua," in: *Proc. of Topology Conf.*, Univ. Texas, Austin, 1980, to appear.

13. Moise, E. E., "An indecomposable plane continuum which is homeomorphic to each of its nondegenerate subcontinua," *Trans. Amer. Math. Soc. 63* (1948), 581-594.

14. Sieklucki, K., "On a class of plane acyclic continua with the fixed point property," *Fund. Math. 63*(1968), 257-278.

HOMOGENEOUS 1-DIMENSIONAL CONTINUA

C. E. Burgess

University of Utah, Salt Lake City

Attempts to classify homogeneous plane continua began about 1920 when Knaster and Kuratowski [20] asked whether every such continuum is a simple closed curve. (A continuum M is defined to be homogeneous if for any two points $p,q \in M$ there is a homeomorphism of M onto itself that carries p to q.) While we know now of two more homogeneous plane continua (the pseudo-arc [4] [24] and the circle of pseudo-arcs [7]), the classification problem has not yet been completed. However, some work in recent years considerably narrows the search for any additional such continua. In this note, I wish to offer a brief summary of this important recent work, mainly by Charles Hagopian, Wayne Lewis, and James Rogers, and with a bit of wild speculation, suggest how I think the classification problem may turn out, both for plane continua and for 1-dimensional continua. A brief history of earlier work on homogeneous plane continua can be found in the paper in which Bing and Jones [7] described the circle of pseudo-arcs.

My own interest in this topic began almost thirty years ago as a result of some conversations with Burton Jones. I am indebted to him for his continuing encouragement and influence that began, for me, when I was a student in some of his classes at the University of Texas at Austin.

1. *Homogeneous plane continua.* Considerable progress toward a classification of these continua has occurred during the last twelve years with the following results:

ISBN 0-12-481820-X

(a) Fearnley [12] and Rogers [25] showed that the pseudo-circle [5] is not homogeneous. Alternative proofs, as special cases of more general theorems, have been given recently by Lewis [22] and Rogers [27].

(b) Lewis [21] showed that the pseudo-arc is the only tree-like homogeneous continuum for which there is an upper bound on the number of junction links in tree-coverings describing the continuum.

(c) Rogers [28] showed that every homogeneous continuum that separates the plane is decomposable. Hence, by Jones' classification theorem [18], any such continuum is a circle of homogeneous nonseparating continua.

(d) Hagopian [14] showed, in 1976, that every homogeneous continuum that does not separate the plane is hereditarily indecomposable [17]. (At the Auburn Topology Conference in March 1969, Jones presented an outline of a more lengthy proof of this theorem. This, of course, generalized his earlier result that every such continuum is indecomposable.)

Hagopian [13] identified the following useful consequence of a theorem about transformation groups that was proved by Effros [11] in 1965:

If M *is a homogeneous continuum and* $\varepsilon > 0$, *then there is a positive number* δ *such that if* $p,q \in M$ *and* $\rho(p,q) > \delta$, *some homeomorphism of* M *onto itself carries* p *to* q *and moves no point more than a distance* ε.

This theorem was used in the theorems mentioned in (b), (c), and (d) above. Jones [19] further illustrated its use with an alternative proof of Bing's theorem [6] that a simple closed curve is the only homogeneous plane continuum that contains an arc. (Ungar [30] had first noticed that Effros' theorem is applicable to theorems about homogeneous continua when he proved that every 2-homogeneous continuum is locally connected.)

With known examples and results, including those mentioned in (a), (b), (c), and (d) above, homogeneous plane continua can now be classified into the following five types, although examples of Type 4 and of Type 5 are not known to exist.

Type 1. Simple closed curves.

Type 2. Pseudo-arcs.

Type 3. Circles of pseudo-arcs.

Type 4. Homogeneous hereditarily indecomposable continua that do not separate the plane and that cannot be described with tree-coverings with an upper bound on the number of junction elements.

Type 5. Circles of continua of Type 4. (Assuming that continua of Type 4 exist, this fifth type is suggested by Jones' classification [18] of homogeneous decomposable plane continua.)

The plane E^2 can be filled with a continuous collection G of pseudo-arcs [1] [23]. Under the projection map P for this decomposition, $P(E^2) = E^2$ and the inverse of each 1-dimensional continuum is 1-dimensional. Thus, for any simple closed curve J in E^2, $P^{-1}(J)$ is a circle of pseudo-arcs [7]. Notice also that the inverse of each pseudo-arc is a pseudo-arc. This suggests that if there is a plane continuum of Type 4, then it might be possible to split the continua of this type (and also those of Type 5) into two types--those that contain pseudo-arcs and those that do not. Such a refinement of classifications is suggested by the following question:

If M *is a homogeneous continuum in* E^2, *and* P *is the projection map described above, then is the continuum* $P^{-1}(M)$ *homogeneous?*

2. *Homogeneous 1-dimensional continua.* The results described in Section 1 for plane continua, along with some recent results mentioned below, suggest that it might be possible to classify homogeneous 1-dimensional continua into the following nine types:

Type 1. Solenoids.

Type 2. Pseudo-arcs.

Type 3. Universal 1-dimensional curves.

Type 4. Solenoids of pseudo-arcs.

Type 5. Universal curve solenoids.

Type 6. Universal curves of pseudo-arcs.

Type 7. Universal curve solenoids of pseudo-arcs.

Type 8. Homogeneous tree-like continua that are not pseudo-arcs.

Type 9. Replace "pseudo-arcs" in Types 4, 6, and 7 with "continua of Type 8."

There are known examples of homogeneous continua of Types 1-5, but there are no known examples of homogeneous continua of Types 6-9. Van Dantzig [10] described solenoids and proved that they are homogeneous. (In this classification, we consider simple closed curves to be solenoids.) Anderson showed that the universal 1-dimensional curve is homogeneous [2] and that it and the circle are the only locally connected homogeneous 1-dimensional continua [3]. Rogers [26] described homogeneous "solenoids of pseudo-arcs" by extending the method Bing and Jones [7] used to describe "circles of pseudo arcs." However, it is not yet known whether, for each solenoid, all corresponding solenoids of pseudo-arcs are homeomorphic. Hagopian and Rogers [16] showed that every homogeneous circle-like continuum must be either a pseudo-arc, a solenoid, or a solenoid of pseudo-arcs. (This extended a previous such result for homogeneous circle-like continua in the plane [8].) Hagopian [15] characterized solenoids as homogeneous continua for which every proper subcontinuum is an arc. Rogers [27] recently obtained the more general result that a homogeneous 1-dimensional continuum is a solenoid if it is atriodic and contains an arc. It follows from a recent result by Lewis [22] that a homogeneous continuum is a pseudo-arc if each of its proper subcontinua is a pseudo-arc. Case [9] used inverse limits of the universal curve, in a manner similar to the way solenoids are described as inverse limits of the circle, to obtain "universal curve solenoids" that are homogeneous.

By considering $E^2 \times I$, we can use the decomposition of E^2 into a continuous collection of pseudo-arcs [23] to obtain a decomposition of E^3 into a continuous collection of pseudo-arcs so that the decomposition space is E^3. Under this decomposition, some 2-dimensional continua would project to 1-dimensional continua. It would be interesting to know the answer to the following question:

Is there a continuous decomposition G *of* E^3 *into pseudo-arcs such that* (1) $E^3/G = E^3$ *and* (2) *the inverse, under the projection map* P, *of each 1-dimensional continuum is 1-dimensional?*

With such a decomposition, we could consider the following question:

If M *is a homogeneous 1-dimensional continuum in* E^3, *and* P *is the projection map described above, then is* $P^{-1}(M)$ *homogeneous?*

With an affirmative answer to this question, we could wonder whether every homogeneous 1-dimensional continuum would be one of the first five types or the inverse of such a continuum under the projection map P. Rogers' recent theorem [29] about decompositions of homogeneous decomposable continua should be helpful with this. Involved in this, of course, is the question whether there is a continuum of Type 8.

Anyone who has worked on the classification problems for homogeneous continua knows that some of the problems and questions suggested here will not be simple to resolve. However, with the progress that has been made in the decade of the 1970's, I think that we can hope to be much closer to a suitable classification within another decade. It is my own impression, perhaps on the basis of wishful thinking, that there are no continua of Type 8 or 9 and that there are no homogeneous plane continua other than those that are already known.

Added in proof. I heard very recently that Oversteegen and Tymchatyn have distributed preprints of a paper in which they show that every homogeneous non-separating plane continuum has span zero. While every chainable continuum has span zero [31], it is not known whether a tree-like continuum must be chainable if it has span zero. An affirmative answer, along with the result claimed by Oversteegen and Tymchatyn, would complete the classification of homogeneous plane continua as the three that are already known.

REFERENCES

1. Anderson, R. D., *Pathological continua and decompositions*, Summary of Lectures and Seminars, Summer Institute on Set Theoretic Topology, University of Wisconsin, 1955, (revised 1958), 81-83.

2. ______________, "A characterization of the universal curve and a proof of its homogeneity," *Ann. of Math. (2)*67(1958), 313-324.

3. ______________, "One-dimensional continuous curves and a homogeneity theorem," *Ann. of Math. (2)*68(1958), 1-16.

4. Bing, R. H., "A homogeneous indecomposable plane continuum," *Duke Math. J. 15*(1948), 729-742.

5. ______________, "Concerning hereditarily indecomposable continua," *Pacific J. Math. 1*(1951), 43-51.

6. ______________, "A simple closed curve is the only bounded plane continuum that contains an arc," *Canad. J. Math. 12*(1960), 209-230.

7. Bing, R. H. and F. B. Jones, "Another homogeneous plane continuum," *Trans. Amer. Math. Soc. 90*(1959), 171-192.

8. Burgess, C. E., "A characterization of homogeneous plane continua that are circularly chainable," *Bull. Amer. Math. Soc. 75*(1969), 1354-1356.

9. Case, J. H. "Another 1-dimensional homogeneous continuum which contains an arc," *Pacific J. Math. 11*(1961), 455-469.

10. van Dantzig, D., "Ueber topologisch homogene Kontinua," *Fund. Math. 15*(1930), 102-125.

11. Effros, Edward G., "Transformation groups and C*-algebras," *Ann. of Math. (2)*(1965), 38-55.

12. Fearnley, L., "The pseudo-circle is not homogeneous," *Bull. Amer. Math. Soc. 75*(1969), 554-558.

13. Hagopian, Charles L., "Homogeneous plane continua," *Houston J. Math. 1*(1975), 35-41.

14. ______________, "Indecomposable homogeneous plane continua are hereditarily indecomposable," *Trans. Amer. Math. Soc. 224*(1976), 339-350.

15. ______________, "A characterization of solenoids," *Pacific J. Math. 68*(1977), 425-435.

16. Hagopian, Charles L., and J. T. Rogers, Jr., "A classification of homogeneous circle-like continua," *Houston J. Math.* *3*(1977), 471-474.

17. Jones, F. B. , "Certain homogeneous unicoherent indecomposable continua," *Proc. Amer. Math. Soc.* *2*(1951), 855-859.

18. __________, "On a certain type of homogeneous plane continuum," *Proc. Amer. Math. Soc.* *6*(1955), 735-740.

19. Jones, F. B., 'Use of a new technique in homogeneous continua," *Houston J. Math.* *1*(1975), 57-61.

20. Knaster, B., and C. Kuratowski, "Probléme 2," *Fund. Math.* *1*(1920), 223.

21. Lewis, Wayne, *Homogeneous tree-like continua.* (Manuscript)

22. __________, *Almost chainable homogeneous continua are chainable.* (Manuscript)

23. Lewis, Wayne and John J. Walsh, "A continuous decomposition of the plane into pseudo-arc," *Houston J. Math.* *4*(1978), 209-222.

24. Moise, E. E., "A note on the pseudo-arc," *Trans. Amer. Math. Soc.* *64* (1949), 57-58.

25. Rogers, J. T., Jr., "The pseudo-circle is not homogeneous," *Trans. Amer. Math. Soc.* *148*(1970), 417-428.

26. __________, "Solenoids of psuedo-arcs," *Houston J. Math.* *3* (1977), 531-537.

27. __________, "Completely regular mappings and homogeneous aposyndetic continua," *Canad. J. Math.*

28. __________, *Homogeneous separating plane continua are decomposable.* (Manuscript).

29. __________, *Decompositions of homogeneous continua.* (Manuscript).

30. Ungar, G. S., "On all kinds of homogeneity," *Trans. Amer. Math. Soc.* *212*(1975), 393-400.

31. Lelek, A., 'Disjoint mappings and the span of spaces," *Fund. Math.* *55*(1964), 199-214.

THE LEFSCHETZ THEOREM FOR SELF-MAPS OF COMPACTA

J. Dugundji

University of Southern California, Los Angeles

The Lefschetz fixed-point theorem is a powerful algebraic tool in fixed-point theory; with various modifications, it quickly yields many of the currently used fixed-point theorems of Analysis and Topology. Given first by Lefschetz for manifolds, it was extended by Hopf to all finite polyhedra; its extension to compact ANR follows very quickly by a simple device discussed here later on. Extensions to arbitrary compacta have not been considered until recently, primarily because there are known examples [1],[12] of compacta, even in $\mathbb{R}^3$, having self-maps satisfying the hypothesis of the Lefschetz theorem, but not its conclusion. Nevertheless, it was first pointed out by Borsuk [3] that it is fruitful to concentrate directly on the maps themselves, rather than on the type of the space, and he showed [3] that there is a rather general, and fairly extensive, class of self-maps of compacta for which the Lefschetz theorem remains valid. This paper is expository, to describe the results found in [3] and their extensions given in [6].

We begin with the terminology needed to formulate the Lefschetz theorem. Letting Q = rationals, recall that on the category of compacta and homotopy classes of maps, there are [8] various functors $H_*: X \mapsto H_*(X,Q) = \{H_0(X,Q), H_1(X,Q), \ldots\}$ to the category of graded Q-vector spaces, such as the singular, and the Čech, homology functors; the details of their construction does not concern us here, but it turns out that $H_0(X,Q) \approx Q$

ISBN 0-12-481820-X

when X is connected (pathwise for the singular theory, topologically for the Čech); we shall assume all spaces connected in all that follows. The functoriality implies that each $f:X\to Y$ induces a linear transformation of graded Q-vector spaces, with homotopic maps inducing the same transformation; schematically, the topological scheme $f:X\to Y$ has the algebraic analog

$$\begin{array}{ccccc} X & H_*(X,Q) & = \{H_0(X,Q), & H_1(X,Q), & \ldots\} \\ f\downarrow \; : & f_*\downarrow & \quad f_0\downarrow & f_1\downarrow & \\ Y & H_*(Y,Q) & = \{H_0(Y,Q), & H_1(Y,Q), & \ldots\} \end{array}$$

where each f_i is a linear transformation.

All ordinary homology theories agree on the subcategory of finite polyhedra [8]; moreover, each finite polyhedron K has homology of finite type (i.e., all $H_i(K,Q)$ are finite-dimensional, and almost all are zero) so, for an $f:K\to K$ the trace $tr(f_i)$ of each $f_i:H_i(K,Q)\to H_i(K,Q)$ is well defined. The Lefschetz theorem states: Let K be a finite polyhedron. Given any $f:K\to K$, let $\lambda(f) = \sum_{i=0} (-1)^i tr(f_i)$. Then $\lambda(f)$ is an integer, and $\lambda(f) \neq 0$ implies that f has a fixed point.

This gives a sufficient condition for the existence of a fixed point. Note that if the hypothesis is valid for any one member of a homology class, then it is valid for all members of that homology class. In particular, homotopic maps being homologous, the theorem can, at best, detect that an $f:K\to K$ has a fixed point only in those cases where every map homotopic to f also has a fixed point; thus, e.g., it cannot detect that $id:S^1\to S^1$ has a fixed point (in fact, $\lambda(id) = 0$) because a small rotation has none. In the converse direction, even if every map in a homotopy class has a fixed point, the Lefschetz theorem may not be able to detect that without some additional hypotheses on K (such as that K is a 1-connected manifold). In any case, as the statement of the theorem shows, it is the homology behavior of the map, rather than its homotopy behavior, that is decisive: for example, the map $f = (\text{Hopf}\vee\text{constant}):S^3\vee S^2 \to S^2\vee S^3$ is not null-homotopic, but it is homologically trivial and, because $\lambda(f) = 1$, it has a fixed point.

Though a map $f:K\to K$ may not have a fixed point, some iterate f^n may (e.g., the antipodal map $S^n\to S^n$); a fixed point of some iterate f^n is called a periodic point of f and n its period, a fixed point of f being a periodic point of period 1. There are several algebraic refinements of the Lefschetz theorem that detect the existence of periodic points. For example, [4],[7]: Let $K(f_i) = \bigcup_{n=0}^{\infty} \ker f_i^n$; if $\sum_{i=0} (-1)^i \dim[H_i(K,Q)/K(f_i)] \neq 0$, then f has a periodic point. Another refinement [7] works with the set of all nonzero eigenvalues $\{\lambda\}$ of f_* : let $e(\lambda)$ (resp. $o(\lambda)$) be the number of times that λ appears in even (resp. odd) dimensions, and let $\tau(f) = \mathrm{card}\{\lambda \mid e(\lambda)-o(\lambda) \neq 0\}$; $\tau(f) = 0$ implies that $\lambda(f) = 0$, but not conversely; in fact, if $\tau(f) \neq 0$, then f has a periodic point of period $\leq \tau(f)$. Neither of these refinements can possibly detect that the antipodal map $\alpha:S^{2n+1}\to S^{2n+1}$ of an odd-dimensional sphere has a periodic point, but they show immediately that $\alpha:S^{2n}\to S^{2n}$ does; more generally, they show that, for any finite polyhedron K with Euler characteristic $\chi(K) \neq 0$, every homeomorphism $h:K\to K$ has a periodic point.

There is a simple technique for extending the Lefschetz theorem to compact ANR spaces X, based on the following property of such spaces: Given any $f:X\to X$ and $\varepsilon > 0$, there is a diagram

$$\begin{array}{ccc} & P & \\ {}^{k}\nearrow & & \searrow^{g} \\ X & \xrightarrow{\quad f \quad} & X \end{array} \qquad (1)$$

where P is a finite polyhedron, $gk \simeq f$ and $d[gk(x),f(x)] < \varepsilon$ for all $x \in X$. Using any homology theory that satisfies the Eilenberg-Steenrod [8] axioms (they all agree on finite polyhedra), the special case $f = \mathrm{id}$ shows that X has homology of finite type, so that $\lambda(f)$ is defined. We next note that

$$\lambda(f) = \lambda(gk)\ , \qquad \text{because } f \simeq gk$$
$$= \lambda(kg) \qquad \text{by commutativity of trace,}$$

so if $\lambda(f) \neq 0$, then $kg:P\to P$ has a fixed point p . From $g(p) = g[kg(p)] = gk[g(p)]$ and $d[g(p),f[g(p)]] = d[gk[gp],f[gp]] < \varepsilon$ follows that $g(p)$ is an ε-fixed point for f . Since $\varepsilon > 0$ is arbitrary, this implies that f has a fixed point: otherwise, by compactness of X , there would be some fixed $\varepsilon_0 > 0$ with $d(x,f(x)) \geq \varepsilon_0$ for all $x \in X$.

Extensions of the Lefschetz theorem to spaces more general than compact ANR have been made, all using variations of the above technique: to arbitrary ANR and compact maps [11], to AANR (in the sense of Gmurczyk [9],[10], and the more general ones of Clapp [5]). Compactness is an essential ingredient: $\tau:x \mapsto x+1:\mathbb{R}^1 \to \mathbb{R}^1$ has no fixed point, yet $\lambda(\tau) = 1$. All the above extensions are concerned with finding a class of spaces for which the Lefschetz theorem is true; it cannot be true for all compacta, since Borsuk [1] has shown that there are acyclic compacta in $\mathbb{R}^3$ admitting homeomorphisms, homotopic to the identity, which do not have fixed points.

It was the idea of Borsuk [3] to shift from studying spaces to studying maps of arbitrary compacta: the problem becomes to determine a general class, or type, of map which will in fact have a fixed point whenever the Lefschetz conditions say it does. The present results [3],[6] lead to a fairly well-determined such class of maps; they also give some information about the types of self-maps of, say, the intersection of a descending sequence of closed discs, that have fixed points.

We will use Čech homology and, to motivate the definitions, we look at the ingredients of the diagram (1) when X is a compactum. At first we will assume that X has Čech homology of finite type, so that $\lambda(f)$ is defined. It is a fact [6] that each such compactum X comes equipped with an $\varepsilon(X) > 0$ such that, for each compactum C , any two maps $f,g:C\to X$ with $\sup\{d(f(c),g(c))\,|\,c\in C\} < \varepsilon(X)$ induce the same homomorphism in Čech homology. Thus, the homotopy factorizations in diagram (1) can be replaced by any sufficiently close (not necessarily homotopic) factorizations through spaces known to satisfy the Lefschetz theorem as, e.g., compact ANR.

To develop such a class of maps, we begin with

Definition. Let X be a compactum. By a Borsuk presentation $\mathcal{B} = \{Z_i \mid i=1,2,\ldots\}$ of X is meant a descending sequence $Z_1 \supset Z_2 \supset \cdots$ of compact ANR such that $\bigcap_1^\infty Z_i = X$.

Since all compacta can be embedded in the Hilbert cube and have arbitrarily small compact ANR neighborhoods there, every compactum has Borsuk presentations.

Definition. Let $\mathcal{B} = \{Z_n\}$ be a Borsuk presentation of X. A map $f:X\to Y$ of compacta is called a $\mathcal{B}$-map if for each $\varepsilon > 0$ there is some Z_n and some $g:Z_n\to Y$ such that $d[g(x),f(x)] < \varepsilon$ for each $x \in X$.

Given a presentation $\mathcal{B}$ of X, the class of $\mathcal{B}$-maps is quite extensive. For any compactum Y, the set of $\mathcal{B}$-maps in the function space Y^X (sup metric) is a nonempty closed subset containing the constant maps; the set of $\mathcal{B}$-maps $= Y^X$ for every Y if and only if $id:X\to Y$ is a $\mathcal{B}$-map (such spaces are called $\mathcal{B}$-spaces, and are essentially the Clapp AANR [5]). Each compact ANR is a $\mathcal{B}$-space, as is every locally connected plane continuum [2]. Given a $\mathcal{B}$-map $f:X\to Y$ and any $g:Y\to Z$, the composition $g\circ f$ is also a $\mathcal{B}$-map; in particular, composition in X^X preserves the $\mathcal{B}$-property.

With these definitions, the entire argument based on the diagram (1) goes through for any given $\mathcal{B}$-map $f:X\to X$ satisfying $\lambda(f) \neq 0$: Choose $\varepsilon(X)$ and use only $\varepsilon < \varepsilon(X)$; for each such ε there is a $g:Z_n\to X$ with $d[gi(x),f(x)] < \varepsilon$ where $i:X\hookrightarrow Z_n$; thus gi and f induce the same homomorphism of $H_*(X,Q)$ so $\lambda(f) = \lambda(gi)$. The same reasoning as before therefore leads [6] to the

THEOREM 1. *Let* X *be a compactum having homology of finite type, and let* $\mathcal{B}$ *be a Borsuk presentation of* X.

(a) *If* $f:X\to X$ *is a* $\mathcal{B}$*-map with* $\lambda(f)\neq 0$, *then* f *has a fixed point.*

(b) *A* $\mathcal{B}$*-map of an acyclic* X *into itself has a fixed point.*

(c) *If* $\chi(X)\neq 0$ *then any homeomorphism* $h:X\to X$ *that is a* $\mathcal{B}$*-map will have a periodic point.*

Observe that the choice of $\mathcal{B}$ is irrelevant; it is the existence of some $\mathcal{B}$ making f into a $\mathcal{B}$-map that is crucial. Using the Leray trace [11], the Lefschetz theorem can be extended [6] to suitable compact maps of arbitrary metric spaces:

COROLLARY. *Let* X *be any metric space and* $f:X\to X$ *such that* $\overline{f(X)}$ *is compact, and has homology of finite type. If* $\overline{f(X)} = \bigcap_{i=0}^{\infty} Z_i$, *where all* $Z_i\subset X$ *and each* Z_i *is a compact ANR, then using Leray traces, the* $\lambda(f)$ *will exist, and* $\lambda(f)\neq 0$ *implies that* f *has a fixed point.*

This places no restriction of f other than that $\overline{f(X)}$ have a Borsuk presentation $\mathcal{B}$ contained in X: for, using $g_n = f|Z_n$ shows that $f|\overline{f(X)}$: $\overline{f(X)}\to\overline{f(X)}$ is a $\mathcal{B}$-map; and since $\overline{f(X)}$ has homology of finite type, the $\lambda(f)$ computed using Leray traces will exist and be equal to $\lambda[f|f(X)]$ computed with the usual trace.

The hypothesis in Theorem 1(a) that X have homology of finite type can be eliminated by a procedure, suggested in [3], that associates a "Lefschetz set" $\Lambda(f)$ with each $\mathcal{B}$-map $f:X\to X$: for each pair of integers n,k, let $A(f;n,k) = \{\lambda(g)\,|\,g:Z_n\to X, d[g(x),f(x)] < \frac{1}{k}$ for all $x\in X\}$ and define $\Lambda(f) = \bigcap_{n=0}^{\infty}[\bigcup_{k=0}^{\infty} A(f;n,k)]$; then

THEOREM 2. *If* X *is any compactum,* $f:X\to X$ *a* $\mathcal{B}$*-map, and* $0\notin\Lambda(f)$, *then* f *has a fixed point.*

The continuous image of an ANR compactum is locally connected; since the compactum X may not be locally connected, the requirement for a $\mathcal{B}$-map that the approximating g_k send Z_k into X seems rather specialized for the purposes of General Topology. Following the approach in [2] we get a class of self-maps of compacta more general than $\mathcal{B}$-maps by requiring that the approximating g_k have values in small Z_n rather than in X ; the diagrams (1) are replaced by diagrams of form

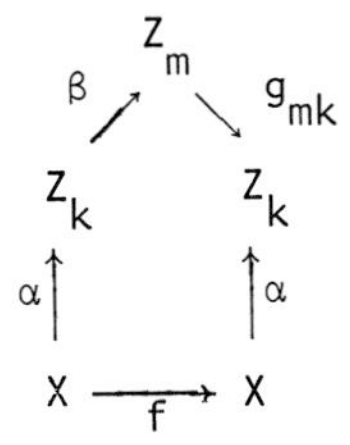

(2)

where α,β are insertions. Precisely:

Definition. Let X be a compactum, $\mathcal{B} = \{Z_k\}$ a Borsuk presentation. An $f:X \to X$ is called an $N\mathcal{B}$-map if there is an extension $\tilde{f}:Z_1 \to Z_1$ of f having the property: for each $\varepsilon > 0$ there is a Z_m and, for each Z_k , a $g_{mk}:Z_m \to Z_k$ such that $d[\tilde{f}(z),g_{mk}(z)] < \varepsilon$ for all $z \in Z_m$.

The existence of an extension $\tilde{f}$ is no restriction: regarding the sets in $\mathcal{B}$ as contained in the Hilbert cube, we can change $\mathcal{B}$ by taking $Z_1 = I^\infty$; the $N\mathcal{B}$-property of f is independent of the particular extension $\tilde{f}$ that is selected [2]. The class of $N\mathcal{B}$-maps is a closed subset of the function space X^X ; it contains the $\mathcal{B}$-maps and the composition of any two $N\mathcal{B}$-maps.

Assume that X has homology of finite type. Given $\varepsilon > 0$, then for each sufficiently large m and all sufficiently large $k > m$, the approximating maps g_{mk} in the diagram (2) will satisfy not only $d[\tilde{f}(z),g_{mk}\beta(z)] < \varepsilon$ but also [3],[6], because of the continuity [8] of the Čech homology theory, $\lambda(f) = \lambda[g_{mk}\beta]$. Thus, if $\lambda(f) \neq 0$ then $g_{mk}\beta$ has a fixed point $z_m \in Z_k \subset Z_m$ and $d[\tilde{f}(z_m),z_m] < \varepsilon$. Taking $\varepsilon \to 0$ and $m \to \infty$, the sequence of fixed points so obtained can be assumed to converge to some $x_0 \in X$; since $\tilde{f}(z_m) \to x_0$ and $\tilde{f}|X = f$, it follows that x_0 is a fixed point for f . This shows that all the statements in Theorem 1 are valid also for the broader class of NB-maps.

REFERENCES

1. Borsuk, K., "Sur un continu acyclique qui se laisse transformer topologiquement en lui-même sans points invariants," *Fund. Math.* *24*(1934), 51-58.

2. ______, "On nearly-extendable maps," *Bull. Ac. Pol. Sc.* *23*(1975), 753-760.

3. ______, "On the Lefschetz-Hopf theorem for nearly extendable maps," *Bull. Ac. Pol. Sc.* *23*(1975), 1273-1279.

4. Bowszyc, C., "On the Euler-Poincare characteristic of a map and the existence of periodic points," *Bull. Ac. Pol. Sc.* *17*(1969), 367-372.

5. Clapp, M. H., "On a generalization of ANR," *Fund. Math.* *40*(1971), 117-130.

6. Dugundji, J., "On Borsuk's extension of the Lefschetz-Hopf theorem," *Bull. Ac. Pol. Sc.* *25*(1977), 805-811.

7. Dugundji, J. and A. Granas, *Fixed Point Theory, Vol. I,* Monografie Matematyczne no. 61, Warszawa, 1981 (to appear).

8. Eilenberg, S. and N. Steenrod, *Foundations of Algebraic Topology,* Princeton Univ. Press, New Jersey, 1952.

9. Gmurczyk, A., "On approximative retracts," *Bull. Ac. Pol. Sc.* *16*(1968), 9-14.

10. Granas, A., "Fixed point theorems for approximative ANR's," *Bull. Ac. Pol. Sc.* *16*(1968), 15-19.

11. ______, "Generalizing the Hopf-Lefschetz fixed-point theorem for non-compact ANR's," *Proc. Symp. Infinite-dimensional Topology,* Baton Rouge, Louisiana, 1967.

12. Kinoshita, S., "On some contractible continua without the fixed point property," *Fund. Math.* *40*(1953), 96-98.

SOME REMARKS ON FIXED POINT THEORY

Eldon Dyer

Graduate School & University Center
City University of New York, New York

This article is intended as a brief survey of selected topics in fixed point theory. We hope to convey some sense of the historical development of the subject as well as some of its current directions. There is also a bibliography which may guide the interested reader further into this important area.

We present first a set of examples; they reappear subsequently as relevant theorems are discussed. A rough division of fixed point theory can be made by first considering those aspects of the subject that appear as homotopy invariants of the maps involved, principally dealing with the Lefschetz index theory, and then by considering more rigid types of structure; for example, the Smith theory. There is also an enriched variant of the Lefschetz index due to Atiyah and Bott which we discuss briefly. The article is concluded with a discussion of a few applications.

The author greatly appreciates the assistance of his colleague Professor A. T. Vasquez in the preparation of the material included here.

I. EXAMPLES

A. (Knill [19]). Let A be a half line spiraling to a circle. For example, let $r:[0,\infty\rangle \to [0,1\rangle$ be a homeomorphism and take A to be the graph of $\rho(\theta) = r(\theta)\cdot\exp(i\theta)$ for $0 \leq \theta < \infty$ together with the unit circle $\rho = 1$. Then $X = TA$, the cone of A, does not have the fixed

ISBN 0-12-481820-X

point property; that is, there is a continuous $f:X \to X$ such that for every $x \in X$, $x \neq f(x)$.

Of course X is contractible to its cone point; so the fixed point property is not an invariant of homotopy type. Moreover, by selecting the cone properly, for example to be $(0,0,1)$ in R^3 with A regarded as lying in $R^2 \times 0 \subset R^3$, it can be seen that X is the intersection of a decreasing sequence of closed 3-cells in R^3.

The corresponding question in R^2 is a classical unsolved problem. That is, suppose $Y \subset R^2$ is the intersection of a decreasing sequence of closed 2-cells in R^2. Does Y have the fixed point property?

B. (Lopez [20]). Let $P_n(C)$ denote the complex n-dimensional projective space (of real dimension $2n$). It is obtained from $C^{n+1}-0$ by identifying the points on each complex line through 0 to a single point. Specific imbeddings $P_n(C) \subset P_{n+1}(C)$ can be obtained from the inclusions $C^{n+1} \times 0 \subset C^{n+1} \times C$. The space $P_1(C)$ is readily seen to be homeomorphic to the 2-sphere S^2.

Let $a,b \in S^2 \times S^2$ and form a quotient space Z of the disjoint union

$$P_2(C) + S^2 \times S^2 + P_4(C) + \Sigma P_8(C)$$

by identifying

$$S^2 = P_1(C) \subset P_2(C) \quad \text{with} \quad S^2 \times b \quad \text{in} \quad S^2 \times S^2$$

$$S^2 = P_1(C) \subset P_4(C) \quad \text{with} \quad a \times S^2 \quad \text{in} \quad S^2 \times S^2$$

and (a,b) with any point of $\Sigma P_8(C)$, where $\Sigma P_8(C)$ denotes the suspension of $P_8(C)$.

The space Z is triangulable and has the fixed point property. But the space $Z \times I$ does not have the fixed point property. This is another example showing that the fixed point property is not an invariant of homotopy type. Although not so easily visualized as example A, it has the advantage of having no local pathology. Also note that both Z and I

have the fixed point property and their product does not.

C. (Bredon [5] and Husseini [17]). Let $m > n$ and $f:S^m \to S^n$ be a suspension. Form the space $C(f)$ by adjoining the unit (m+1)-disk D^{m+1} to S^n by f ; that is, form the quotient space of the disjoint union $D^{m+1} + S^n$ by identifying each $x \in S^m \subset D^{m+1}$ with $f(x)$.

For m and n having the same parity, the space $C(f)$ has the fixed point property if and only if $0 \neq [f] \in \pi_m(S^n)$. But if $f':S^m \to S^n$ is another such map and the order of $[f']$ in $\pi_m(S^n)$ is prime to that of $[f]$, then the product $C(f) \times C(f')$ does not have the fixed point property. By a careful selection of m, n, f , and f' , these examples can be thickened to manifolds with boundary and then doubled to yield closed manifolds M and M' , each having the fixed point property but whose product does not.

D. (Floyd [13], Conner and Floyd [7], Floyd and Richardson [14] and Kister [18]). This is a class of examples of finite groups of homeomorphisms on some space with anomalous fixed or stationary point phenomena. The basic idea in each of these examples is to find an action appropriate for the type of phenomenon sought on a simplicial complex, then to imbed the complex in a suitably high dimensional Euclidean space and extend the action to a regular neighborhood of the imbedded complex. An application of Whitehead's regular neighborhood theorem, or one of its refinements, leads to the example.

There is an action of the icosohedral group on a closed n-cell, for some $n \leq 876$, such that no point is left fixed by every element of the group. There is an action of Z_6 , the cyclic group of order six, on a sphere S^n , for some $n \leq 41$, such that the fixed point set of the generator is nonempty but fails to have the homology of a sphere. For each Z_c , with c any integer other than a power of a prime, and any $n \geq 8$, there is a smooth action of Z_c on R^n such that no point is left fixed by the generator of Z_c .

E. (Conner and Floyd [8]). Let $n = pq$ with $p>q>1$ relatively prime. Let $S \subset P_2(C)$ be the set of all equivalence classes of elements of the form $z = (z_1,z_2,z_3) \in C^3$ with z having norm 1 and with $z_1^n + z_2^n + z_3^n = 0$. Let λ be a primitive n^{th} root of 1 . Define $T:S \to S$ by $z_1,z_2,z_3 \mapsto z_1,\lambda^p z_2,\lambda^q z_3$ and $\tau:S \to S$ by $z_1,z_2,z_3 \mapsto \lambda z_1,z_2,z_3$. Then $T^n = 1_S = \tau^n$ and $T\tau = \tau T$. Let $M = S/T$. The space M is a Riemann surface of genus $(p-1)\cdot(q-1)/2$. Since τ commutes with T , it coinduces $\tau^*:M \to M$. The map τ^* has exactly one fixed point and $(\tau^*)^n = 1_M$.

F. (Atiyah and Bott [2]). Let $V(a)$ be the subspace of C^n of all solutions $z = (z_1,\ldots,z_n)$ of the equation

$$z_1^{a_1}+\cdots+z_n^{a_n} = 0$$

for $a = (a_1,\ldots,a_n)$, and let $\Sigma(a)$ be the intersection of $V(a)$ with the unit sphere S^{2n-1} in C^n . For $a = (a_1,\ldots,a_{2m}) = (2,2,\ldots,2,k)$ with k odd, $\Sigma(a)$ is homeomorphic to S^{4m-3} . The involution $T:C^n \to C^n$ given by $z_i \mapsto -z_i$ for $1 \le i \le n-1$ and $z_n \mapsto z_n$ induces an involution $T:\Sigma(a) \to \Sigma(a)$. For $m \ge 2$, the involution T on $\Sigma(a) = S^{4m-3}$ with $k = 3$, is not isomorphic to the standard antipodal map.

II. FIXED POINTS DETECTED BY ALGEBRAIC INVARIANTS

A. Lefschetz Theory. The theorem of Lefschetz in this section is a homological device for inferring that a continuous function $f:P \to P$ has a fixed point. Conditions of sufficient strength are imposed on P so that homology has a chance of measuring the relevant phenomena. The results of Lefschetz furnish sufficient but not necessary conditions for a self-map f to have a fixed point. We return to the more sensitive analysis regarding necessity in the next section.

For a finite dimensional vector space V over a field F and an endomorphism $\phi:V\to V$, there is an element $tr(\phi)$ of F, the trace of ϕ, given by choosing a basis $v_1,v_2,\ldots,v_n$ of V, writing the matrix $\phi(v_i) = \sum_j a_i^j v_j$ of ϕ with respect to this basis, and letting $tr(\phi) = \sum_i a_i^i$. The sum so obtained is independent of choice of basis. Trace has the following easily verifiable properties:

(i) if $\phi:V\to V$ and $\psi:W\to W$, then

$$tr(\phi+\psi) = tr(\phi)+tr(\psi)$$

$$tr(\phi\otimes\psi) = tr(\phi)\cdot tr(\psi) ;$$

(ii) if $W\subset V$ and $\phi:V\to V$ satisfies $\phi|W:W\to W$, letting $\hat{\phi}:V/W\to V/W$ be the coinduced endomorphism,

$$tr(\phi) = tr(\phi|W)+tr(\hat{\phi}) ;$$

(iii) if $\phi:V\to W$ and $\psi:W\to V$, then

$$tr(\phi\psi) = tr(\psi\phi) .$$

Suppose that K is a finite simplicial complex and $f:K\to K$ is a simplicial mapping having the property that for each simplex σ of K, σ and $f(\sigma)$ are disjoint. Let F denote a field and $C_i(K)$ denote the group of i-chains of K with coefficients in F. $C_i(K)$ is a finite dimensional vector space over F; in fact, we can take as a basis for $C_i(K)$ the set $\{1\cdot\sigma^i\}$ for all i-simplices of K. With this basis it is clear that the assumption on f implies

$$(*) \qquad tr(C_i(f)) = 0 \qquad \text{for all } i$$

where $C_i(f):C_i(K)\to C_i(K)$ is the chain homomorphism induced by f.

For any simplicial map $g:K\to K$ we have the induced homomorphisms on chains, cycles, boundaries, and homology

$$C_i(g):C_i(K)\to C_i(K)$$

$$Z_i(g):Z_i(K)\to Z_i(K)$$

$$B_i(g):B_i(K)\to B_i(K)$$

$$H_i(g):H_i(K) \to H_i(K)$$

with $C_i(g)|Z_i(K) = Z_i(g)$ and $Z_i(g)|B_i(K) = B_i(g)$. Also by definition we have

$$\widehat{C_i(g)} = B_{i-1}(g):B_{i-1}(K) \to B_{i-1}(K)$$

$$\widehat{Z_i(g)} = H_i(g):H_i(K) \to H_i(K) \ .$$

Thus by property (ii) above

$$\begin{aligned} tr(C_i(g)) &= tr(Z_i(g))+tr(B_{i-1}(g)) \\ &= tr(B_i(g))+tr(H_i(g))+tr(B_{i-1}(g)) \ . \end{aligned}$$

Also $B_{-1}(K) = 0$ and for $n = \dim K$, $B_n(K) = 0$. Thus,

$$L(g) = \sum_i (-1)^i tr(C_i(g)) = \sum_i (-1)^i tr(H_i(g)) \ .$$

This element $L(g) \in F$ is called the Lefschetz Number of the map g for coefficients in F . (Note that for $g = 1_K:K\to K$ the identity function, $L(1_K)$ is just the Euler character of K .)

For our map $f:K\to K$ above, we had the conclusion (*) that $tr(C_i(f)) = 0$. Thus we conclude that if K is a finite simplicial complex, $f:K\to K$ is a simplicial map such that σ and $f(\sigma)$ are disjoint for every simplex σ of K , then for every field F , $L(f) = 0$.

From this observation, standard facts about the barycentric subdivision of a simplicial complex (in particular that diameters of simplices tend to 0 under iterated barycentric subdivision), and the usual arguments of topological invariance of homology of polyhedra, an easy indirect argument leads to the

LEFSCHETZ INDEX THEOREM. *If* K *is a compact polyhedron,* $f:K\to K$ *is a continuous function,* F *is a field, and* $L(f) = \sum_i (-1)^i tr(H_i(f)) \neq 0$, *then* f *has a fixed point.*

There is an important generalization of this result which utilizes the notion of absolute neighborhood retract (ANR) due to Borsuk. A compact metric space is said to be an ANR if it is a retract of some neighborhood of it in each separable metric space containing it. (There is a considerable amount of literature on this and related classes of spaces.) Of particular interest for our present discussion is the fact that for any compact metric ANR space X, continuous $f:X\to X$, and $\varepsilon > 0$, there exist a compact polyhedron P and two continuous functions $\phi:X\to P$ and $\psi:P\to X$ such that the functions f and $\psi(\phi)$ are homotopic and for every $x \in X$, $d(f(x),\psi(\phi x)) < \varepsilon$.

Let X be a compact metric ANR and $f:X\to X$ be a continuous function without any fixed point. Then there is a positive number ε such that for every $x \in X$, $d(x,f(x)) > \varepsilon$. Let P, $\phi:X\to P$, and $\psi:P\to X$ be as described in the previous paragraph for this X, f and ε. Note that the composite $P \xrightarrow{\psi} X \xrightarrow{\phi} P$ has no fixed point; for if $y = \phi(\psi y)$ then $d(\psi y,f(\psi y)) \leq d(\psi y,\psi(\phi(\psi y))) + d(\psi(\phi(\psi y)),f(\psi y)) < \varepsilon$, which is not possible. Thus by the Lefschetz Index Theorem, $L(\phi(\psi)) = 0$. By property (iii) above we have $L(\psi(\phi)) = 0$. However, since f and $\psi(\phi)$ are homotopic, it is also true that $L(f) = 0$. Hence we have the following corollary of the Lefschetz Index Theorem.

COROLLARY. *If* X *is a compact metric ANR and* $f:X\to X$ *is a continuous function, then* $L(f) \neq 0$ *implies that* f *has a fixed point.*

(Insofar as the author is aware, the specific argument above is due to J. Dugundji.)

While the above have been formulated for homology with coefficients in a field, it follows from the fact that

$$H^i(X;F) \approx \mathrm{Hom}(H_i(X;F),F)$$

for F a field that $L(f) = \Sigma(-1)^i \mathrm{tr}(H^i(f;F))$ and that corresponding conclusions can be reached using cohomology.

As an example of the utility of cohomology in this context, let us consider a continuous self-map $f:P_n(C) \to P_n(C)$ on a complex projective space. With integer coefficients the cohomology of $P_n(C)$ is a graded polynomial ring on a generator β in dimension 2 with $\beta^{n+1} = 0$. There is an integer b such that $f^*\beta = b\cdot\beta$ in $H^2(P_n(C);Z)$. Using the ring structure, one can conclude that $L(f) = (1+b+b^2+\cdots+b^n)\cdot 1$ in F for any field F. Since it can happen in the field of rational numbers that $L(f) = 0$ only if $b = -1$ and n is odd, we conclude that for other b and n, f has a fixed point.

For n odd there is a fixed point free self-map on $P_n(C)$. Regard $P_{2k-1}(C)$ as the quotient space of $C^k \oplus C^k - 0$ by the complex lines in $C^k \oplus C^k$ through 0. Define the self-map $\tilde{f}$ on $C^k \oplus C^k$ by $x_1,x_2 \mapsto -\bar{x}_2,\bar{x}_1$. Then $\tilde{f}$ coinduces a continuous function $f;P_{2k-1}(C) \to P_{2k-1}(C)$. Were f to have a fixed point, then for some $0 \neq x_1,x_2$ in $C^k \oplus C^k$ and some $0 \neq w$ in C, $w\tilde{f}(x_1,x_2) = x_1,x_2$. Examination shows this implies the absurdity $w\bar{w} = -1$.

An interesting variant occurs if we consider self-maps f defined on quaternionic projective spaces $P_n(H)$. Their integral cohomology has a ring structure similar to that of complex projective space, with a generator β in dimension 4 instead of 2 as in the complex case. Identical analysis shows then that for $f^*\beta = b\cdot\beta$, the map f has a fixed point unless n is odd and $b = -1$.

For $n = 1$, $P_n(H)$ is the 4-sphere and of course the antipodal map on S^4 is fixed point free. For other odd values of n, it turns out that b cannot be -1; thus, for all $n \geq 2$, $P_n(H)$ has the fixed point property. To see this (following an argument of G. Bredon) we use coefficients Z_3 and the Steenrod reduced power $\mathcal{P}^1$. It is known that $\mathcal{P}^1\beta = \beta^2$ and that $\mathcal{P}^1 f^* = f^*\mathcal{P}^1$. Thus we have

$$f^*\beta^2 = f^*\mathcal{P}^1\beta = \mathcal{P}^1 f^*\beta = \mathcal{P}^1 b\cdot\beta = b\cdot\beta^2$$
$$f^*\beta^2 = (f^*\beta)\cup(f^*\beta) = (b\cdot\beta)\cup(b\cdot\beta) = b^2\cdot\beta^2 .$$

Thus $b \equiv b^2 (3)$ and in particular $b \neq -1$.

B. Local Indices - The Hopf Formula. Any self-map f on a finite simplicial complex K is homotopic to a self-map f' on K with the fixed point set of f' finite and with each fixed point of f' in a simplex of K which is not a face of another simplex of K. If x is such a fixed point and σ^n the simplex containing it, we can assume that x is the only fixed point of f' within σ and compare f' to the identity function, using the linear coordinates within σ. Thus we consider

$$(1_\sigma - f') : \sigma, \sigma - x \to R^n, R^n - 0 .$$

In integral homology in dimension n, this induces a homomorphism $g: Z \to Z$ and we let $\deg(f',x) = (-1)^n g(1)$. H. Hopf showed in [16] that

$$L(f) = L(f') = \sum_{x \in Fix(f')} \deg(f',x) .$$

The above formulation is inadequate in several respects, and the reader is referred to [9] or [6] for more details and more complete development of these ideas. The "local index" and its relation with the Lefschetz index have been subjects of considerable interest and study. We shall indicate briefly two modern lines of development below.

C. Atiyah-Bott Index Theory. This theory provides a refinement of the Lefschetz-Hopf formula of the previous paragraph in the setting of smooth manifolds. Let X be a compact, C^∞-manifold and $\pi: TX \to X$ be the tangent bundle of X. Let E be a smooth complex vector bundle over X, $\Gamma(E)$ be the space of smooth cross-sections of E, and $d: \Gamma(E) \to \Gamma(F)$ be a differential operator (a linear map given locally by a matrix of partial differential operators). If d is of order k, the terms of order k define a bundle map, the symbol, $\sigma_k(d): \pi^* E \to \pi^* F$, where $\pi^* E$ is the vector bundle over TX induced by $\pi: TX \to X$. (See [2] for exact definitions.)

A sequence $\cdots \to \Gamma(E^k) \xrightarrow{d_k} \Gamma(E^{k+1}) \to \cdots$ of complex vector bundles over X, only finitely many nonzero, and differential operators is called an elliptic complex provided

(i) $d_{k+1}(d_k) = 0$ for every k ;

(ii) the sequence $\cdots \to \pi^* E^k \xrightarrow{\sigma(d_k)} \pi^* E^{k+1} \to \cdots$ is exact over the complement of the 0-section of TX .

One defines cohomology $H^k(\Gamma E)$ of an elliptic complex in the usual manner. An endomorphism $T:\Gamma E \to \Gamma E$ is of course required to commute with the d_k and induces endomorphisms $H^k(T):H^k(\Gamma E) \to H^k(\Gamma E)$ for all k . In analogy with the Lefschetz number, let

$$L(T) = \sum_k (-1)^k \cdot tr(H^k(T)) .$$

Let $f:X \to X$ be a C^∞ self-map with graph transversal to the diagonal in $X \times X$. (This implies that the fixed point set of f is finite.) A lifting ϕ of f to the elliptic complex ΓE is a family of bundle morphisms $\phi^k:f^*E^k \to E^k$ such that for T^k defined to be the composite $\Gamma(E^k) \xrightarrow{f^*} \Gamma(f^*E^k) \xrightarrow{\Gamma\phi^k} \Gamma(E^k)$ we have $T_{k+1}(d_k) = d_k(T_k)$.

With these definitions Atiyah and Bott prove the

THEOREM. *Let* ΓE *be an elliptic complex over* X , $f:X \to X$ *be a smooth map with graph transversal to the diagonal,* ϕ *be a lifting of* f , *and* $T:\Gamma E \to \Gamma E$ *be the endomorphism induced by* ϕ . *Then* $L(T) = \sum_{x \in Fix(f)} \nu(x)$ *where* $\nu(x) = \Sigma(-1)^k \cdot tr(\phi_x^k)/|\det(1-df_x)|$.

This theorem and its proof are quite deep. There are a number of applications. For the de Rham complex of exterior powers on a smooth manifold, it is the classical Lefschetz-Hopf Index Theorem.

For a holomorphic $f:X \to X$ with X a complex manifold, the exterior powers of differential forms are bifiltered by dz_k and $d\bar{z}_k$ and the corresponding conclusions have an added degree of sensitivity reflecting the underlying complex structure. This is known as the Wood's Hole Fixed Point Theorem.

Further specific cases yield precise descriptions for isometries on Riemannian manifolds---the sort of thing occurring in transformation group situations. As an example one obtains the

THEOREM. *Let* X *be a compact, connected, oriented* C^{∞} *manifold of positive dimension and* $f:X\to X$ *be a smooth function satisfying* $f^{p^j} = 1_X$ *for some* j *and odd prime* p . *Then* f *does not have just one fixed point.*

(This theorem has also been proved by P. E. Conner and E. E. Floyd by very different techniques [8].) See also Example 1.E.

D. Vertical Fixed Point Theory. This is a relatively new line of development with major advances due to A. Dold [10], and J. C. Becker and D. H. Gottlieb [3]. One considers a commuting diagram

$$\begin{array}{ccc} X & \xrightarrow{f} & X \\ & {\scriptstyle p}\searrow \quad \swarrow{\scriptstyle p} & \\ & B & \end{array}$$

of spaces and maps. The fixed point set of f , Fix f , is then also a space over B . For each $b \in B$, we have $f_b:X_b \to X_b$ where $X_b = p^{-1}b$ and f_b is the restriction of f . Then $\mathrm{Fix}\, f = \cup \mathrm{Fix}\, f_b$. Thus we have a continuously varying family of maps f_b (and of spaces X_b) and fixed point sets Fix f_b parametrized over the space B . Of course, for B a singleton space, this is just ordinary fixed point theory. Analysis of the more general problem involves extraordinary cohomology theories over B and leads to results of a somewhat familiar form. However, this approach also leads to results on stable fiber homotopy equivalence and to another verfication of "the" Adams conjecture.

III. FIXED POINT PROPERTIES OF HOMOTOPY CLASSES OF MAPPINGS

Since the homomorphism $H_n(f):H_n(X) \to H_n(Y)$ induced by the continuous

$f:X \to Y$ is dependent only on the homotopy class of f, it is clear that for a self-map $f:X \to X$, the Lefschetz number $L(f)$ is dependent only on the homotopy class of f. Thus for a compact metric ANR space X and self-map $f:X \to X$, if $L(f) \neq 0$ and $g:X \to X$ is homotopic to f, then g has a fixed point. It is natural to consider whether the hypothesis that every self-map f of a given homotopy class has a fixed point implies that $L(f)$ be nonzero in some field. There are conditions under which this is true. We discuss this area in the present section.

Recall that the fundamental groupoid of a space X is the category πX with objects the points of X and whose morphisms $x \to y$ are homotopy classes of paths $u:I \to X$ with $u0 = x$ and $u1 = y$, the homotopies keeping endpoints fixed. Composition in the category is given by addition of paths. Each morphism has an inverse. The set of morphisms of πX from x to x with composition in the category as product is the fundamental group $\pi_1(X,x)$ at x.

For a continuous self-map $f:X \to X$ there is the induced functor $\pi f: \pi X \to \pi X$. Consider the subcategory $\mathrm{Fix}(\pi f)$ of objects and morphisms left fixed by πf. The category $\mathrm{Fix}(\pi f)$ is of course the sum of its components. (A component of a category A containing the object a of A is the full subcategory of A of all objects b of A joined by a finite string $a \to a_1 \leftarrow a_2 \to a_3 \leftarrow \cdots \to b$ of morphisms of A.) Since every morphism of πX has an inverse, in considering components of $\mathrm{Fix}(\pi f)$ we need only consider morphisms of the form $a \to b$.

Suppose that X is a polyhedron and $f:X \to X$ has only a finite number of fixed points. Then $\mathrm{Fix}(\pi f)$ has only a finite number of components. In a component C of $\mathrm{Fix}(\pi f)$ there is a finite number of points $x_1, \ldots, x_n$. We call C an essential component if $\Sigma \deg(f, x_i) \neq 0$. And define the Nielsen number of f, $N(f)$, to be the number of essential components of $\mathrm{Fix}(\pi f)$. It can be shown that if $g:X \to X$ also has only finitely many fixed points and g is homotopic to f, then $N(g) = N(f)$ Since every continuous self-map $f:X\to X$ of a polyhedron is homotopic to some $f':X\to X$ with only finitely many fixed points, we see by the above assertion that we can attach an integer $N(f)$ to any self-map of a polyhedron by selecting any such f', and that $N(f)$ is an invariant of the homotopy class of f. Notice that f has at least $N(f)$ fixed points.

There are combinatorial conditions on a triangulation of the polyhedron X, due essentially to F. Wecken [23] (see also [6]), which allow us to alter f by a homotopy so as to remove those components of $\mathrm{Fix}(\pi f)$ which are not essential and to coalesce each of the essential ones into a single fixed point. Thus in the presence of these conditions, a given self-map $f:X\to X$ of a polyhedron is homotopic to a map with exactly $N(f)$ fixed points. Triangulable manifolds always satisfy the Wecken conditions.

Determination of the Nielsen number is of course nontrivial, but there are situations in which it can be related with the Lefschetz number. Not surprisingly these involve the fundamental group, and certain subgroups of it.

In order to describe the relevant subgroups and their properties, we first discuss some generalities. Any continuous $p:X\to B$ has a factorization $X \xrightarrow{q} \Gamma(p) \xrightarrow{c} B$ in which $\Gamma(p)$ is the quotient space of X obtained by identifying each path component of each subspace $p^{-1}b$ of X to a point. In the event that p is a fibration, X is an ANR, and each inclusion $b \to B$ is a cofibration, then q is again a fibration and c is a covering map. The fibers of q are all path connected. For each $x \in X$, this determines a subgroup $C(x)$ of $\pi_1(B,px)$ by $\pi_1 c:\pi_1(\Gamma(p),qx) \subset \pi_1(B,px)$.

For any ANR space X and space Y, the function

$$ev\wedge pr_2:\langle X,Y\rangle\times X \to Y\times X$$

given by $f,x \mapsto fx,x$, where $\langle X,Y\rangle$ is the suitably topologized space of continuous functions $X \to Y$, is a fibration. If $X = Y$, then $X \times X$ is an ANR and each inclusion $x,x' \to X \times X$ is a cofibration. For $x_0 \in X$, let $Z \xrightarrow{p} X \times x_0$ be the restriction $ev \wedge pr_2$ to $X \times x_0$ and factor it $Z \xrightarrow{q} \Gamma(p) \xrightarrow{c} X$ as in the previous paragraph. Then for each $f: X \to X$, the pair f,x_0 is a point of Z and we obtain the subgroup

$$C(f,x_0) = T(f,x_0) \subset \pi_1(X,fx_0) ,$$

called the Chiang group. For $f = 1_X$ and $x_0 \in X$, the subgroup $C(1_X,x_0)$ of $\pi_1(X,x_0)$ is called the Gottlieb group, $G(X,x_0)$. Thus

$$G(X,x_0) = T(1_X,x_0) \subset \pi_1(X,x_0) .$$

From usual properties of covering spaces, a number of properties of these subgroups are easily established. Assume that X is path-connected. For any $f \in \langle X,X \rangle$ and $x_0 \in X$, there is $g \in \langle X,X \rangle$ with f homotopic to g and $gx_0 = x_0$. A choice of homotopy $\alpha: f \simeq g$ yields an isomorphism $\alpha_{\#}: T(g,x_0) \approx T(f,x_0)$. While a different homotopy from f to g will yield a different isomorphism, the two isomorphisms differ at most by a conjugation in $\pi_1(X,x_0)$.

If for $f,x_0 \in \langle X,X \rangle \times X$ we choose $g \in \langle X,X \rangle$ with f homotopic to g and $gx_0 = x_0$, it can be seen that

$$G(X,x_0) \subset T(g,x_0) \subset \pi_1(X,x_0) .$$

From this inclusion, properties of the Chiang group can sometimes be deduced from properties of the Gottlieb group. Examples of known properties of the Gottlieb group (see [15]) are given in the next two paragraphs.

For $x_0 \in X$, let $ev: \langle X,X \rangle \to X$ be evaluation at x_0; i.e., $ev(f) = fx_0$. If $x_0 \to X$ is a cofibration, then ev is a fibration. Let X be a path-connected polyhedron, and regard $1_X \in \langle X,X \rangle$ as the basepoint. Then using composition in $\langle X,X \rangle$ as "product," both $\langle X,X \rangle$ and $F = ev^{-1}(x_0) \subset \langle X,X \rangle$ are monoids with unit and the inclusion $F \to \langle X,X \rangle$ is a monoid morphism. These monoids have classifying spaces and the inclusion induces a continuous function $B_F \to B_{\langle X,X \rangle}$. The homotopy fiber of this function is the space X . In the homotopy sequence of the fibration the homomorphism $\pi_2 B\langle X,X\rangle \to \pi_1 X$ can be identified with

$$\pi_1 ev:\pi_1(\langle X,X\rangle,1_X) \to \pi_1(X,x_0) .$$

In this way we see that $\mathrm{Im}\ \pi_1(ev)$ lies in the center of $\pi_1(X,x_0)$. It follows that the Gottlieb group $G(X,x_0)$ lies in the center of $\pi_1(X,x_0)$.

If X itself is a path-connected H-space with unit x_0 , an easy argument shows that $G(X,x_0) = \pi_1(X,x_0)$. If X is an aspherical space, $X = K(\pi,1)$, then $G(X,x_0)$ is in fact the center of π .

The important fact about the Chiang group $T(f,x_0)$ is contained in the following.

THEOREM. *If* $T(f,x_0) = \pi_1(X,x_0)$ *and* $L(f) = 0$, *then* $N(f) = 0$.

Note in particular, that if X is a polyhedron satisfying the combinatorial conditions of Wecken, and if for $f:X\to X$ we have $L(f) = 0$ and $T(f,x_0) = \pi_1(X,x_0)$, then f is homotopic to some $g:X\to X$ with no fixed point.

More specifically we find in view of the previous remarks on the Gottlieb group that $T(f,x_0) = \pi_1(X,x_0)$ if any of the following conditions is satisfied.

(i) X is simply connected;

(ii) X is a path-connected H-space;

(iii) X is aspherical and $\mathrm{Im}\ \pi_1 f$ is in the center of $\pi_1 X$.

For more detail on this area we refer to the paper of D. H. Gottlieb [15] and to the monograph of R. F. Brown [6].

IV. SMITH THEORY

No discussion of fixed point theory can fail to mention the major discoveries of P. A. Smith in this subject. In a monumental sequence of papers written about 1930, Smith developed a homological description of the fixed point set of a self-map $f:X\to X$ satisfying the condition that

$f^p = 1_X$ for some prime p. His techniques anticipated and made major advances in areas involving spectral sequences, homological algebra, and cohomology operations, areas which did not receive definitive treatment for roughly twenty years after Smith achieved his breakthroughs. This author believes the level of innovation exhibited in this work of Smith's to be one of the most impressive achievements in topology.

Our brief discussion of Smith theory is based on the approach of R. G. Swan [22]. Let G be a finite group acting as a group of homeomorphisms on a space X ; assume the action is such that each point of X is left fixed by every element of G or by no element of G. This will be the case if G has no nontrivial proper subgroup.

Let $u:E \to K(G,1)$ be a universal principal right G-bundle. (In our case this just means that E is a universal covering space of $K(G,1)$.) Form the quotient space $E \times_G X$ by identifying eg,x with e,gx for all e,x in $E \times X$ and g in G, and define $r:E \times_G X \to K(G,1)$ by $r(eg,x) = u(e)$. The function r is well defined and is in fact a fibration with fiber the space X. Let $L \subset X$ be the subspace of those points of X left fixed by G and define $j:K(G,1)\times L \to E \times_G X$ by $k,\ell \mapsto [e,\ell]$ where e is any point of E satisfying $ue = k$. Since $eg,\ell \sim e,g\ell = e,\ell$, the choice of $e \in u^{-1}k$ is irrelevant and j is well defined. The diagram

$$\begin{array}{ccc} K(G,1)\times L & \xrightarrow{\;j\;} & E \times_G X \\ & {}_{pr_1}\searrow \quad \swarrow_{r} & \\ & K(G,1) & \end{array}$$

commutes.

If A is a subspace of X left invariant by the action of G on X, then the function r can be regarded as a relative fibration with relative fiber X,A, and the diagram above can be expanded to the following diagram

$$\begin{array}{ccc} K(G,1)\times L, K(G,1)\times(L\cap A) & \xrightarrow{\ j\ } & E \underset{G}{\times} X, E \underset{G}{\times} A \\ & {\scriptstyle pr_1}\searrow \qquad \swarrow{\scriptstyle r} & \\ & K(G,1) & \end{array}$$

of relative fibrations.

There are various spectral sequences for such fibrations r ; we expect for example

$$E_2^{p,q} = H^p(G;H^q(X,A))$$

and

$$E_\infty \sim H^*(E \underset{G}{\times} X, E \underset{G}{\times} A)$$

Spectral sequences are particularly gentle in their treatment of Euler characters.

We would like to assert that $H^*(j)$ is an isomorphism; this would enable us to complete our analysis of the spectral sequence and deduce many of the theorems of Smith. A glance at the triple

$$E \underset{G}{\times} A \to E \underset{G}{\times} A \cup K(G,1)\times L \to E \underset{G}{\times} X$$

and the excision isomorphism

$$H^*(E \underset{G}{\times} A \cup K(G,1)\times L, E \underset{G}{\times} A) \approx H^*(K(G,1)\times L, K(G,1)\times(L\cap A))$$

shows that $H^*(j)$ is an isomorphism if and only if

$$H^*(E \underset{G}{\times} X, E \underset{G}{\times} A \cup K(G,1)\times L) = 0 .$$

An examination of examples will convince the reader that the latter is not true in general.

In [22] R. G. Swan shows how to avoid this difficulty through the systematic use of Tate cohomology. Then the analysis proceeds roughly as outlined above. A number of (slight extensions of) results of Smith theory then follow. We mention a few of these.

THEOREM A. *Let* X *be a paracompact Hausdorff space such that every open covering has a finite dimensional refinement. If* $G = Z_p$ *for* p *a prime and*

$$H^i(X,A) = \begin{cases} 0 & i \neq m \\ Z_p & i = m \end{cases}$$

then for some r , $-1 \leq r \leq m$,

$$H^i(L, L \cap A; Z_p) = \begin{cases} 0 & i \neq r \\ Z_p & i = r \end{cases} .$$

(-1 *occurs only if* $A = \emptyset$ *and reduced cohomology is used in both cases.)*

THEOREM B. *If also* Z_p *acts trivially on* $H^*(X,A;Z)$, *then for all* k

$$\sum_{\substack{j \geq k \\ j-k \text{ even}}} \dim H^j(L, L \cap A; Z_p) \leq \sum_{\substack{j \geq k \\ j-k \text{ even}}} \dim H^j(X,A;Z_p) .$$

Moreover,

$$\chi(L, L \cap A; Z_p) = \chi(X,A;Z_p) .$$

THEOREM C. *If* X *is cohomologically locally connected over* Z_p , *then so is* L .

Notice for example that Theorem A implies that a cyclic group of prime order cannot act on a sphere so as to leave just one point fixed; equivalently, it cannot act freely on Euclidean space. The structure is such that by successive factorizations, we can draw the same conclusion for cyclic groups of prime-power order. However, for composite orders we have the last mentioned example in I.D. (Compare also with the second theorem in II.C.)

V. RELATED STRUCTURES AND APPLICATIONS

A. The Weil Formulas. The formulas in this section relate the Lefschetz numbers of the iterates of a self-map $f:X\to X$ with a certain rational function. In this setting they are rather easy to establish. However, analogous formulas were conjectured by A. Weil (see [25]) to hold for nonsingular algebraic varieties over fields of any characteristic. These Weil conjectures led to major developments in algebraic geometry and have recently been fully verified by Deligne.

Let V be a finite dimensional vector space over a field k and $\phi:V\to V$ be a linear map. Let

$$P(\phi,z) = \det(I-z\phi)$$

be the characteristic polynomial of ϕ in the variable z. If $\dim V = 0$, we set $P(\phi,z) = 1$.

Let X be a path-connected space with $H^*(X;k)$ finite dimensional. Let $f:X\to X$ be a self-map and set

$$P_i(z) = P(H^i(f;k),z) .$$

Define the rational function

$$\zeta_f(z) = \frac{P_1(z)\cdot P_3(z)\cdot P_5(z)\cdot \cdots}{P_0(z)\cdot P_2(z)\cdot P_4(z)\cdot \cdots}$$

THEOREM 1. *With the above hypotheses,*

$$\frac{d}{dz}\log \zeta_f(z) = \sum_1^\infty L(f^\nu)z^{\nu-1} \quad in \quad k([z]) .$$

Proof. The argument is just a formal manipulation. We can assume that k is algebraically closed and put ϕ in Jordan canonical form with diagonal entries $\lambda_1,\ldots,\lambda_r$. Then

$$P(\phi,z) = \prod_1^r (1-\lambda_i z)$$

$$\mathrm{tr}(\phi^\nu) = \sum_1^r \lambda_i^\nu .$$

Thus we have

$$-\frac{d}{dz}\log P(\phi,z) = \sum_{1}^{r} \frac{\lambda_i}{1-\lambda_i z}$$
$$= \sum_{i=1}^{r} \lambda_i \sum_{\nu=0}^{\infty} (\lambda_i z)^{\nu}$$
$$= \sum_{\nu=0}^{\infty} \left(\sum_{i=1}^{r} \lambda_i^{\nu+1} \right) z^{\nu}$$
$$= \sum_{\nu=1}^{\infty} \mathrm{tr}(\phi^{\nu}) z^{\nu-1}$$

Hence,

$$\sum_{\nu=1}^{\infty} L(f^{\nu}) z^{\nu-1} = \sum_{\text{even } i} -\frac{d}{dz}\log P_i(z) - \sum_{\text{odd } i} -\frac{d}{dz}\log P_i(z)$$
$$= \frac{d}{dz}\log \frac{\prod_{\text{odd } i} P_i(z)}{\prod_{\text{even } i} P_i(z)}$$
$$= \frac{d}{dz}\log \zeta_f(z) . \qquad \blacksquare$$

Consider next a compact oriented n-manifold M with orientation class $[M] \in H_n(M;Z)$. For a self-map $f:M \to M$ the degree d of f is the number satisfying $f_*[M] = d \cdot [M]$.

THEOREM 2. *For the self-map* $f:M \to M$ *of degree* d *on the compact oriented n-manifold* M,

(a) if n *is even, then*

$$\zeta_f(1/zd) = d^{\chi/2} \cdot z^{\chi} \cdot \zeta_f(z)$$

where χ *is the Euler character of* M,

(b) and if n *is odd, then*

$$\zeta_f(1/zd) \cdot \zeta_f(z) = \frac{\Pi \det H^{2i+1}(f)}{\Pi \det H^{2i}(f)} .$$

Sketch of Proof. Poincaré duality $D:H^i \approx H_{n-i}$ is given by $Du = u \cap [M]$. In general, $f_*(f^* u \cap [M]) = u \cap f_*[M] = d(u \cap [M])$. Thus,

$$\{D^{-1} H_{n-i}(f) D\} H^i(f) = d \cdot \mathrm{Id}: H^i(M) \to H^i(M) .$$

Hence if $P(H^i(f),z) = \prod_{j=1}^{r_i} (1-\lambda_{j,i}z)$, then $P(D^{-1}H_{n-i}(f)D,z) = \prod_{j=1}^{r_i} (1-\frac{d}{\lambda_{j,i}}\cdot z)$ and so $P(H^{n-i}(f),z) = \prod_{j=1}^{r_i} (1-\frac{d}{\lambda_{j,i}}\cdot z)$. Also $P(H^n(f),z) = 1-dz$. The remainder of the verification is just a formal exercise. ▯

COROLLARY (of Thm. 1). *Let* X *be a compact metric ANR with* $\chi(X) \neq 0$ *and* $f:X\to X$ *be a self-map such that for some field* k , $H^i(f)$ *is an isomorphism for all* i . *Then for some integer* ν , f^ν *has a fixed point.*

(In fact F. B. Fuller showed by a more careful analysis that $\nu \leq \mathrm{Max}(\sum_{i\ \mathrm{odd}} \beta_i, \sum_{i\ \mathrm{even}} \beta_i)$ where β_i is the dimension of $H^i(X,k)$.)

Proof. Suppose no f^ν has a fixed point. Then $L(f^\nu)$ is zero for all ν and so $\zeta_f(z) = 1$. Thus

$$P_0(z)\cdot P_2(z)\cdot P_4(z)\cdot \cdots = P_1(z)\cdot P_3(z)\cdot P_5(z)\cdot \cdots .$$

Since the $H^i(f)$ are isomorphisms, each $P_i(z)$ is a polynomial of degree β_i . Thus $\beta_0+\beta_2+\beta_4+\cdots = \beta_1+\beta_3+\beta_5+\cdots$ and $\chi(X) = 0$. ▯

We note the hypotheses of this corollary hold for any orientable closed manifold M with $\chi(M) \neq 0$ and any self-map $f:M\to M$ of nonzero degree. By way of contrast, consider an irrational rotation of a circle.

B. Nielsen's Theorem. A classical result stated by J. Nielsen is that if S is an orientable surface and $t:S\to S$ is an orientation preserving homeomorphism such that for some n , t^n is homotopic to the identity, then t itself is homotopic to a homeomorphism u such that u^n is the identity.

We outline here an argument using Smith theory for this result in the case n is a prime power; the argument is due to A. M. Macbeath [21].

Let S be of genus g . Let T be the Teichmuller space of complex structures on S_g with two identified if there is a biholomorphic map from one to the other which is homotopic to the identity. The space T is homeomorphic to Euclidean space R^{6g-6}. The homeomorphism t induces a homeomorphism $\bar{t}:T \to T$; and since $t^n \simeq 1_{S_g}$, $\bar{t}^n = 1_T$. Thus for n a prime power we conclude by Smith theory that for some analytic structure Σ on S_g, Σ and $t\Sigma$ are equal in T; i.e., there is a biholomorphic u with $u\Sigma = t\Sigma$ and $ut^{-1} \simeq 1_{S_g}$. Hence $u \simeq t$ and $u^n \simeq t^n \simeq 1_{S_g}$. But u^n is also biholomorphic and so $u^n = 1_{S_g}$.

The author does not know if there is a correct proof of the Nielsen assertion for composite n in print. A much stronger assertion has been announced recently, but is not yet in print.

C. Maximal Tori in a Lie Group. A. Weil gave [24] a fixed point argument for the theorem of E. Cartan that each element of a compact, connected Lie group lies in some conjugate of any maximal torus T in the group. The argument proceeds as follows. Let $N(T)$ be the normalizer of T in G. It has only finitely many path-components and the one containing the unit is T itself.

Let $g \in G$ and define $f:G/T \to G/T$ by $f(xT) = gxT$. Upon showing that $L(f) \neq 0$, we can conclude that f has a fixed point; i.e., for some $x \in G$, $xT = gxT$ and so $g \in xTx^{-1}$. Using the path-connectedness of G, we have a path from g to some generator g_0 of T. By this path we obtain a homotopy from f to f_0, defined using g_0 in place of g. Then $L(f_0) = L(f)$. An explicit computation in the tangent space of $(G/T)_e$ shows that $L(f_0) = |N(T)/T| > 0$. (For details, see pp. 89-92 of [1].)

The list of applications of fixed point theorems goes on and on. Among them, a proof of the Hurwitz theorem that the group of analytic automorphisms on a compact Riemann surface of genus greater than one is finite, W. S. Massey's verification of a conjecture of H. Whitney on the possible normal bundles of nonorientable surfaces imbedded in S^4, results of Rohlin and others on representing 2-dimensional homology classes in 4-manifolds by imbedded surfaces, and results of Milnor on compact groups of diffeomorphisms on a homology sphere with exactly two fixed points, the action being free elsewhere.

There is a major part of fixed point theory omitted in these remarks, that pertaining to functional analysis. It is the author's understanding that J. Dugundji is preparing a book which treats these aspects of the subject, as well as the sort of material alluded to here. Much of the material contained herein is treated more fully in the book by R. F. Brown [6]. Also we would like to mention the well prepared survey article by E. Fadell [12].

Respectfully dedicated to Professor F. B. Jones

REFERENCES

1. Adams, J. F., *Lectures On Lie Groups*, W. A. Benjamin, Inc., Mass., 1969.

2. Atiyah, M. F. and R. Bott, "A Lefschetz fixed point formula for elliptic complexes, I, II," *Annals Math. 86*(1967), 374-407; *88*(1968), 451-491.

3. Becker, J. C. and D. H. Gottlieb, "Transfer maps for fibrations and duality," *Comp. Math. 33*(1976), 107-133.

4. Becker, J. C. and R. E. Schultz, "Fixed point indices and left invariant framings," in: *Geometric Applications of Homotopy Theory*, Lecture Notes in Math. No. 657, Springer-Verlag, New York, 1978, 1-31.

5. Bredon, G., "Some examples for the fixed-point property," *Pacific J. Math. 38*(1971), 571-575.

6. Brown, R. F., *The Lefschetz Fixed Point Theorem*, Scott, Foresman & Co. Calif., 1971.

7. Conner, P. E. and E. E. Floyd, "On the construction of periodic maps without fixed points," *Proc. Amer. Math. Soc. 10*(1959), 354-360.

8. ______, "Maps of odd period," *Annals Math. 84*(1966), 132-156.

9. Dold, A., "Fixed point index and fixed point theorem for Euclidean neighborhood retracts," *Topology 4*(1965), 1-8.

10. Dold, A., "The fixed point index of fibre-preserving maps," *Inventiones Math.* *25*(1974), 281-297.

11. ______, "The fixed point transfer," *Math. Z.* *148*(1976), 215-244.

12. Fadell, E., "Recent results in the fixed point theory of continuous maps," *Bull. Amer. Math. Soc.* *76*(1970), 10-29.

13. Floyd, E. E., "Fixed point sets of compact abelian Lie groups of transformations," *Annals Math.* *66*(1957), 30-35.

14. Floyd, E. E. and R. W. Richardson, "An action of a finite group on an n-cell without stationary points," *Bull. Amer. Math. Soc.* *65*(1959), 73-76.

15. Gottlieb, D. H., "On fibre spaces and the evaluation map," *Annals Math.* *87*(1968), 42-55.

16. Hopf, H., "Über die algebraische Anzahl von Fixpunkten," *Math. Z.* *29* (1929), 493-524.

17. Husseini, S. Y., "The products of manifolds with the f.p.p. need not have the f.p.p.," *Bull. Amer. Math. Soc.* *81*(1975), 441-442.

18. Kister, J. M., "Examples of periodic maps on Euclidean spaces without fixed points," *Bull. Amer. Math. Soc.* *67*(1961), 471-474.

19. Knill, R. J., "Cones, products and fixed points," *Fund. Math.* *60*(1967), 35-46.

20. Lopez, W., "An example in the fixed point theory of polyhedra," *Bull. Amer. Math. Soc.* *73*(1967), 922-924.

21. Macbeath, A. M., "On a theorem of J. Nielsen," *Quarterly J. Math.* *13* (1962), 235-236.

22. Swan, R. G., "A new method in fixed point theory," *Comm. Math. Helv.* *34*(1960), 1-16.

23. Wecken, F., "Fixpunktklassen, I, II, III," *Math. Ann.* *117*(1941), 659-671; *118*(1942), 216-234 and 544-577.

24. Weil, A., *Demonstration Topologique d'un Théorème Fondamental de Cartan*, C. R. Acad. Sci. No. 200, Paris, 1935, 518-520.

25. ______, "Solutions of equations in finite fields," *Bull. Amer. Math. Soc.* *55*(1949), 497-508.

WEAKLY CONFLUENT MAPS ON TREES

C. A. Eberhart
J. B. Fugate

University of Kentucky, Lexington

The purpose of this paper is to announce some theorems which give sufficient conditions for a tree-like continuum to have the fixed point property. The principal result is Theorem 7, which extends theorems of Hamilton [H], McCord [M], and En-Nashef [E]. We begin with some definitions.

A *continuum* is a compact connected metric space. A *tree* is a continuum which is the union of a finite collection of arcs and contains no simple closed curve. A *simple triod* is a tree which is the union of three arcs having an end point of each as the common part of any two.

We will give three equivalent definitions for a continuum M to be tree-like:

1) For each $\varepsilon > 0$, there is a finite open cover $\mathcal{U}$ of M, mesh $\mathcal{U} < \varepsilon$, and nerve $\mathcal{U}$ is a tree.
2) For each $\varepsilon > 0$, there is a tree T and a surjective map $f:M \to T$ so that for each $t \in T$, diam $f^{-1}(t) < \varepsilon$. (Such a map is called an ε-*map* (cf. [M-S]).)
3) There is a sequence of trees T_i and surjective bonding maps $\pi_{ij}:T_j \to T_i$ so that $M = \text{Inv lim}(T_i, \pi_{ij})$.

For example, a ray limiting on a simple triod is simple-triod like. See Figure 1. The union of a convergent sequence of vertical segments, connected by a horizontal segment, is tree-like, but is not like any given tree. See Figure 2.

ISBN 0-12-481820-X

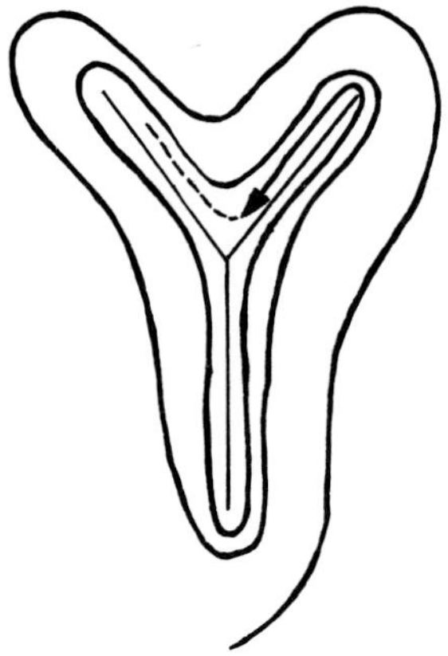

Fig. 1

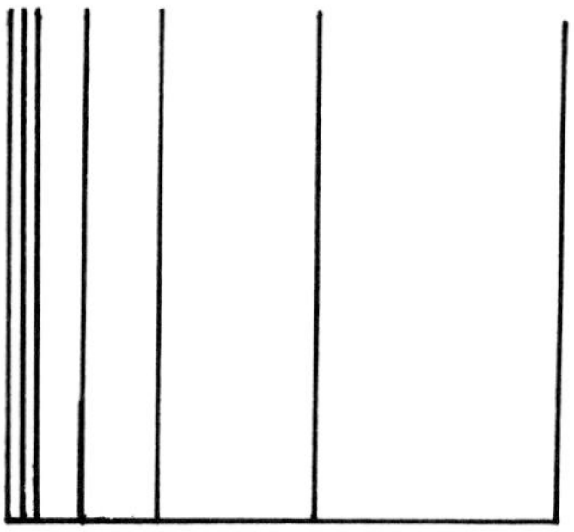

Fig. 2

A continuum which is *arc-like* (also called *chainable*) is obviously tree-like.

R. H. Bing asked [B] whether tree-like continua have the *fixed point property* (abbreviated fpp); i.e., if M is tree-like and $f:M \to M$ is a map, then there is a point $p \in M$ so that $f(p) = p$. We are primarily interested in two of the many theorems which give affirmative answers to special cases of Bing's question.

Hamilton's Theorem [H]. Arc-like continua have fpp .

McCord's Theorem [M]. If each bonding map π_{ij} is open, then M has fpp .

David Bellamy [Be] answered Bing's question by contructing a tree-like continuum which admits a fixed-point-free map. However, his example does not answer any of the following questions:

a) Must a planar tree-like continuum have fpp ?

b) Must an hereditarily indecomposable tree-like continuum have fpp ?

c) If S is a tree and M is S-like, must M have fpp?

One way to prove Hamilton's theorem is to use the following result.

Incidence Point Property for Arcs. If M is a continuum, $f,g:M \to [0,1]$ are maps and f is surjection, then there is a point $p \in M$ so that $f(p) = g(p)$.

Proof of Hamilton's Theorem. Let d be a metric for M and $h:M \to M$ be a map. Since M is compact, it is enough to show that for each $\varepsilon > 0$, there is an $m \in M$ so that $d(m,h(m)) < \varepsilon$. Since M is tree-like, using definition 2), we obtain an ε-map $P:M \to [0,1]$. Then P and Ph must have an incidence point m, so $d(m,h(m)) < \varepsilon$.

This argument suggests that we can obtain a fixed point theorem for tree-like continua by first proving an incidence point theorem for trees. The following example shows, however, that we cannot merely replace "arc" by "tree" in the Incidence Point Theorem for arcs.

Example 1. A pair of self maps of a simple triod which have no incidence point. $f(a) = a'$, $g(a) = a''$, etc. Both f and g are piecewise

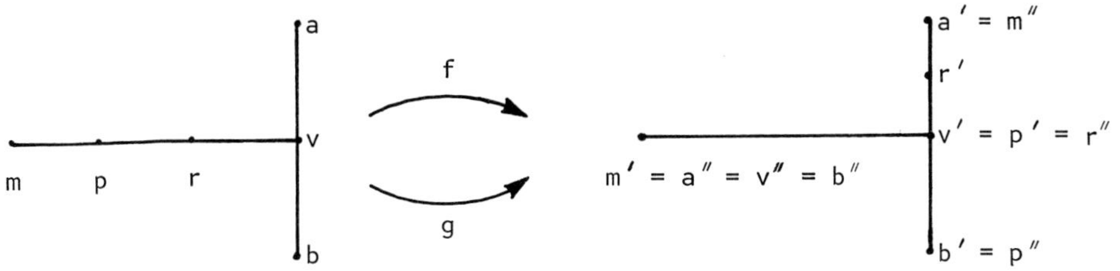

linear.

Thus, if we are to prove an incidence point theorem for trees, we need to restrict the maps. An appropriate restriction is given by the following pair of definitions.

DEFINITION. A map $f:X \to Y$ is *confluent (weakly confluent)* provided that if L is a subcontinuum of Y then every (some) component of $f^{-1}[L]$ is mapped onto L.

We combine several easily-established or well known ([C],[W]) results in the following theorem.

THEOREM 1. *Suppose that* X *and* Y *are continua and* $f:X \to Y$ *is a surjective map. If* f *is either monotone or open, then* f *is confluent. If* f *is either confluent or a retraction, or* Y *is an arc, then* f *is weakly confluent.*

Since maps of continua onto arcs must be weakly confluent, any arc-like continuum is an inverse limit of trees with weakly confluent bonding maps. Thus, McCord's theorem and Hamilton's theorem provided partial affirmative answers to a question of Howard Cook.

Cook's Question. Suppose that $M = \text{Inv} \lim(T_i, \pi_{ij})$, each T_i is a tree and each bonding map π_{ij} is weakly confluent. Must M have fpp ?

The remainder of the paper will discuss our answer to this question. The main results are Theorem 4 and Theorem 7. Before presenting these results, we want to recast our ideas in the language of universal mappings.

DEFINITION. A mapping $f: X \to Y$ is *universal* provided that if $g: X \to Y$ is a map, then f and g have an incidence point; i.e., there is an $x \in X$ so that $f(x) = g(x)$.

Clearly, universal maps must be surjections. A space has fpp if and only if the identity is universal. The Incidence Point Theorem for arcs says that any map of a continuum onto an arc is universal. Hamilton's theorem is then a corollary of the following theorem of Holsztynski [H].

THEOREM 2. *If the continuum* $M = \text{Inv} \lim(K_i, \pi_{ij})$, *each* K_i *is a polyhedron and each bonding map* π_{ij} *is universal, then* M *has* fpp.

Using this result, we can approach Cook's question by asking if a weakly confluent map between trees is universal. We first obtain a special case.

THEOREM 3. *If* T *is a tree and* $f: T \to T$ *is weakly confluent, then* f *is universal.*

Using this and Theorem 2, we obtain a fixed point theorem.

THEOREM 4. *If* T *is a tree,* $M = \text{Inv} \lim(T_i, \pi_{ij})$, *each* $T_i = T$ *and each* π_{ij} *is weakly confluent, then* M *has* fpp.

The next example shows that if we allow, in Theorem 3, the domain and range to be different trees, then the resulting statement is not true.

[D] Dyer, E., "A fixed point theorem," *Proc. Amer. Math. Soc.* *7*(1956), 662-672.

[E] En-Nashef, B., "Coincidence and fixed point theorems for functions on generalized trees," *Bull. Acad. Polon. Sci. Ser. Sci. Math. Astron. Phys.* *22*(1974), 943-948.

[H] Hamilton, O. H., "A fixed point theorem for pseudo arcs and certain other metric continua," *Proc. Amer. Math. Soc.* *2*(1951), 173-174.

[Ho] Holsztynski, W., "Universal mappings and fixed point theorems," *Bull. Acad. Polon. Sci.* *15*(1967), 433-438.

[M] McCord, M. C., "Inverse limit sequences with covering maps," *Trans. Amer. Math. Soc.* *114*(1965), 197-209.

[M-S] Mardesic, Sibe and Jack Segal, "ε-maps onto polyhedra," *Trans. Amer. Math. Soc.* *109*(1963), 146-164.

[W] Whyburn, G. T., *Analytic Topology*, Amer. Math. Soc., Rhode Island, 1942.

POLISH TRANSFORMATION GROUPS AND CLASSIFICATION PROBLEMS

Edward G. Effros*

University of California, Los Angeles

I. INTRODUCTION

In many instances, the classification of a family of mathematical objects simply amounts to describing the orbits of an appropriate transformation group. For the latter we shall use the notation (G,X) , where G is a group, X is a set, and we are given a map

(1.1) $$\phi: G \times X \to X : (s,x) \to s(x)$$

satisfying the usual relations. We let X/G denote the corresponding orbit space.

To illustrate our point, consider the following examples of orbit spaces:

(1) X/G = the similarity classes of $n \times n$ complex matrices. Here we let

X = the complex $n \times n$ matrices A

$G = GL(n,\mathbb{C})$ = the invertible elements T in X

$\phi(T,A) = TAT^{-1}$.

(2) X/G = the isomorphism classes of n-dimensional complex algebras:

X = the bilinear maps $M : \mathbb{C}^n \times \mathbb{C}^n \to \mathbb{C}^n$

$G = GL(n,\mathbb{C})$

$\phi(T,M)(a,b) = TM(T^{-1}a,T^{-1}b) \qquad (a,b \in \mathbb{C}^n)$

*This material is based upon work supported by the National Science Foundation under Grant No. MCS 80-02991.

ISBN 0-12-481820-X

(3) X/G = the isomorphism classes of the countably infinite discrete groups:

X = all maps $M : \mathbb{N} \times \mathbb{N} \to \mathbb{N}$ satisfying the group axioms

G = all bijections $T : \mathbb{N} \to \mathbb{N}$

$$\phi(T,M)(a,b) = TM(T^{-1}a, T^{-1}b) \qquad (a,b \in \mathbb{N})$$

(4) X/G = the unitary equivalence classes of irreducible unitary representations of a countable discrete group G on a given separable Hilbert space H:

X = all irreducible unitary representations ρ of G on a given Hilbert space H

$G = U(H)$ = all unitary maps $V : H \to H$

$$\phi(V,\rho)(g) = V\rho(g)V^{-1}$$

(5) X/G = the embeddings $h : S^2 \to S^3$ modulo the self-homeomorphisms of S^3:

X = all embeddings h of S^2 into S^3

G = the group of self-homeomorphisms $g : S^3 \to S^3$

$$\phi(g,h) = g \circ h .$$

In all of the above examples, X and G may be provided with natural Hausdorff topologies with respect to which (1.1) is continuous. In 1961, Glimm showed that if G and X and both locally compact and second countable, then a number of desirable "smoothness" conditions that might be satisfied by the given action are equivalent [11] (see below). Thus there is a fundamental dichotomy for such systems. However, as may be seen in examples (3) - (5) above, the transformation groups arising in classification theory are often not locally compact. This was particularly frustrating in the case of (4), where Glimm had already proved the existence of a similar dichotomy by using C^*-algebraic methods [10]. In fact the latter paper, in which he proved Mackey's celebrated "smooth dual if and only if type I" Conjecture [15,p.75], naturally led Glimm to the transformation group results.

In [8] the author succeeded in generalizing Glimm's theory to transformation groups (G,X) where G and X are Polish, i.e., metrizable by separable complete metrics. This in particular provided a transformation group proof of Mackey's Conjecture, and as we will see in §3, has had some interesting applications to topology as well.

An unfortunate aspect of [8] is that to obtain a complete generalization of Glimm's Theorem, it was necessary to impose two increasingly restrictive hypotheses on the system (G,X), unimaginatively called "conditions C and D". It was shown in [8] (using example (3) above) that the generalization was false if one did not impose some additional hypothesis. Nonetheless, the author felt that these conditions were somewhat cumbersome, and difficult to verify in certain situations. In §2 we will show that conditions C and D may both be replaced by the more natural hypothesis that the equivalence relation be an F_σ set. In §3 we consider how this improved result can be applied to examples related to those above. In particular we give an example for C^* - algebras (see (4') in §3) for which the F_σ condition holds, but condition D remains unverified.

II. THE F_σ CONDITION AND ITS IMPLICATIONS

Given a transformation group (G,X), we define the *G-equivalence relation* $R_G \subseteq X \times X$ by

$$R_G = \{(x,y) \in X \times X : y = gx \text{ for some } g \in G\} .$$

We recall that a subset of a topological space is said to be G_δ (respectively, F_σ) if it is a countable intersection of open sets (respectively, a countable union of closed sets). We will say that (G,X) is a smooth transformation group if it satisfies condition (7) in the following theorem.

THEOREM. *Let* (G,X) *be a Polish topological transformation group. Then the following are equivalent:*

(1) *For each* $x \in X$, *the map* $gG_x \to gx$ of G/G_x *onto* Gx *is a homeomorphism.*

(2) *Each orbit is second category in itself.*

(3) *Each orbit is* G_δ *in* X.

(4) X/G *is* T_0.

If in addition, $R = R(G,X)$ *is an* F_σ *subset of* $X \times X$, *these conditions are equivalent to each of the following:*

(5) *Each orbit is locally closed.*

(6) *The quotient topology on* X/G *generates the quotient Borel structure.*

(7) *The Borel structure on* X/G *is countably separated.*

(8) X/G *has no non-trivial atoms.*

(9) X *has no non-trivial ergodic measures.*

(10) X/G *is almost Hausdroff.*

(11) X/G *has a standard Borel structure.*

(12) *The quotient map* $\pi : X \to X/G$ *has a Borel transversal.*

(13) *The quotient map* $\pi : X \to X/G$ *has a Borel cross-section.*

(14) R *is a* G_δ *subset of* $X \times X$.

Proof. The equivalence of (1) - (4) was proved in [8, Theorem 2.1]. Now assume that R is an F_σ set. Since it is not clear that this implies the hypothesis condition C of [8, Theorem 2.6], in which the equivalence of (1) - (9) was proved, it is necessary to check each step of that demonstration. It is still true that each orbit is F_σ since the homeomorphism $y \to (x,y)$ of X into $X \times X$, carries Gx onto $(\{x\} \times X) \cap R$. Thus (2) $\Rightarrow$ (5) follows as before.

The only other application of condition C in the proof of equivalence for (1) - (9) occurred in (9) $\Rightarrow$ (4), in the disguise of Lemma 2.8.

In order to replace the latter, define a continuous map θ of $G \times X$ into $X \times X$ by

(2.1) $$\theta(s,x) = (sx,x) .$$

It is clear that $\theta(G \times X) = R$. If $s \in G$, define homeomorphisms f_s and g_s of $G \times X$ and $X \times X$, respectively, by

$$f_s(t,x) = (st,x) ,$$
$$g_s(y,x) = (sy,x) .$$

Then $\theta \circ f_s = g_s \circ \theta$. Let R_n be closed subsets of $X \times X$ with $R = \bigcup R_n$. Then $S_n = \theta^{-1}(R_n)$ is closed in $G \times X$, and $G \times X = \bigcup S_n$. Since $G \times X$ is of second category, one of the sets S_n must have interior. If $(s^{-1},x) \in \operatorname{int} S_n$, then $(e,x) \in \operatorname{int} f_s(S_n)$, and we have that $\theta(f_s(S_n)) = g_s(R_n)$ is closed in $X \times X$. Thus we may select a neighborhood M of e and an open set P in X for which the closure of $\theta(M \times P)$ lies in R . If $x \in P$ and Q_m is a basis of open sets at x , we claim that

(A) $$\bigcap_{m=1}^{\infty} \mathrm{Cl}[MQ_m] \subseteq Gx .$$

If y lies in the intersection, there exist sequences s_m in M and x_m in Q_m with $s_m x_m$ converging to y . It follows that $(s_m x_m, x_m) = \theta(s_m,x_m)$ converges to (y,x) . Since eventually x_m is in P , we have that $(y,x) \in R$, i.e., $y \in Gx$.

We modify the proof of (9) $\Rightarrow$ (4) as follows. After the reduction to the case $X = \mathrm{Cl}\pi^{-1}(p)$, we choose P and M as above. Since $\pi^{-1}(p)$ is dense in X , we may select $x \in \pi^{-1}(p)$ with $x \in P$. In the subsequent induction, we let $P(\emptyset) = P$. Lemma 2.8 was used in [8] to show that having selected $g(0),\ldots,g(n)$ in G and open sets $P(i_1,\ldots,i_k)$, $0 \leq k \leq n$, one then has for any sequence $1 \leq k_1 < \ldots < k_r \leq n$, and basis R_m at x ,

$$\text{(B)} \qquad \mathrm{Cl}\left[\bigcap_{m=1}^{\infty} Mg(k_1)\cdots g(k_r)R_m\right] \subseteq Gx\ .$$

We instead proceed as follows. By hypothesis $x \in P(0,\ldots,0_n)$, and

$$\begin{aligned} g(k_1)\cdots g(k_r)\ P(0,\ldots,0_n) &\subseteq g(k_1)\cdots g(k_r)\ P(0,\ldots,0_{k_r}) \\ &\subseteq g(k_1)\cdots g(k_{r-1})\ P(0,\ldots,1_{k_r}) \\ &\subseteq g(k_1)\cdots g(k_{r-1})\ P(0,\ldots,0_{k_{r-1}}) \\ &\subseteq \cdots \\ &\subseteq P(\emptyset) = P\ . \end{aligned}$$

Thus $y = g(k_1)\cdots g(k_r)\ x \in P$, and since $Q_m = g(k_1)\cdots g(k_r)R_m$ is a basis at y, we may apply (A). (B) is an immediate consequence, and the remainder of (9) $\Rightarrow$ (4) proceeds as before.

Turning to (4) $\Rightarrow$ (14'), let H_n be a countable basis of open sets for X/G (see [8, Lemma 2.3]). Assuming that X/G is T_0, one has for $p,q \in X/G$ that $p = q$ if and only if for each n, $p \in H_n$ if and only if $q \in H_n$. Letting $H_n' = \pi^{-1}(H_n)$ it follows that

$$R = \bigcap_n [H_n' \times H_n'] \cup [(X - H_n') \times (X - H_n')]\ .$$

Since closed sets in a separable metric space are G_δ, it follows that R is G_δ.

We next prove that (14') $\Rightarrow$ (10). If (G,X) satisfies (14), then R is both an F_σ and a G_δ subset of X. It follows that if F is a non-empty closed subset of X/G, then $(G,\pi^{-1}(F))$ is a Polish transformation group, and the corresponding equivalence relation $R \cap [\pi^{-1}(F) \times \pi^{-1}(F)]$ is also both G_δ and F_σ. Changing notation, it suffices to show that X/G has a non-empty, open Hausdorff subset.

Let $R = \cup R_n$, with R_n closed subsets of $X \times X$. Since R is A G_δ set, it is itself a Polish space, and thus of second category. It follows that one of the sets R_n has relative interior in R. If

$(s^{-1}x,x) \in \mathrm{int}_R R_n$, then since g_s is a homeomorphism of $X \times X$ with $g_s(R) = R$, $(x,x) \in \mathrm{int}_R g_s(R_n)$, and $Q = g_s(R_n)$ is also closed in $X \times X$. Let P be an open neighborhood of x in X with $(P \times P) \cap R \subseteq Q$. The restricted equivalence relation on P is

$$R_P = (P \times P) \cap R = (P \times P) \cap Q ,$$

and it is relatively closed in $P \times P$. It follows that P/R_P is Hausdorff (see [14, p.98]). Since π is an open mapping, a subset H of $\pi(P)$ is open if and only if $P \cap \pi^{-1}(H)$ is open, i.e., the spaces $\pi(P)$ and P/R_P are homeomorphic. $\pi(P)$ is the desired open Hausdorff subset of X/R .

The remaining implications proceed as in [8].

Remarks:

1. If (G,X) is a second countable locally compact transformation group, it is easy to give a direct proof that $R(G,X)$ is F_σ . Since $G \times X$ is K_σ , i.e., a countable union of compact sets and θ is continuous (see (2.1)), $R(G,X)$ is also K_σ , hence F_σ .

2. Condition D of [8] implies the existence of a neighborhood M of e such that (A) is true for all x in X . Equivalently, we have that if

$$R_M = \{(sx,x) : x \in X,\ s \in M\} = \theta(M \times X) ,$$

(see (2.1)) the closure of R_M is contained in R . The open sets tM, $t \in G$ cover G , hence there is a countable subcover t_nM . The sets

$$R_n = g_{t_n}(R_M) = \{(t_nsx,x) : x \in X,\ s \in M\}$$

are also closed, and $R = \cup R_n$. Thus condition D implies that R is F_σ .

3. It should be noted (14') is not the same as (14) in [8].

4. In contrast with [8], we used the joint continuity of $(g,x) \to gx$.

5. John Burgess [5] has recently shown that $(7) \Longleftrightarrow (11) \Longleftrightarrow (12)$

even if R is not an F_σ set. It would be tempting to conjecture that in this general context, these conditions are also equivalent to (8) (it was shown in [8] that these conditions are not equivalent to (4)).

III. SOME APPLICATIONS

So illustrate Theorem 2.1, let us consider the examples of §1. Since G and X are locally compact in (1) and (2), we have from Remark 1 in §2 that $R = R(G,X)$ is F_σ. In both cases the action is defined algebraically. It is essentially a theorem of Chevalley that such systems must be smooth (see [7, p. 183 bottom]). The points in X/G are explicitly described by the Jordan canonical form in example (1). To see that X/G is not T_1 even when $n = 2$, we note that the matrices $\begin{bmatrix} 1 & 0 \\ \varepsilon & 1 \end{bmatrix}$ are similar for all $\varepsilon > 0$, but converge to the non-similar-matrix $\begin{bmatrix} 1 & 0 \\ 0 & 1 \end{bmatrix}$ as $\varepsilon \to 0$. An explicit description of X/G in Example (2) seems to be unobtainable even for relatively low dimensions (see [19]).

Turning to Example (3), it was shown in [8, §3B] that (G,X) is Polish. In this case $R = R(G,X)$ is not F_σ. To see this let $Y \subseteq X$ be the abelian groups of exponent 4. As pointed out in [8], Y is a closed invariant subset of X such that Y/G is Borel isomorphic to a countable discrete set, but is not T_0. It follows from Theorem 2.1 that $R(G,Y) = R(G,X) \cap (Y \times Y)$ is not an F_σ subset of $Y \times Y$, and thus $R(G,X)$ is not an F_σ subset of $X \times X$.

It was shown in [8, Lemma 4.1] that (G,X) in Example (4) is a Polish transformation groups satisfying condition D and thus it has an F_σ-equivalence relation. On the other hand, the latter follows directly from [6, Lemma 3.7.3]. Let us consider a closely related example of considerable interest to C^*-algebraists:

(4') X/G = the unitary equivalence classes of irreducible representations of a separable, unital C^*-algebra A, where

$X = ES(A)$ = the extremal (pure) states p on A
$G = U(A)$ = the group of unitaries $u \in A$

$\phi(u,p)(a) = p(u^*au)$.

G and X are Polish in the relative topologies induced from the norm and weak* topologies on A and A^*, respectively (see [4, Corollary 1.2.3], [12, pp. 118-119]), and it is a simple matter to verify that $\phi : G \times X \to X$ is continuous. We have been unable to prove that (G,X) satisfies condition D. Nonetheless we have recently verified that the unitary equivalence relation $R = R(G,X)$ is an F_σ subset of $X \times X$ [9]. This is a simple consequence of the fact observed by Alfsen and Schultz [1], that if p,q are distinct pure states in the set of all states $S(A)$, then $(p,q) \in R$ if and only if the face $F(p,q)$ spanned by p and q is not just the line segment $[p,q]$ from p to q. If K is any metrizable compact convex set, then the set

$$Q = \{(p,q) \in K \times K : F(p,q) = [p,q]\}$$

is G_δ in $K \times K$ (this generalizes the fact that

$$E(K) = \{(p \in K : F(p,p) = \{p\}\}$$

in G_δ in K). Thus

$$R = E(K) \times E(K) \setminus Q$$

is relatively F_σ in $E(K) \times E(K)$.

The applications of Theorem 2.1 to topology were initially discovered by Hagopian [13]. Specifically he was concerned with a compact metric space X.

Letting G be the group of all homeomorphisms of X onto itself with the compact open topology, (G,X) is a Polish transformation group(see [20]. X is said to be *homogeneous* if for some $x \in X$ (and thus for all $x \in X$), $Gx = X$, If this is the case, it follows from Theorem 2.1 that the map $G \to Gx : s \to s(x)$ is an open mapping of G onto X. In particular, given an open neighborhood M of 1 in G, let N be an open neighborhood of 1 with $N^{-1} = N$ and $N^2 \subseteq M$. Then the sets Nx

form an open covering of X. Letting d be the metric on X, there exists a $\delta > 0$ such that for all $x,y \in X$, $d(x,y) < \delta$ implies $x,y \in N_z$ for some z, and thus $x \in M_y$. If we let D be the metric on G, we conclude that for all $\varepsilon > 0$, there exists a $\delta > 0$ such that $d(x,y) < \delta$ implies $y = g(x)$ for some $g \in G$ with $D(g,1) < \varepsilon$. This latter condition is known as "microtransitivity", and the fact that it is implied by transitivity has proved to be quite useful in the theory of homogeneous continua [3], [16], [17], [18].

It would be of some interest to see how Theorem 2.1 might be applied to Example (5) in §1. This has been studied by Ric Ancel [2]. We do not know if the corresponding equivalence relation is F_σ.

REFERENCES

1. Alfsen, E., and F. Shultz, "State spaces of Jordan algebras," *Acta Math.* *140*(1978), 155-190.

2. Ancel, R., "An alternate proof of Effros' Theorem," to appear.

3. Bellamy D., and L. Lum, "Simple closed curves in arcwise connected continua," to appear.

4. Browder, A., *Introduction to Function Algebras*, Benjamin, New York, 1969.

5. Burgess, J. "A selection theorem for group actions," *Pacific J. Math* *80*(1979), 333-336.

6. Dixmier, J., *Les C* -algèbras et leurs Représentations*, Gauthier-Villars, Paris, 1964.

7. ___________, "Représentations induites holomorphes des groupes résolubles algebriques," *Bull. Soc. Math. Fr.* *94*(1966), 181- 206.

8. Effros, E., "Transformation groups and C^*-algebras," *Ann. Math.* *81* (1965), 38-55.

9. ___________, "The structure theory of C^*-algebras, some old and new problems," to appear.

10. Glimm, J., "Type I C^*-algebras," *Ann. Math* *73*(1961), 572-612.

11. ___________, "Locally compact transformation groups," *Trans. AMS* *101* (1961), 124-138.

12. Godement, R., "Sur la theorie des representations unitaires," *Ann. of*

Math. *53*(1951), 68-124.

13. Hagopian, C., "Homeogeneous plane continua," *Houston J. Math 1*(1975), 35-41.

14. Hurewicz, W., and H. Wallman, *Dimension Theory*, Princeton University Press, Princeton, 1948.

15. Kelley, J., *General Topology*, van Nostrand, New York, 1952.

16. Mackey, G., "The Theory of Group Representations," *mimeo. notes*, University of Chicago, 1955.

17. Rogers, J., "Decompositions of homogeneous continua," to appear.

18. __________, "Completely regular mappings and homogeneous, aposyndetic continua," to appear.

19. __________, "Homeogeneous, separating plane continua are decomposable," to appear.

20. Chung-Yun Chao, "Infinitely many non-isomorphic nilpotent algebras," *Proc. AMS 24*(1970), 126-133.

ALMOST CONTINUOUS RETRACTS

B. D. Garrett

University of Alabama, University

All spaces herein are intended to be separable and metric. Notation and terminology are generally standard (I is the closed unit interval $[0,1]$, I^n is the n-cube, R is the real line, etc.). A continuum is a compact connected space and an arcwise connected space is nondegenerate. Whenever $M \subset (X \times Y)$, then $p(M)$ denotes the projection of M into the first factor X ; if $S \subset X$ then $M|S$ is the restriction of M to S or $\{(x,y)$ in $X \times Y : (x,y)$ is in M and x is in $S\}$.

By a ϕ retraction of X onto Y is meant a function f of type-ϕ such that $f: X \to Y$, $Y \subset X$ and $f(x) = x$, for x in Y . Then Y is called a ϕ retract of X .

Stallings defined the almost continuous functions in [18]. He was concerned with a gap in an argument from Hamilton's paper [8] in which Hamilton was answering J. Nash's question. Nash asked, does a connectivity function from an n-cell into itself leave a point fixed? A function $f: X \to Y$ is a connectivity function if, whenever C is a connected subset of X , then $f|C$ is connected, and f is almost continuous provided each open set of $X \times Y$ which contains f also contains a continuous function from X to Y .

In relation to almost continuous retracts, the most important parts of Stallings paper are Theorem 3, Proposition 1 and Question 10. From Theorem 3 of [18] we have

ISBN 0-12-481820-X

THEOREM 1. *If* X *has the fixed point property and* $f:X\to X$ *is almost continuous, then* f *leaves some point fixed.*

THEOREM 2 (Stallings' Prop. 1). *If the function* $f:X\to Y$ *is almost continuous and the function* $g:Y\to Z$ *is continuous, then* $gf:X\to Z$ *is almost continuous.*

In Question 10 of [18] Stallings describes an almost continuous retract from a 2-cell into itself and asserts that, based on these two theorems and the Brouwer Fixed Point Theorem, an almost continuous retract has the fixed point property. This assertion has been questioned [12][13] because of the application of Theorem 2 above in this situation. However, in [6] it is shown that for cases such as Stallings considered in his Question 10, Stallings assertion is valid. The problem in such application of Theorem 2 is with just which product spaces contain the open sets from the definition of almost continuity. By the statement of Theorem 2, and by Stallings proof, the open sets from the definition of almost continuity and which contain gf are in $X\times Z$ and not in $X\times X$. As has been observed in [12], [13], and [5], $gf:X\to X$ may be almost continuous, while $Z\subset X$ and $gf:X\to Z$ is not almost continuous. It is true that if $Z\subset X$ and $f:X\to Z$ is almost continuous, then $f:X\to X$ is almost continuous.

Because of the difficulty in applying Theorem 2, most of the work done on almost continuous retracts has been done using Hoyle's definition [10] which considers only the case where $M\subset X$ is an almost continuous retract of X provided there is an almost continuous function $f:X\to M$ and $f(x) = x$, for each x in M. Definition 1 below states the definition for almost continuity with emphasis upon the product space containing the open sets in question, and Definition 2, using that emphasis allows for consideration of both the sense used in Stallings' Question 10 and the more restricted sense as defined by Hoyle.

DEFINITION 1. The function f with domain X is said to be *almost continuous relative to* $X \times Y$ provided $f \subset (X \times Y)$ and each open set of $X \times Y$ containing f contains a continuous function $g: X \to Y$. (The notation $f: X \xrightarrow{X \times W} Y$ means that f is a function from X to Y which is almost continuous relative to $X \times W$.)

DEFINITION 2. Suppose each of X, Y, and M is a space with $M \subset X$ and $M \subset Y$. Then M is an *almost continuous retract of* X *relative to* $X \times Y$ means that there is an almost continuous function $f: X \to M$ relative to $X \times Y$ such that $f(x) = x$ for x in M. The function f is called an *almost continuous retraction of* X *onto* M relative to $X \times Y$.

In [18] Stallings described a certain closed subspace of I^2 which would intersect each continuous function $f: I \to I$ and whose complement therefore, if it contained a function $g: I \to I$, would be an open set containing no continuous function $f: I \to I$. The suggestion was that such a technique might be used to find a connectivity function from I to I which could not be almost continuous. Roberts [16], Cornette [3], and Jones and Thomas [11] each used such a technique to show the existence of a connectivity function from I to I which is not almost continuous relative to I^2. In [4], Garrett and Kellum, following Roberts, named such subsets of a product space "blocking sets." Using these blocking sets, it is possible under "nice" conditions to construct functions which are either not almost continuous or are almost continuous. Most of the results which are known about almost continuous retracts depend upon such a construction.

DEFINITION 3. The subset M of $X \times Y$ is a *blocking set of* $X \times Y$ provided M is closed in $X \times Y$, $(X \times Y) - M$ contains a function $f: X \to Y$, and whenever $g: X \to Y$ is a continuous function, g intersects M. A blocking set of $X \times Y$ is an *irreducible* (or minimal) *blocking* set of $X \times Y$ if no proper subset of M is a blocking set of $X \times Y$.

THEOREM 3. *If each of* X *and* Y *is a continuum and* M *is a blocking set of* $X \times Y$, *then* M *contains an irreducible blocking set of* $X \times Y$.

THEOREM 4. *In order that the function* $f: X \to Y$ *be almost continuous, it is necessary and sufficient that* f *intersect each irreducible blocking set of* $X \times Y$.

Whenever the domain projection of blocking sets (or some particular subset of the domain projection of blocking sets) must be at least as numerous as the collection of blocking sets, then transfinite induction can be used to construct functions intersecting each blocking set (or subset of a blocking set). Such techniques are used in [4],[12]-[15], and [17].

THEOREM 5 [17]. *If the subset* M *of* I^n *is either arcwise connected or an almost continuous retract of* I^n, *relative to* $I^n \times M$, *then for each irreducible blocking set* K *of* $I^n \times M$, $\overline{p(K)}$ *is connected.*

THEOREM 6 [1],[17]. *If* M *is an arcwise connected subcontinuum of* I^n, *then* M *is an almost continuous retract of* I^n, *relative to* $I^n \times M$, *if and only if, for each blocking set* K *of* $I^n \times M$, *with* $p(G) \subset M$, K *intersects the diagonal of* $M \times M$.

In Question 10 of [18], Stallings proposed the use of almost continuous retracts in attacking the question of whether or not each nonseparating plane continuum has the fixed point property. More generally, in his introduction, he suggested that noncontinuous functions could be used for such a purpose. In [9] Hildebrand and Sanderson consider connectivity retracts and ask if each nonseparating plane continuum is a connectivity retract of a disk. Cornette [3] showed that a connectivity retract of a unicoherent Peano continuum is a unicoherent Peano continuum, thereby ending that line of attack. Kellum in [12]-[14] showed that the almost continuous retracts offered hope.

THEOREM 7 [12]. *There exists a noncompact and nonarcwise connected subset* M *of* I^2 *which, relative to* $I^2 \times M$ *, is an almost continuous retract of* I^2 *.*

THEOREM 8 [13]. *There exists an arcwise connected but nonlocally connected subcontinuum* M *of* I^2 *which, relative to* $I^2 \times M$ *, is an almost continuous retract of* I^2 *.*

THEOREM 9 [14]. *There exists a subcontinuum* M *of* I^2 *which is not locally connected and is not arcwise connected, but which, relative to* $I^2 \times M$ *, is an almost continuous retract of* I^2 *.*

Since in these cases almost continuity is considered relative to $I^2 \times f(x)$, the difficulties with Stallings, Proposition 1, mentioned above, did not interfere. From the following theorem we have that each such "retraction" will have the fixed point property.

THEOREM 10 [18]. *If* M *is an almost continuous retract of* I^n *relative to* $I^n \times M$ *, then* M *has the fixed point property.*

Since at the time these last few results were being discovered, no way of getting by the problems of using Stallings Proposition 1 seemed available [12],[13], and since there seemed no need to emphasize the relative product space which determined almost continuity, Kellum stated Theorem 11 below as follows: An almost continuous retract of a Peano continuum must be almost arcwise connected. This did seem to end the hope of using almost continuous retracts for such attacks on the more familiar fixed point problems.

DEFINITION 4. The space X is *almost arcwise connected* if for each pair of open sets in X there is an arc in X intersecting both of them.

THEOREM 11 [12],[13]. *If* M *is an almost continuous retract of the Peano continuum* X *, relative to* $X \times M$ *, then* M *is almost arcwise connected. In particular, a pseudo-arc* A *, cannot be an almost continuous* *retract of disk* D *, relative to* D×A .

For "nice" spaces, such as n-cells, Stallings assertion from Question 10 of [18] can be seen to be true by using the following.

THEOREM 12 [6]. *Suppose each of* X, Y, Z *, and* W *is a space, either* $W \subset Y$ *or* $Y \subset W$ *, the function* $f:X \to Y$ *is almost continuous relative to* X×W *, and the continuous function* $g:f(X) \to Z$ *can be extended to a continuous* $G:W \to Z$ *. Then* $gf:X \to Z$ *is almost continuous relative to* X×Z .

THEOREM 13 [6],[18]. *If the continuum* M *is an almost continuous retract of* I^n *, relative to* $I^n \times I^n$ *, then* M *has the fixed point property.*

Contrasting Theorem 11,

THEOREM 14 [6]. *There is an almost continuous retract of* I^2 *which is not almost arcwise connected.*

If one looks for limitations to such an attack on the problem of which nonseparating plane continua have the fixed point property, then looking at restrictions on the types of images that are possible might be an indication. Cornette's result for connectivity retracts [3] has about ended the hope for obtaining any results with them and Kellum's result (Theorem 11) seemed to eliminate the almost continuous retracts. In [12] he states that with a simple modification of the proof of 3.16 from [9], it can be shown that if M is a subset of the interior of I^n and is an almost continuous retract of I^n , relative to $I^n \times M$, then M does not separate I^n . Recently Rosen showed me an argument which is a combination of that from [9] and Theorem 12 with which he showed that, if M is an almost continuous retract of I^2 , relative to $I^2 \times I^2$, which is a subset of the interior of I^2 , then M does not separate I^2 . (This argument did not generalize to I^n .) So far as I know, the first example for an almost continuous image of a Peano continuum which is not almost arcwise connected was in [5]; Theorem 14 above, from [6] gives another example. Kellum has recently shown that whenever the subcontinuum M of I^2 is the intersection of disk, then M is the image of an almost continuous $f: I^2 \xrightarrow{I^2 \times I^2} M$. These results are encouraging in that continua which are almost continuous retracts of a disk do not separate the plane and *may* be of just about any type.

There are other fixed point problems than the plane continuum fixed point problem which can be attacked with almost continuous retracts. At the March 1980 annual Spring Topology Conference in Birmingham, I presented an argument for Theorem 15 (and almost had time to finish one for Theorem 16). These give a partial result to Question 7 of [2].

THEOREM 15. *If the continuum* M *is an almost continuous retract of the disk* X, *relative to* $X \times X$, *the continuum* N *is an almost continuous retract of the disk* Y *relative to* $Y \times Y$, *and* $X \cap Y = M \cap N$ *is an arc, then the continuum* $M \cup N$ *is an almost continuous retract of* $X \cup Y$ *relative to the space* $(X \cup Y) \times (X \cup Y)$.

THEOREM 16. *If the continuum* M *is an almost continuous retract of a disk* X, *relative to* $X \times X$, *and* D *is a disk such that* $M \cap D$ *is an arc, then* $M \cup D$ *has the fixed point property.*

Proofs for some of these theorems do not appear in print (Theorems 12-16). Theorem 12 can be proved by altering the argument given by Stallings in [18] for Proposition 1. To prove Theorem 12, replace g_* of Stallings' argument with G of the statement of Theorem 12 and the argument goes through. Theorem 12, the Brouwer Fixed Point Theorem, and Theorem 1 can be used to prove Theorem 13. Theorem 16 follows from Theorem 15 and Theorem 1, where D in Theorem 16 is N in Theorem 15 and $D = N = Y$. A proof of Theorem 16 can be developed using a method similar to that given in the example below which combines almost continuous retractions in order to prove Theorem 15.

EXAMPLE. With $I = [-1,1]$, for a compact set that is not almost arcwise connected, take M to be the set of all points and limit points in I^2 of $q:(I-0) \to R$, where $q(x) = \sin \frac{1}{x}$. Define also $X_1 = [-1,0]\times I$, $X_2 = [0,1]\times I$, $M_1 = M \cap X_1$, $M_2 = M \cap X_2$, and $A = M_1 \cap M_2$.

From [4], M_1 is an almost continuous retract of X_1 relative to $X_1 \times M_1$ and M_2 is an almost continuous retract of X_2 relative to $X_2 \times M_2$. Then there exist an almost continuous retraction $f_1 : X_1 \to M_1$ relative to $X_1 \times M_1$ and an almost continuous retraction $f_2 : X_2 \to M_2$ relative to $X_2 \times M_2$; in both cases, it is easily shown that the second factor in the relative product space can be replaced by I^2 [1]. Denote by f the function from I^2 to I^2 which is the union of f_1 and f_2, and denote by U an open set of $I^2 \times I^2$ containing f.

For some positive number $\varepsilon > 0$, there is a finite collection of open ε-balls in I^4 covering $f|_A$, whose union is called W, and such that each point of $\bar{W} \cap I^4$ is in U. Suppose $s = d(f|_A, \bar{W}-W)$ and suppose $c > 0$ is less than $\sqrt{\frac{1}{2}}s$. Because $[-c,c]\times I = C$ is a 2-cell and is homeomorphic to $Id(C)$, the set L which is the set of all points (x,y) in I^4 with x in C and $d((x,y), Id(C)) < \sqrt{\frac{1}{2}}s$ is homeomorphic to $C \times I^2$. Since, for each point p of L, $d(p, f|_A) < s$, the set L is a subset of $W \cap U$.

Define V to be the set formed by adding to $U|_{I^2-A}$ the set of all interior points relative to I^4 of the set L which are in $A \times I^2$. From Definition 1 there are continuous functions $g_1 : X_1 \to I^2$ and $g_2 : X_2 \to I^2$ lying in V and therefore in U. Using the uniform continuity of these functions, with $r = d(g_1|_A \cup g_2|_A, Bd_{I^4}(L))$, there is a positive number $e < c$ such that, if p is a point of $[-e,e] \times I$, either $d(g_1(p), g_1|_A \cup g_2|_A) < r$ or $d(g_2(p), g_1|_A \cup g_2|_A) < r$. Now we have that $g_1|_{-e\times I}$ and $g_2|_{e\times I}$ are subsets of L and we define $h:(-e\times I) \cup (e\times I) \to I^2$ to be the union of these two sets. For $E = [-e,e] \times I$, the subset $L|_E$ of L is homeomorphic to $E\times I^2$. Using this homeomorphism, h can be extended to a continuous function $H:E \to I^2$, which is g_1 on $-e\times I$ and is g_2 on $e\times I$, and each point of which is a point of L.

Each of the functions $g_1|_{[-1,e]\times I}$, H, and $g_2|_{[e,1]\times I}$ is a continuous function and the function $g:I^2 \to I^2$, which is their union, is a continuous function which is a subset of U. This shows that $f:I^2 \to M$, where $f|_M = Id(M)$ is almost continuous, and, relative to $I^2\times I^2$, M is an almost continuous retract of I^2.

REFERENCES

1. Alexander, D. L. and B. D. Garrett, "Blocking sets and almost continuous functions," preprint.

2. Bing, R. H., "The elusive fixed point property," *Amer. Math. Monthly* *76*(1969), 119-132.

3. Cornette, J. L., "Connectivity functions and images of Peano continua," *Fund. Math.* *75*(1966), 184-192.

4. Garrett, B. D. and K. R. Kellum, "Almost continuous real functions," *Proc. Amer. Math. Soc.* *33*(1972), 181-184.

5. Garrett, B. D., "Almost continuous images of Peano continua," preprint.

6. ______, "Almost continuity," preprint.

7. Hagopian, C. L., "Almost arcwise connected continua without dense arc components," preprint.

8. Hamilton, O. H., "Fixed points for certain noncontinuous functions," *Proc. Amer. Math. Soc.* *8*(1957), 750-756.

9. Hildebrand, S. K. and D. E. Sanderson, "Connectivity functions and retracts," *Fund. Math.* *57*(1965), 237-245.

10. Hoyle, H. B., "Connectivity maps and almost continuous functions," *Duke Math. J.* *37*(1970), 671-680.

11. Jones, F. B. and E. S. Thomas, Jr., "Connected G_δ graphs," *Duke Math. J.* *33*(1966), 341-345.

12. Kellum, K. R., "Noncontinuous retracts," *Studies In Topology*, Academic Press, New York, 1975, 255-261.

13. ______, "On a question of Borsuk concerning noncontinuous retracts I," *Fund. Math.* *87*(1975), 89-92.

14. Kellum, K. R., "On a question of Borsuk concerning noncontinuous retracts II," *Fund. Math.* *92*(1976), 135-140.

15. ______, "The equivalence of absolute almost continuous retracts and ε-absolute retracts," *Fund. Math.* *96*(1977), 229-235.

16. Roberts, J. H., "Zero-dimensional sets blocking connectivity functions," *Fund. Math.* *57*(1965), 173-179.

17. Rosen, H., "Connected projections of blocking sets of $I^n \times M$," *Proc. Amer. Math. Soc.*, to appear.

18. Stallings, J., "Fixed point theorems for connectivity maps," *Fund. Math.* *47*(1959), 249-263.

λ-CONNECTED PRODUCTS

Charles L. Hagopian+

California State University, Sacramento

In 1948 F. Burton Jones [2] proved that the product of any two continua is aposyndetic.* Such a product may fail to be arcwise connected. In fact, the product of two continua is arcwise connected only if both factors are arcwise connected. However, the following question is open.

Question. Must the product of two continua be λ-connected?

DEFINITION (Kuratowski [4, p. 262]). A continuum M is of *type* λ if M is irreducible and each indecomposable subcontinuum of M is a continuum of condensation.

If a continuum M is of type λ, then M admits a monotone upper semi-continuous decomposition to an arc with the property that each element of the decomposition has void interior relative to M.

DEFINITION (Knaster and Mazurkiewicz [3][1]). A continuum M is *λ-connected* if for every pair p,q of points of M, there exists a continuum of type λ in M that is irreducible between p and q.

One might hope to answer this question by showing that every aposyndetic continuum is λ-connected and then applying Jones's product theorem. This is impossible. The continuum pictured below is aposyndetic and not

+The author was partially supported by a grant from the National Science Foundation, Grant No. MCS79-16811.

*A *continuum* is a nondegenerate compact connected metric space. See Jones's article in these Proceedings for the definition of aposyndesis.

ISBN 0-12-481820-X

λ-connected. It consists of two copies of Knaster's indecomposable continuum [5, p. 204] and a collection of intervals joining the corresponding endpoints of all semicircles.

The following theorem may be of interest to those who are looking for a product that is not λ-connected.

THEOREM. *The product of any two hereditarily indecomposable continua is* λ-*connected.*

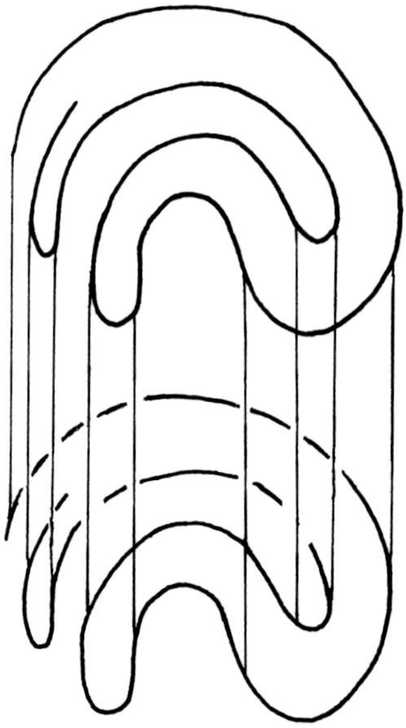

REFERENCES

1. Hagopian, C. L., "Mapping theorems for plane continua," *Topology Proc.* *3*(1978), 117-122.

2. Jones, F. B., "Concerning non-aposyndetic continua," *Amer. J. Math.* *70* (1948), 403-413.

3. Knaster, B. and S. Mazurkiewicz, "Sur un probleme concernant les transformations continues," *Fund. Math.* *21*(1933), 85-90.

4. Kuratowski, C., "Theorie des continus irreductibles entre deux points II," *Fund. Math.* *10*(1927), 225-276.

5. ______, *Topology, Vol. 2*, 3d ed., Monografie Mat., TOM 21, PWN, Warsaw, 1961; English transl., Academic Press, New York; PWN, Warsaw, 1968.

A CLASSIFICATION THEOREM IN TOPOLOGY

V. Kannan

School of Mathematics & CIS
University of Hyderabad
Hyderabad, India

We consider the class $\mathcal{C}$ of all countable topological spaces admitting a complete metric and having derived length $\leq \omega$. It is a result of Sierpinski and also of DeGroot that this class contains exactly c homeomorphic types, where c is the cardinality of the continuum. We here describe a method of associating some number-invariants for members in this class. The main theorem asserts that these invariants completely determine the space up to homeomorphism.

I. INTRODUCTION

The problem of classifying topological spaces up to homeomorphism, by means of certain associated invariants, is a well known one. Usually, in many known theorems, the invariants are not enough to specify the spaces uniquely. But, for certain nice classes of spaces, a complete classification has been obtained. For example, [M-S] obtained a complete set of invariants (as ordinal pairs) for the class of countable compact Hausdorff spaces. DeGroot [G] extended the result to countable locally compact Hausdorff spaces. By a result of [S], the problem becomes trivial, for the class of countable perfect metrizable spaces. By a result of [G] and [S], no finite-tuple of countable ordinals can suffice, when we consider the class $\mathcal{C}$ of countable spaces admitting a complete metric, even if we

ISBN 0-12-481820-X

consider only spaces of derived length $\leq \omega$. Our theorem shows that sequences of finite ordinals, extracted as topological invariants, do the job in this class. Our invariants are not only global but also local in the sense that they may be defined for each point of the space. Our theorem thus describes a sequence of topological properties of points of spaces in the class C in a definite order and asserts that there is a homeomorphism taking x to y if and only if x and y have the same properties in our list.

II. A DESCRIPTION OF THE INVARIANTS

We write a sequence of finite sequences as follows:

$$a_{0,1};b_{0,1} \ ; \ a_{1,1},a_{1,2};b_{1,1},b_{1,2} \ ;$$

$$a_{2,1},a_{2,2},\ldots,a_{2,t_2};b_{2,1},b_{2,2},\ldots,b_{2,t_2} \ ; \ \cdots$$

$$\cdots$$

$$a_{n,1},a_{n,2},\ldots,a_{n,t_n};b_{n,1},b_{n,2},\ldots,b_{n,t_n} \ ; \ \cdots .$$

Here, t_n is a positive integer, arising naturally out of some combinatorial questions in topology, and is independent of the space for which the invariants are computed. Further $t_0 = 1$ and $t_1 = 2$. The numbers $a_{i,j}$ and $b_{i,j}$ vary with the space X. Each $a_{i,j}$ is a cardinal number, taking one of the values between 0 and $\aleph_0$. Each $b_{i,j}$ takes one of the two values, 0 or 1. These are defined as below. For each space X in C and each pair (i,j) in the set D of all pairs (i,j) of nonnegative integers with $1 \leq j \leq t_i$, we define the sets $A_{i,j}(X)$ recursively by induction as follows:

$A_{0,1}(X)$ = the set of all isolated points of X.

If $A_{i,j}(X)$ has been already defined for all $i < n$ and $1 \leq j \leq t_i$, then to define $A_{n,j}(X)$, we let $D_n = \{(i,j):0\leq i<n;1\leq j\leq t_i\}$ and define a function g ($=g_n$) from $A_n X D_n$ to $\{0,1,2\}$ as below. Here A_n is the

set of all points of X having derived length n . This is defined as follows:

Let X be a topological space and let $x \in X$. Then we define the *derived length* of X at x (denoted by $\delta(X,x)$) recursively as follows:

$$\begin{cases} \delta(X,x) = 0 \text{ if and only if } x \text{ is isolated in } X ; \\ \quad \text{for any ordinal number } \alpha , \\ \delta(X,x) = \alpha+1 \text{ if and only if } x \text{ is isolated in the subset } \{y \in X: \\ \quad \delta(X,y) \not\leq \alpha\} ; \\ \delta(X,x) \text{ is never a limit ordinal.} \end{cases}$$

X is scattered if and only if X has a derived length at each of its points. In this case the derived length of X is defined as $\delta(X) = \sup\{\delta(X,x) : x \in X\}$. This is also known as the Cantor-Bendixon length of X .

If $x \in A_n$ and $(i,j) \in D_n$, then

$$g(x,(i,j)) = \begin{cases} 0 & \text{if } A_{i,j} \cup \{x\} \text{ is discrete,} \\ 1 & \text{if } A_{i,j} \cup \{x\} \text{ is not discrete, but there is a relative neighborhood of } x \text{ with compact closure,} \\ 2 & \text{otherwise.} \end{cases}$$

This function induces a partition of A_n thus: two points x_1 and x_2 are put in the same class if and only if $g(x_1,(i,j)) = g(x_2,(i,j))$ for every (i,j) in D_n . The number of these partition classes (that are nontrivial for at least one space X) is found to be t_n . There is a definite total ordering on these classes, since we have a natural total order both on D_n and on $\{0,1,2\}$. Now let $A_{n,j}(X) = \{x \in A_n : x$ is in the jth class above$\}$. This completes the inductive definitions of the sets $A_{n,j}$. We finally let $a_{n,j}(X)$ = the cardinality of $A_{n,j}(X)$ and $b_{n,j}(X)$ = 0 or 1 according as the closure of $A_{n,j}(X)$ is compact or not.

III. THE STATEMENTS OF RESULTS

THEOREM 1. *Let* X *and* Y *be two spaces in* C . *Then they are homeomorphic if and only if* $a_{i,j}(X) = a_{i,j}(Y)$ *and* $b_{i,j}(X) = b_{i,j}(Y)$ *hold for every* (i,j) *in* D .

THEOREM 2. *Let* X *be a space in* C *and let* $x,y \in X$. *Then the following are equivalent:*

a) There is a self-homeomorphism h *of* X *such that* $h(x) = y$.

b) There is a pair (r,s) *in* D *such that both* x *and* y *belong to* $A_{r,s}(X)$.

THEOREM 3. *Let* X *be a space in* C , *let* F *be a discrete closed subset of* X *and let* Θ *be any permutation of* F . *Then* Θ *can be extended to a self-homeomorphism of* X *if and only if* $\Theta(F \cap A_{r,s}) = F \cap A_{r,s}$ *holds for every* (r,s) *in* D .

A brief outline of the Proofs. We consider the well-ordered set $0 < 0^* < 1 < 1^* < 2 < 2^* < \cdots < n < n^* < n+1 < \cdots$. We say that a space X in C is of type n , if it has a unique point having derived length n . It is said to be of type n^* if it is not of type n , and if $\delta(X) = n$. Our proof of Theorem 1 is based on induction on the type of X .

If X is any space in C , we construct in our proof a sequence $X_1, X_2, \ldots, X_n, \ldots$ of spaces in C such that

a) the invariants $a_{i,j}(X_n)$ and $b_{i,j}(X_n)$ are completely determined by the invariants of X (and do not depend on any other property of X);

b) each X_n is of a strictly lower type than X ; and

c) X is uniquely constructible from this sequence $X_1, X_2, \ldots, X_n, \ldots$ in a well defined way.

When this is done, the invariants of X determine those of X_n's (because of a)); these in turn, by induction hypothesis, determine the spaces X_n completely (because of b)); these spaces X_n then determine X uniquely (because of c)). The first step of the inductive proof is furnished by the following proposition (known to many in the field):

"If X is of type 1, then X should be homeomorphic to one and only one of the following three spaces: i) The well-ordered space $\omega+1$, ii) The well-ordered space $\omega\cdot 2$ and iii) The space of rational numbers with all but one point made isolated."

It is in fact this proposition that has motivated and directed the author to guess at the present system of invariants.

The disjoint topological sum and the sequential sum are the two construction processes involved in c) above.

If $x \in X \in \mathcal{C}$ and if V is a neighborhood of x, then we define standardness of V in terms of the invariants of the subspaces V and V^c. We prove the existence of standard neighborhoods at each point. We use them to construct the above sequence $X_1, X_2, \ldots, X_n, \ldots$ and also to obtain later results.

IV. SOME REMARKS

We also prove that no proper subcollection of the above collection of invariants is sufficient to determine the space uniquely. We further indicate other types of consequences of the theorems, among which we give here a sample: A space X in $\mathcal{C}$ can have at most t_n types of points of derived length n; further t_n is the smallest number with this property.

Acknowledgement. The author thanks Professor M. Venkataraman for some helpful discussions and encouragement. He is glad to dedicate this article to Professor F. B. Jones, whose stimulating lecture at Madurai University (in 1968) is still remembered by the author.

REFERENCES

[G] DeGroot, J., "Topological classification of all closed countable and continuous classification of all countable point sets," *Indag. Math.* *8*(1946), 11-17.

[K] Kannan, V., "Ordinal invariants in topology," a monograph. (To appear.)

[K-R] Kannan, V. and M. Rajagopalan, "On scattered spaces," *Proc. Amer. Math. Soc.* *43*(1974), 402-408.

[M-S] Mazurkiewicz and W. Sierpinski, "Contribution a la topologie des ensembles dénombrables," *Fund. Math.* *1*(1920), 17-27.

[S] Sierpinski, W., "Sur une propriete topologique des ensembles dénombrables denses en soi," *Fund. Math.* *1*(1920), 11-16.

LAWSON SEMILATTICES WITH BIALGEBRAIC CONGRUENCE LATTICES

Garr S. Lystad
Albert R. Stralka

University of California, Riverside
Texas Instruments, Texas

This paper is concerned with lattices of closed congruences on compact topological semilattices. As usual in such discussions, topological constructions such as the standard dimension-raising map from the Cantor set onto the closed unit interval play important roles. However, in this treatment most of the topology will be hidden behind an algebraic facade, which is provided by recently developed duality theories. Our original intent was to discuss the lattice of closed semilattice congruences on compact Boolean algebras (compact Boolean algebras are, of course, nothing more than products of the two element discrete chain). However, we were able to extend our scope a bit to include families of lattices similar to those used by Kahn and others in some aspects of computer programming language theory. Our main result stated rather loosely is: Compact topological semilattices which are similar to compact Boolean algebras have lattices of closed semilattice congruences which possess very strong convergence properties. In particular, these congruence lattices can be made into compact topological lattices in a very natural way.

1. Definitions and Background

Throughout this collection of definitions, we shall assume that L is a lattice. For an element a of L and a subset A of L, $\downarrow a = \{x \in L : x \leq a\}$ and $\downarrow A = \bigcup\{\downarrow a : a \in A\}$. The sets $\uparrow a$ and $\uparrow A$ are defined

ISBN 0-12-481820-X

dually. The element a is way below b in L , written $a \ll b$, if whenever D is a directed set in L with $\sup D \geq b$, there is an element $d \in D$ such that $d \geq a$. The set of elements way below b in L is denoted by $\Downarrow b$. If $a \ll a$ then it is said to be compact. The set of compact elements of L is denoted by k(L). L is *continuous* if it is complete and for each $x \in L$, $x = \sup \Downarrow x$. It is said to be *algebraic* if it is complete and for each $x \in L$, $x = \sup(\downarrow x \cap k(L))$.

The category of continuous lattices and maps between such objects which preserve arbitrary infs and directed sups will be denoted by $\mathcal{CL}$. Within $\mathcal{CL}$, the subcategory of all algebraic lattices and $\mathcal{CL}$-morphisms between such objects will be denoted by $\mathcal{a}$.

Algebraic and continuous lattices have naturally defined compact Hausdorff topologies. In our discussion we will only need the topologies for algebraic lattices. For the algebraic lattice A , $\mathcal{T}(A)$ will be the topology on A generated by declaring the sets $\uparrow x$ to be open and closed for every $x \in k(A)$. This topology will be compact, Hausdorff and totally disconnected. With respect to $\mathcal{T}(A)$, the meet operation of A is continuous but not necessarily the join operation. Thus $(A,\mathcal{T}(A),\wedge)$ is a compact (Hausdorff) topological semilattice with identity, or maximal, element. Moreover, all such semilattices can be obtained in this way. (Note that if $(S,\mathcal{T},\wedge)$ is a compact topological semilattice with identity element, then it becomes a lattice when $x \vee y$ is defined to be the smallest element of nonempty set $\uparrow x \cap \uparrow y$.) A lattice A is *bialgebraic* if both A and A^{op} (A^{op} is A with the reverse order) are algebraic lattices. If, in addition, the two topologies $\mathcal{T}(A)$ and $\mathcal{T}(A^{op})$ coincide, then A is said to be a *coordinated bialgebraic lattice*. From Lystad [5] we know that a bialgebraic lattice can be made into a compact topological lattice if and only if it is coordinated. The canonical uncoordinated bialgebraic lattice is the following:

EXAMPLE 1.1. Let M_∞ be the set consisting of the natural numbers

N plus two additional symbols $\bot$ and $\top$. Declare $\bot < n < \top$ for every $n \in N$. With the order generated by these inequalities, L becomes a bialgebraic lattice. By a result of Stepp [7], L cannot be imbedded in a compact topological lattice by a lattice map.

Although the definition we are about to give for congruence is not standard, in our context it will create no difficulties. For a category $\mathcal{C}$ and a $\mathcal{C}$-morphism $\phi: S \to T$, we let $[\phi]$ be the relation on S associated with ϕ , i.e., $[\phi] = \{(s_1,s_2) \in S\times S : \phi(s_1) = \phi(s_2)\}$. Then any subset of $S\times S$ which can be obtained from a $\mathcal{C}$-morphism in this way is called a $\mathcal{C}$-*congruence* on S . When ordered by set theoretic inclusion (in the cases in which we are interested), the family of $\mathcal{C}$-congruences on S , denoted by $\Theta(S,\mathcal{C})$, forms a complete lattice.

For $S \in \mathcal{a}$ we see that two congruence lattices are of significance, $\Theta(S,\mathcal{a})$ and $\Theta(S,\mathcal{CL})$. From [4] we know that $\Theta(S,\mathcal{a}) = \Theta(S,\mathcal{CL})$ if and only if there is not a $\mathcal{CL}$-surmorphism from S onto I , the closed unit interval of the real line with its usual order.

For a complete discussion of continuous lattices, we recommend the soon to be published Compendium [2].

2. Congruences and Subobjects of $\mathcal{a}$

While proofs of the next result appear in the literature, we include our own proof both for the sake of completeness and because we feel that it is simpler than the other proofs.

PROPOSITION 2.1 *If* $A \in \mathcal{a}$ *then* $\Theta(A,\mathcal{a})$ *is a co-algebraic lattice (i.e.,* $\Theta(A^{op},\mathcal{a})$ *is algebraic).*

Proof. Define ch(A) to be the family of all congruences associated with $\mathcal{CL}$-surmorphisms onto the two element chain. From various sources, we know that ch(A) generates $\Theta(A,\mathcal{a})$ (cf. [3] or [2]). Since ch(A) constitutes the family of co-atoms of $\Theta(A,\mathcal{a})$, to complete our proof we need only that each member of ch(A) is co-compact ([1], p. 15). To this end suppose that $[\lambda] \in ch(A)$ and suppose that $\{[\phi_\gamma]:\gamma \in \Gamma\}$ is a downward directed family of congruences in $\Theta(A,\mathcal{a})$ whose meet is below $[\lambda]$. Because $ch(A)$ generates $\Theta(A,\mathcal{a})$ and finite intersections of members of $ch(A)$ have finite range, we may assume that each of the congruences $[\phi_\gamma]$ has finite range. Let $x = \inf\lambda^{-1}(1)$ and for each $\gamma \in \Gamma$ let $x_\gamma = \inf\phi_\gamma^{-1}\circ\phi_\gamma(x)$. Then each x_γ belongs to $k(A)$ and since $\bigcap\{[\Theta_\gamma]:\gamma \in \Gamma\} \subseteq [\lambda]$, it follows that $\sup\{x_\gamma:\gamma \in \Gamma\} = x$. But x also belongs to $k(A)$ and the set $\{x_\gamma:\gamma \in \Gamma\}$ is directed, hence $x = x_\gamma$ for some $\gamma \in \Gamma$ and it then follows that $[\phi_\gamma] \subseteq [\lambda]$. □

DEFINTION 2.2. Let $(S,*)$ be a semilattice with identity element e (e.e., $e*x = x$ for all $x \in S$). By $\mathcal{O}(S,*)$ we shall mean the lattice of all $*$ subsemilattices of S which contain e. When ordered by inclusion, $\mathcal{O}(S,*)$ becomes a lattice.

The duality theory of [3] provides a direct relationship between congruences and subsemilattices.

LEMMA 2.3. *If* $A \in \mathcal{a}$ *then* $\Theta(A,\mathcal{a})$ *is isomorphic with* $\mathcal{O}(k(A),\vee)^{op}$.

3. Two Classes of Semilattices

At this point we can introduce two classes of abstract semilattices. Let $\mathcal{P}$ denote the category of meet semilattices and meet preserving maps.

DEFINITION 3.1. Let $\mathcal{P}_n$ denote the category of lattices in which every proper filter and every bounded ideal is principal. Let $\mathcal{P}_f$ denote the category of lattices in which every principal ideal is finite. The morphisms for both $\mathcal{P}_n$ and $\mathcal{P}_f$ are the $\mathcal{P}$-morphisms between such lattices.

We will now list a collection of facts about $\mathcal{P}_n$ and $\mathcal{P}_f$ which are rather easily verifiable. Two elementary examples will appear in our discussion. We define U to be the chain of natural numbers with the added symbol τ. We declare $\tau > n$ for every natural number n. The chain D is taken to be U^{op}. D is the most elementary example of a convergent decreasing sequence, while U is the most elementary example of a convergent increasing sequence.

(3.2) The lattice S belongs to P_n if and only if S has a smallest element and S has neither U nor D as a P-quotient.

(3.3) The P-quotients of objects of P_n and P_f belong to P_n and P_f respectively.

(3.4) P_f is a subcategory of P_n .

(3.5) P_n is closed relative to arbitary products, whereas P_f is only closed relative to finite products.

With the facts cited above and the results from [6], we have the following two results:

(3.6) If $S \in P_n$ then $\Theta((S,\wedge),P)$ is bialgebraic.

(3.7) If $S \in P_f$ then $\Theta((S,\wedge),P)$ is a coordinated bialgebraic lattice.

For any set A let F(A) be the lattice of all finite subsets of A . Then with the operations of set theoretic union and intersection, F(A) belongs to P_f . When one forms the ideal completion of F(A) , the compact Boolean algebra 2^A is created.

LEMMA 3.8. *If* $S \in P_n$ *and* $A \in \mathcal{O}(S,\vee)$, *then* $A \in P_n$.

Proof. First suppose that B is an ideal of A which is bounded in S . Then $\downarrow B$ is a bounded ideal of S . As such it is principal. Hence there is $b \in B$ such that $\downarrow b = \downarrow B$ and $\downarrow b \cap A = B$. As one consequence of this face we see that every bounded ideal of A is principal. A second use of this fact will give us a meet operation for A . Suppose that $a,b \in A$. Then $\downarrow(a \wedge b)$ A is an ideal of A (here $a \wedge b$ is taken to be the meet of a and b in S) which is bounded in S . Hence by the fact above, it must have a generator, say c . When the meet of a and b in A is defined to be c we will have obtained a valid meet operation for A .

Finally, let F be any filter in A. Since meets in A are less than or equal to the corresponding meets in S, we see that $\uparrow F$ is also a filter in S. Thus it has a generator which must belong to F. Hence $A \in P_n$. □

With Lemma 2.3 we were able to point out a relationship between congruences and subsemilattices in $\mathcal{a}$. We will now establish a similar relationship in P_n.

PROPOSITION 3.9 *If* $S \in P_n$ then $\Theta((S,\wedge),P)$ *is isomorphic with* $\mathcal{O}(S,\vee)^{op}$.

Proof. Let $A \in \mathcal{O}(S,\vee)$. From Lemma 3.8, A belongs to P_n. Define $\alpha(A)$ by setting $x \simeq y$ in $\alpha(A)$ if and only if $\downarrow x \cap A = \downarrow y \cap A$. Clearly $\alpha(A)$ is a member of $\Theta((S,\wedge),P)$. Now, suppose that $s \in S$. By the first part of Lemma 3.8 we have $\downarrow s \cap A = \downarrow a \cap A$ for some $a \in A$. Thus $\alpha(A)$ defines a retraction of S onto A.

On the other hand, suppose that $[\Theta] \in \Theta((S,\wedge),P)$. From ([6], Prop. 2.7) we know that there is a sup preserving map $\Psi:\Theta(S) \to S$ given by $\Psi(x) = \inf\phi^{-1}(x)$. Then $\Psi \circ \phi(S)$ is a member of $\mathcal{O}(S,\vee)$ which is isomorphic with $\phi(S)$. It then follows that $\Theta((S,\wedge),P)$ and $\mathcal{O}(S,\vee)^{op}$ are isomorphic. □

THEOREM 3.10. *Suppose that* A *is an algebraic lattice such that* k(A) *is a sublattice of* A *and it belongs to* P_n. *Then* $\Theta(A,\mathcal{a})$ *coincides with* $\Theta(A,CL)$ *and is isomorphic to* $\Theta((k(A),\ ),P)$. *Consequently, it is then bialgebraic.*

Proof. To show that $\Theta(A,\mathcal{a})$ and $\Theta(A,CL)$ coincide is equivalent to showing that there is no CL surmorphism from A onto I. Suppose on the contrary that there is such a morphism $\phi:A \to I$. Then since every element of A is a supremum of members of k(A) and since ϕ preserves directed sups, it follows that every element of I must be a supremum of a directed subset from $\phi(k(A))$, i.e., $\phi(k(A))$ is a dense subset of I. Thus $\phi(k(A)))$ must have U as a quotient and so must k(A).

This is contrary to the membership requirements of P_n . Thus $\Theta(A,\alpha) = \Theta(A,CL)$.

From Lemma 2.1 we see that $\Theta(A,CL)$ is co-algebraic and then from 2.1 and 3.9 we see that it is algebraic (the congruence lattice of an abstract semilattice is always algebraic). This completes our proof.

Applying Theorem 3.10 and (3.7), we have

COROLLARY 3.11. *If* A *is an algebraic lattice such that* k(A) *is a sublattice of* A *which belongs to* P_f *, then* $\Theta(A,CL)$ *is a coordinated bialgebraic lattice.*

We can expand our collection of objects in A which have bialgebraic or corrdinated bialgebraic congruence lattices by applying the following easily derived result.

PROPOSITION 3.12. *Let* $S \in CL$ *and suppose that* $S = {\downarrow}x \cup {\uparrow}x$ *for some* $x \in S$. *Then* $\Theta(S,CL)$ *is isomorphic to* $\Theta({\downarrow}x,CL)\times\Theta({\uparrow}x,CL)$.

As an illustration of the use of Proposition 3.12, let S be the object of α formed by attaching the maximal element of U to the minimal element of 2^2 , i.e.,

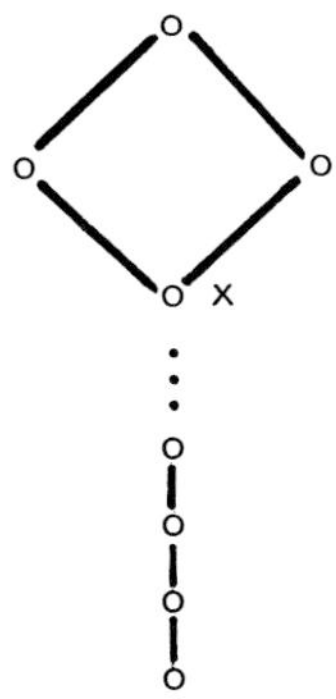

Then $\Theta(S,CL) = \Theta({\uparrow}x,CL)\times\Theta({\downarrow}x,CL) = \Theta(2^2,CL)\times 2^{\omega}$.

REFERENCES

1. Crawley, P. and R.P. Dilworth, *The Algebraic Theory of Lattices*, Prentice Hall, New Jersey, 1973.

2. Gierz, G., K. H. Hofmann, K. Kermel, J. D. Lawson, M. Mislove, and D. S. Scott, *A Compendium of Continuous Lattices*, Springer Verlag, Berlin, Heidelberg, New York, 1980.

3. Hofmann, K. H., M. Mislove and A. Stralka, *The Pontryagin Duality of Compact 0-Dimensional Semilattices and Its Applications*, Lecture Notes in Math. 396, Springer Verlag, Heidelberg, 1974.

4. ______, "Dimension raising maps in topological algebra," *Math. Zeit.* 135(1973), 1-36.

5. Lystad, G. S., *Compact Zero-Dimensional Semilattices*, Dissertation, University of California, Riverside, 1978.

6. Lystad, G. S. and A. Stralka, "Semilattices having bialgebraic congruence lattices," *Pac. J. Math.* 85(1979), 131-143.

7. Stepp, J.W., "Semilattices which are embeddable in topological lattices," *J. London Math. Soc.* 7(1973) 76-82.

A SURVEY OF CYCLIC ELEMENT THEORY AND RECENT DEVELOPMENTS

Byron L. McAllister

Montana State University, Bozeman

I. OUTLINE

An important technique for understanding the structure of a topological space X involves choosing a property P and breaking X up into subsets each of which has property P, usually "maximally" so that there is little or no trace of property P in the relative structure i.e. the way these subsets fit back together to form X. Perhaps the ultimate example of the successful use of this technique is seen when X is a locally connected space and P is the property of being connected. In this case if we know the topological structure of all the chosen subsets (here called connected components) then we know the topological structure of the entire space X. When X is not locally connected, even knowledge of the structure of all components as well as of the topology of the decomposition (quotient) space may require considerable additional detail before we can say we know *all* the structure of X. But even then the technique can sometimes "tell all," and it usually "tells enough," so that it is quite valuable. Thus the study of topological structure is partly reduced (entirely reduced, in the locally connected case) to the study of connected and of totally disconnected spaces.

The cyclic element theory may be thought of as a method of analyzing the structure of connected spaces by the same general technique, taking

ISBN 0-12-481820-X

property P as *cyclic connectedness*. The theory was first developed in about 1926, by Gordon T. Whyburn for continuous curves in the plane, and extended by Whyburn (using theorems of Ayres) to arbitrary metric Peano continua by 1928. (Older citations will often be omitted. They can be found in my 1966 historical paper [9].) (Peano continua are the same as continuous curves. A continuous curve was defined by Jordan as any continuous image of a closed interval of real numbers. The name change came about as the result of Peano's proof that such a continuum need not be one-dimensional.)

A subset M of a Peano continuum X is said to be *cyclicly connected* provided that for each two (distinct) points of M, there is a simple closed curve in M that includes both points. Whyburn defined a *true* (or *non-degenerate*) *cyclic element* of X as any subcontinuum M of X that is maximal with respect to being cyclicly connected. The cut points and end points of X were also defined (as singleton sets) to be cyclic elements of X.

For example if M is the union of two circles C_1 and C_2 tangent at a point p then M has three cyclic elements, C_1, C_2, and the point of tangency, $\{p\}$. Note that cyclic elements need not be disjoint even in such simple cases as this one, and thus such convenient tools as the theory of upper semi-continuous decomposition are not immediately available. Nevertheless, it is easy to convince oneself of the "acyclicity of the relative structure," and Whyburn built an extensive theory to confirm this. It is worth a brief interruption to describe the tools used to do so.

Whyburn called any connected set that is the union of a collection of cyclic elements an H-set. (The name probably comes simply from Whyburn's using H for one such set. H-sets turned out to be the same as the arc-curves previously defined by W. L. Ayres, and it is possible that the name A-set, used by Whyburn to denote an H-set that is actually a subcontinuum of X derives from the first letter of *arc-curve* or of *Ayres.)* An H-set H minimal with respect to containing two given cyclic elements C_1 and C_2 is called a *cyclic chain* with C_1 and C_2 as *endelements* of H. Then X is always "cyclic chainwise connected," defined analogously to arc-wise connectedness, and the acyclicity of the relative structure is reflected in the fact that the cyclic chain with C_1 and C_2 (cyclic elements of X) as endelements is unique.

(I must interrupt to apologize to the memory of Professor Whyburn for saying in my paper [9] that he had characterized a *node* of X as an end-point or a true cyclic element C of X such that the complement of the interior of C is connected. What Whyburn referred to [18] was not the interior of C but the set of *internal* points of C, by which he meant those points of C that are not cut points of X. Since the set of cut points of X that belong to C can be dense in C, easy examples reveal my mis-formulation for what it is.)

Cyclic element theory succeeded remarkably well in the goal of analyzing the structure of connected spaces. The many applications developed by Whyburn, Ayres, Kuratowski, Borsuk, Kelley, Schweigert, and others include, for a small sample, the following:

The fixed point property is cyclicly extensible and reducible. (A property is cyclicly extensible if its possession by every cyclic element of X implies its possession by X; a property is cyclicly reducible if whenever X has the property, so do all the cyclic elements of X.)

The property of being an absolute neighborhood retract is cyclicly extensible and reducible. So is unicoherence.

Those spaces that are obtainable as monotone upper semi-continuous decompositions of the sphere are precisely the Peano continua in which every true cyclic element is a 2-sphere.

Each true cyclic element of a Peano continuum C in the plane that is a boundary curve, i.e. such that C is precisely the boundary of a component of its complement, is a simple closed curve.

Every homeomorphism of a Peano continuum into itself has a fixed element (a cyclic element C such that $f(C) \subsetneq C$).

Morrey's application of cyclic element theory to the theory of surfaces is a splendid example of how results developed for one field of mathematics can find use in another. The following sketch of his method may be of interest.

A (path) *surface* is a continuous function f from a disk D into E_3. (Actually a surface is an equivalence class of such functions, but the intention here is only to convey the general ideas. For details see [13], or [4] or other references found in [9]). A difficulty that impeded study of the *area* assigned to the surface f arises out of the fact that f may map portions of D onto the same *geometric* surface in E_3 repeatedly (much as an exponential function may wrap an interval repeatedly about a circle, but with "messier" possibilities.) Morrey's idea was to break up f in the following way. Since the components of the inverse images of points of $f[D]$ under f form a disjoint closed cover of D, they form an upper semi-continuous decomposition M of D; topologized in the usual fashion, M is called the middle space of the light-monotone factorization $f = \ell \circ m$ of f. (That is, $m:D \to M$ is defined by $m(x) = S$ if and only if $x \in S$, and $\ell:M \to E_3$ is defined by $\ell(S) = p$ if and only if for each point x of S, $f(x) = p$, $x \in D$, $S \in M$, $p \in E_3$.)

Now if C is any cyclic element of a Peano continuum M, there is a retraction r_c of M onto C obtained by letting $r_c(x) = x$ for $x \in C$ and, for $x \in M \backslash C$, setting $r_c(x)$ equal to the unique boundary point of K_x that lies in C, where K_x is the component of $M \backslash C$ that includes x. When M is the middle space of the light monotone factorization

$\ell \circ m$ of the (path) surface f, these retractions of M into cyclic elements of M enable one to form "parts" of f of the form $\pi_C = \ell \circ r_C \circ m$, and Morrey showed not only that these functions π_C are, for area purposes, somewaht easier to handle than f is, but that the area of the surface f is equal to the sum of the areas of the parts π_C.

These early developments and more are summarized with extensive references to the literature in [9]. Here we shall be chiefly concerned with generalizations of the theory. These have been of two main kinds: *refinements*, in which cyclic elements are further subdivided, and *extensions* in the sense that a closely analogous theory is sought for less restrictive categories of spaces.

II. REFINEMENTS

A very early refinement was suggested by Wilder. A singleton set of X can be thought of as a closed point set that carries no 0-dimensional cycle, and a true cyclic element turns out to be maximal with respect to not being cut by such a set. Wilder--and Whyburn, who developed the refinement--called a closed point set that carries no essential r-cycle a T_r-set. An E_r set is then a nondegenerate connected set maximal with respect to being cut by no T_r-set. Thus the E_0 sets are the true cyclic elements. (A "practical" - the quotation marks seem appropriate here - application is to large chocolate bars, which are often made with deep grooves, along which the chocolate can be broken. Imagining the chocolate at the bottom of a groove to be only "one point thick" two adjacent "lumps" are disconnected by removing the groove (a T_1 set) but no lump can be broken by a mere "line of chocolate," so that "lumps" are E_1 sets.) References to this theory of Cyclic Elements of Higher Orders and its literature are also given in [9].

A second kind of refinement arose from the desire to improve the generality of the application to area theory. If two lunes (crescent-shaped regions) have only their vertices in common, the resulting figure is cyclicly connected and hence contains only one cyclic element of itself. Yet, for area purposes, a light map from that figure would appear to have as area the sums of the areas of its restrictions to each lune. Thus a refinement that breaks up cyclic elements along finite cuttings was desirable. Hall and Youngs pioneered this approach, which was greatly developed by Neugebauer, using an approach due to Cesari. Others who have contributed to this approach include Kosinsky, Remage, and McAuley. McAuley's approach breaks along not-necessarily-finite cuttings, using the concept of local cyclical connectedness. It was developed by McAuley and the author in [10] and [11]. References to the others mentioned can be found in [9].

III. EXTENSIONS

Weakenings of the restrictive assumptions on the categories of spaces involved began early, for Ayres frequently used *continuous curve* to mean any continuous image of an *open* interval, a metric space more general than a Peano Continuum. F. B. Jones's 1940 paper [7] is another especially interesting step of this kind. Whyburn, too, frequently developed the theory with greatly weakened hypotheses. For example, in Analytic Topology [17] he was often able to drop compactness and to weaken local connectedness to semi-local connectedness. Plunging to extremes, the ultimate weakening of conditions on the spaces involves developing a theory in which the topology plays virtually no role. The most recent work along these lines is that of L. E. Ward Jr., also included in these Conference Proceedings. Ward's approach includes references to the earliest paper of this type, by C. J. Harry, an approach even more general than Ward's due to Radó and

Reichelderfer (which apparently is "over-general" for some purposes) and a more distantly related study by A. D. Wallace. For further background on matters related to this kind of generalization, also see Ward's extensive bibliography in [16].

In the meantime, theories with more structural detail involving less drastic generalizations have also been developed. In 1968 Whyburn [19] initiated a theory of cyclic elements in Hausdorff spaces--in fact for spaces somewhat more general than that but not so weak as general T_1 spaces. He observed, vis-a-vis his weakened T_2 axiom, that "For spaces not satisfying this axiom, little if anything of significance can be proved without restricting the space in some other way," which, however, may not mean so much that *no* structural theory is likely to be developed for such very general spaces as that close analogues of the *classical* cyclic element theory will demand at least (weakened) T_2 or other strong assumptions.

Now, as a matter of fact, a considerable part of the theory for non-T_2 spaces had already been developed as early as 1942 by Albert and Youngs but note that this did, in fact, involve one of the "other strong assumptions," namely local connectedness. As a matter of fact, Albert and Youngs made a reasonably good case for the position that the "right" category of spaces in which to do structure theory of cyclic element type is that of locally connected topological spaces. They did not work the theory out in full, but they got far enough to show that there are things "worth doing" in such general spaces.

A natural refinement of Albert and Youngs's approach was made in 1971 by Spencer Minear [12] who used the fact--apparently not known to Albert and Youngs--that if p is a cut point and $\{p\}$ is not closed, then $\{p\}$ is open. Minear's work developed the theory far enough, for example, to prove for general connected, locally connected topological spaces that unicoherence is both cyclicly extensible and reducible, and, in fact, if the space is also compact, that the fixed point property is also cyclicly extensible and reducible.

In many respects the key paper of recent times is that of B. Lehman [8] who, by assuming *both* the Hausdorff axiom *and* such "strong assumptions" as local connectedness as a blanket hypothesis and compactness or local compactness where needed, was able to succeed, in many respects, in the goal of "rewriting" the cyclic element theory of [17] for locally connected Hausdorff spaces. [8] Though Lehman left open the important question of what properties are cyclicly extensible or reducible, she compensated by producing a good analogue of the cyclic chain approximation theorem, a device for converging to a space X "from inside" by successive adjunction of smaller and smaller simple cyclic chains that leave less and less of X untouched. And she has initiated the extensible/reducible theory in a paper to be published in the Canadian Journal of Mathematics.

A few further papers should also be mentioned. One set of these consists of two papers from 1938 and 1940 [1,2] by V. W. Adkisson and a paper from 1957 [14] by G. R. Strohl, Jr., which ought to have been mentioned in [9], since they include interesting applications of cyclic element thoery, but which I--inexplicably--left out. Since 1965 [when [9] was completed] I have not tried to catch every application of cyclic element theory, but I shall just cite four more "applications papers" that seem to me to be significant; Ralph Bennett (1966) [3], W. Wiley Williams (1975) [20], C. L. Hagopian (1970) [6], and J. L. Cornette (1974) [5].

REFERENCES

1. Adkisson, V. W., "Plane Peanian Continua with unique maps on the sphere and in the plane," *Trans. Amer. Math. Soc. 44* (1938), 58-67.

2. ______, "Extending maps of plane continua," *Duke Math. J. 6* (1940), 216-228.

3. Bennett, R., "Locally connected 2-cell and 2-sphere like continua, *Proc. Amer. Math. Soc. 86* (1957), 297-308.

4. Cesari, L., "Surface Area," *Ann. of Math. Studies 35, Princeton University Press 1956.*

5. Cornette, J. L., "Image of a Hausdorff arc" is cyclically extensible and reducible," *Trans. Amer. Math. Soc., 199* (1974), 253-267.

6. Hagopian, C. L., "Arc-wise connected plane continua," *Topology Conference* (Proc. General Topology Conference, Emory University, Atlanta Ga., 1970), 41-45.

7. Jones, F. B., "Almost cyclic elements and simple links of a continuous curve," *Bull. Amer. Math. Soc. 46* (1940), 775-783.

8. Lehman, B., "Cyclic element theory in connected and locally connected Hausdorff spaces," *Canadian J. Math. 28* (1976), 1032-1050.

9. McAllister, B. L., "Cyclic elements in topology, a history," *Amer. Math. Monthly, 73* (1966), 337-350.

10. __________ and L. F. McAuley, "A new "cyclic" element theory," *Math. Zeitschr., 101* (1967), 152-164.

11. McAuley, L.F., and B. L. McAllister, "A note on cyclic subelement theory--reducibility of local connectedness and local simple connectedness," *Colloq. Math., 24* (1972), 213-218.

12. Minear, Spencer,"On the Structure of Locally Connected Topological Spaces," Thesis, Montana State University, 1971.

13. Radó, T., "Length and Area," *Amer. Math. Soc.,* Colloq. Publ. 30, 1948.

14. Strohl, G. R. Jr., "Peano spaces which are either strong cyclic or two-cyclic," *Trans. Amer. Math. Soc., 86* (1957), 297-308.

15. Ward, L. E., Jr., "Axioms for cut points" *Conference on General Topology Set Theory and Analysis,* Riverside, CA., May 1980 (This volume).

16. __________, "Partially ordered spaces and the Structure of continua," *Conference on Metric Spaces, Structure of Continua, Etc.,* Guilford College, 1980.

17. Whyburn, G. T., "Analytic Topology," *Amer. Math. Soc.,* Colloq. Publ. 32, 1949.

18. __________, "Concerning the structure of a continuous curve," *Amer. J. of Math., 50* (1928), 167-194.

19. __________, "Cut points in general topological spaces," *Proc. Nat. Acad. Sci., U.S.A., 61* (1968), 380-387.

20. Williams, W. W., "Semilattices on Peano continua," *Proc. Amer. Math. Soc., 49* (1975), 495-500.

CERTAIN POINT-LIKE DECOMPOSITIONS OF E^3 WITH 1-DIMENSIONAL IMAGES OF NON-DEGENERATE ELEMENTS

Louis F. McAuley

State University of New York, Binghamton

Edythe P. Woodruff*

Trenton State College, New Jersey

I. INTRODUCTION

We shall be concerned with upper semicontinuous (usc) decompositions G of E^3 using the standard definitions due to either Whyburn [7] (or Moore). As usual, H will denote the collection of non-degenerate elements of G and $P:E^3 \Rightarrow E^3/G$ will denote the usual quotient (or projection) map onto the decomposition space E^3/G. Each element g of H is point-like, i.e., E^3-g is homeomorphic to E^3 - a point.

Consider the following question:

Does there exist a point-like usc decomposition G *of* E^3 *such that (1)* $P(H)$ *has dimension* 1, *(2)* E^3/G *is not homeomorphic to* E^3, *and (3) if* $H' \subset H$, $P(H')$ *has dimension* 0, *and* G' *is an usc decomposition of* E^3 *whose collection of non-degenerate elements is* H', *then* E^3/G' *is homeomorphic to* E^3 *?*

This question was raised recently by Michael Starbird and has been posed in another form by R. J. Daverman. We give affirmative answers to both questions. In fact, we develop a useful technique (or construction) of *threading a set through the collection* H *of non-degenerate elements*.

*Partially supported by NSF Grant MCS-7909542.

ISBN 0-12-481820-X

We use a *simple arc* and a *2-cell* to answer the questions of Starbird and Daverman, respectively, threaded through the collection H in Bing's Dog Bone decomposition. We could have used open arcs in the Armentrout-Bing example [1] so that under the projection map P , the open arcs *and the collection* H is a simple arc (in order to answer Starbird's question). That is, we could have threaded a simple arc through P(H) in the decomposition space.

We believe that this kind of construction has a variety of potential uses. In fact, we intend to pursue the matter further.

II. AN EXAMPLE

We modify Bing's Dog Bone decomposition [3]. Although we assume that the reader is familiar with the construction, we will briefly describe it. The successive stages are obtained by iteration of the construction in Figure 1.

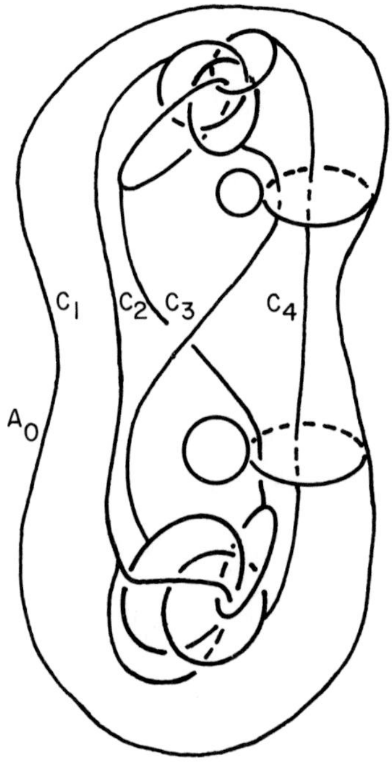

Fig. 1.

The 1-dimensional figures C_i, $1 \le i \le 4$, consist of two disjoint simple closed curves connected by a simple arc. The simple closed curves link as in Figure 1 inside of A_0 ---a 3-cell with two handles. Each C_i is made into a topological copy A_i of A_0 by taking a small tubular neighborhood of C_i so that A_1, A_2, A_3, A_4 are pairwise disjoint and the linking in Figure 1 is preserved. This process is iterated in such a manner that we obtain an usc decomposition G of E^3 such that the collection H of nondegenerate elements of G is a collection of tame arcs (which can be made polygonal). Furthermore, E^3/G is not homeomorphic to E^3 and $P(H)$ has dimension 0.

We modify this Dog Bone construction as follows. Let P be a horizontal plane which separates the two handles of A_0 (assuming that the z-axis runs symmetrically and vertically through the handles). The plane P intersects each A_i in a disk (2-cell). These can be connected by intervals into a chain C_1 of disks and intervals. See Figure 2. We label the disks $D_i = A_i \cap P$. Thus, C_1 consists of disks D_1, D_2, D_3, D_4, and simple

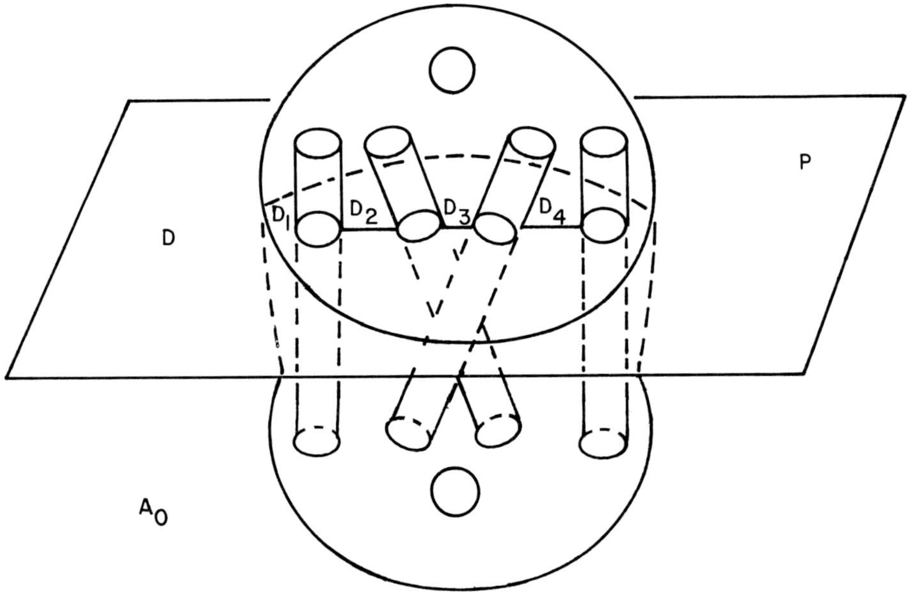

Fig. 2.

arcs a_1, a_2, a_3 . Repeat the procedure in each D_i . That is, we construct simple chains C_{1i} in each D_i consisting of four pairwise disjoint disks and intervals lying in D_i . The four disks constitute the intersection of the four dog bones of the next stage of the construction which lie in A_i . We obtain a chain C_2 consisting of the chains C_{1i} connected by appropriate intervals. See Figure 3.

We construct a sequence $\{C_i\}$ of such chains so that $\bigcap_{i=1}^{\infty} C_i^*$ is a simple arc A which intersects each element h of H in exactly one

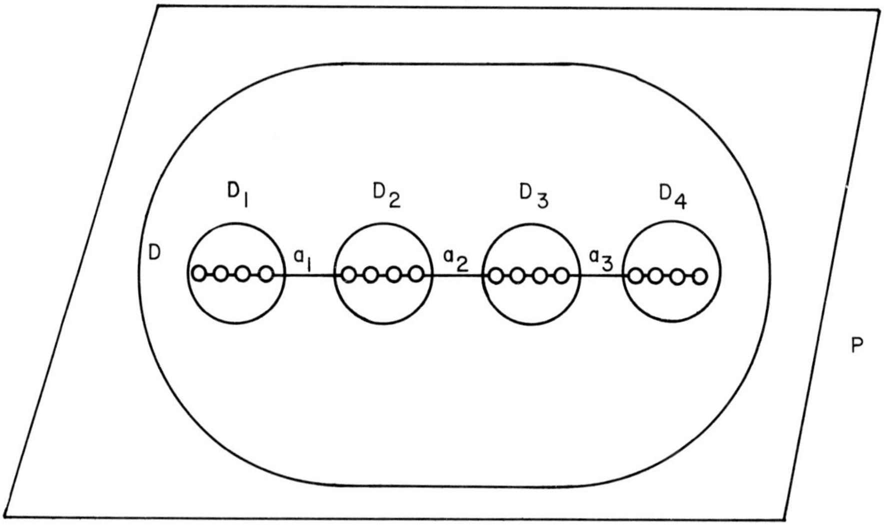

Fig. 3.

point. We could have made A a straight line interval.

Now, for each component of $A-(A\cap H^*)$ which is an open arc (p,q) , let $P^{-1}(P(p))\cup(p,q)\cup P^{-1}(P(q))$ be an element of a collection H_1 . The remaining elements of H_1 are those elements g of H such that $g\cap[p,q]=\emptyset$ where $[p,q]$ is the closure of a component (p,q) of $A-(A\cap H^*)$. We now have a new usc decomposition G_1 of E^3 whose nondegenerate elements form the collection H_1 . Each element of H_1 is point-like with countably many elements being topological letter H's and the remaining elements of H_1 being tame arcs.

III. THE DECOMPOSITION SPACE E^3/G_1 PROVIDES AN AFFIRMATIVE ANSWER TO STARBIRD'S QUESTION

It should be clear that E^3/G_1 is not homeomorphic to E^3. Since the original Dog Bone Space is not homeomorphic to E^3 and its nondegenerate collection H is not shrinkable [6], it follows that this new collection H_1 is not shrinkable. Thus, $E^3/G_1 \neq E^3$ [6].

Clearly, $P(H_1)$ has dimension 1 since $P(H_1)$ is a simple arc.

Now, let H_1' be any subcollection of H_1 such that $P(H_1')$ has dimension 0 and G_1' is an usc decomposition of E^3 with H_1' as the collection of nondegenerate elements. We shall show that G_1' is shrinkable (i.e., H_1' is shrinkable).

Note that $H_1 - H_1'$ contains a collection K consisting of topological H's which is "dense" in H_1. Otherwise, $P(H_1')$ would contain some 1-dimensional set (arc). In any case, the collection C of dog bones *not needed* to define the elements of H_1' has the property that given a dog bone D' used to define H_1, there is a dog bone $D'' \subset D'$ such that $D'' \in C$. That is, there is a "dense" set of dog bones left out of the construction required for H_1'.

LEMMA. *The collection* H_1' *is shrinkable.*

Proof. Suppose that $\varepsilon > 0$ and that U is an open set where $U \subset (H_1')^*$. We shall show that those dog bones needed to define H_1' can be shrunk to sets of diameter less than ε inside U. Choose $N+1$ planes such that (1) each is parallel to the plane P, (2) planes P_1 and P_N divide A_0 into three 3-cells, (3) P_i separates P_{i-1} from P_{i+1} for $1 < i \leq N$ and (4) the part of A_0 between P_i and P_{i+1} has vertical thickness less than $\varepsilon/6$.

STEP 1. The dog bone A_1 of the first stage of the construction of H_1 contains a dog bone $A_{1ij\cdots k}$ that lies in the complement of $(H_1')^*$. If this were not true, then $P(H_1')$ would contain an arc and hence would have dimension 1 .

There is a homeomorphism h_1 of A_0 onto A_0 fixed on ∂A_0 so that $h_1(A_t)$, $t \neq i$, misses one of the planes P_m, $1 \leq m \leq N+1$ and $h_1(A_{1i})$ intersects all of the planes. This is indicated in Figure 4.

In $h_1(A_i)$, we have $h_1(A_{1t})$ for $1 \leq t \leq 4$. Note that $h_1(A_{1i}) \supset h_1(A_{1ij\cdots k})$. There is a homeomorphism h of $h_1(A_1)$ onto itself which is fixed on its boundary such that (letting $h_2 = h \cdot h_1$) $h_2(A_{1it})$ for $t \neq j$ misses one of the planes P_m, $1 \leq m \leq N+1$, while $h_2(A_{1ij})$ intersects all of the planes. See Figure 5.

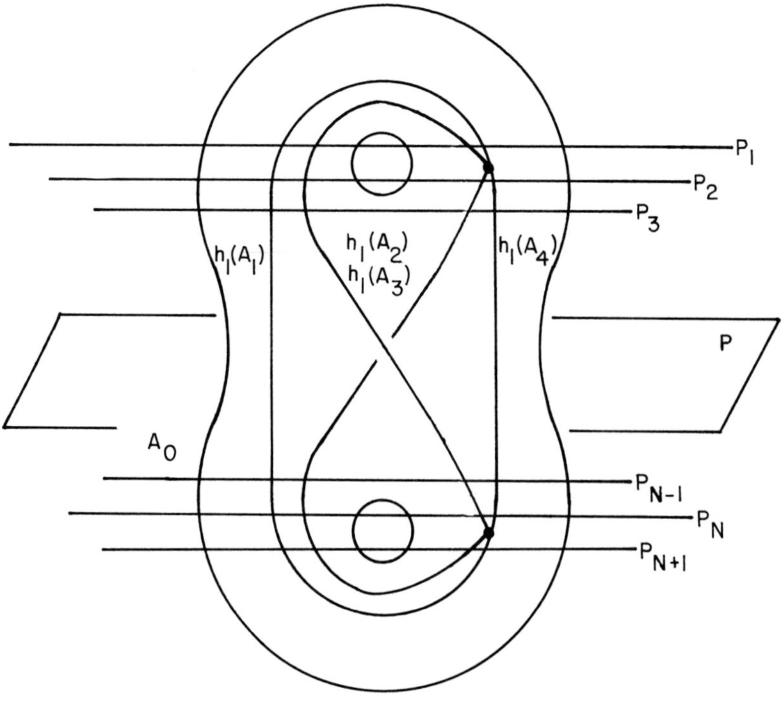

Fig. 4.

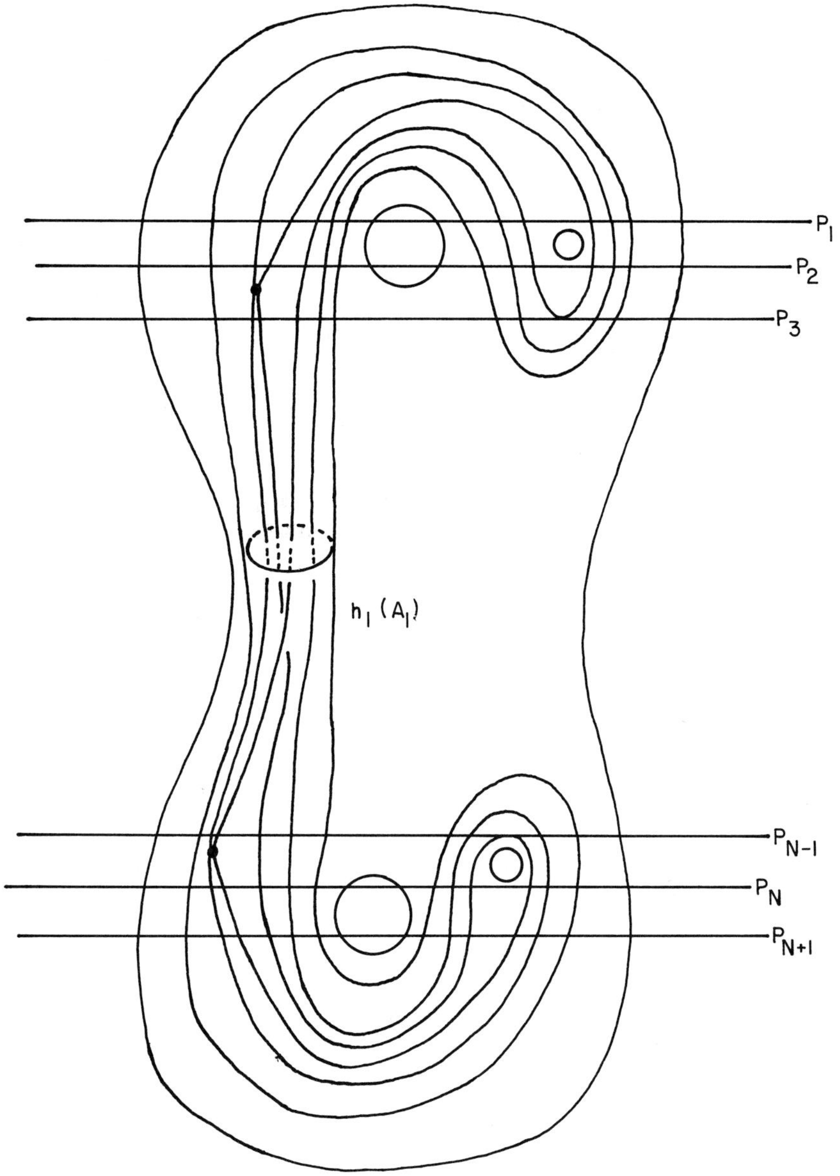

Fig. 5

Next, consider $h_2(A_{1i})$ which contains four dog bones at the third stage of the construction. One contains $h_2(A_{1ij\cdots k})$. Shorten the others so that they miss one of the planes P_m, $1 \leq m \leq N+1$. Continue this process. Finally, we reach a homeomorphic copy of $A_{1ij\cdots k}$ which may be long but the other three $A_{1ij\cdots t}$ for $t = \{1,2,3,4\}-\{k\}$ have homeomorphic copies which miss one of the planes P_m. We can throw away the copy of $A_{1ij\cdots k}$ since $A_{1ij\cdots k}$ lies in the complement of $(H_1')^*$.

At the end of Step 1, we have reached stage S_1 of the construction and we have produced a homeomorphism g_1 (a composition of S_1 homeomorphisms) such that $g_1(A_{su\cdots t})$ misses at least one of the planes P_m, $1 \leq m \leq N+1$, for each $A_{su\cdots t}$ except for $g_1(A_{ij\cdots k})$ which we have thrown away.

STEP 2. For each of the remaining dog bones $g_1(A_{su\cdots t})$ at stage S_1, we repeat Step 1. This is done $4^{S_1}-1$ times. That is, there is $A_{su\cdots t\cdots q}$ in $A_{su\cdots t}$ which lies in the complement of $(H_1')^*$. One of the four dog bones, $A_{su\cdots ti}$, $i \in \{1,2,3,4\}$, contains $A_{su\cdots t\cdots q}$. There is a homeomorphism taking $g_1(A_{su\cdots t})$ onto itself and fixed on its boundary which shrinks $g_1(A_{su\cdots tj})$, $j \neq i$, to miss *two* of the planes P_m, $1 \leq m \leq N+1$. We continue this process as in Step 1 until we reach the stage K_1 where $A_{su\cdots t\cdots q}$ is added in the construction. Again, we throw away a homeomorphic copy of it (composition of the various homeomorphisms).

Repeating Step 1 in this manner $4^{S_1}-1$ times, we reach various stages K_i, $1 \leq i \leq 4^{S_1}-1$. Let $S_2 = \max[K_i]$. Thus, at the end of Step 2, we have reached stage S_2 of the construction and we have produced a homeomorphism g_2 (a composition of homeomorphisms) such that (1) $g_2(A_{r\cdots t})$ misses at least two of the planes P_m, $1 \leq m \leq N+1$ for all $A_{rz\cdots t}$ at stage S_2 for which homeomorphic copies have not been thrown away. Those thrown away lie in the complement of $(H_1')^*$.

We continue until we have completed N steps in this manner. Thus, a homeomorphism g_N is produced such that if D is a dog bone at stage S_N of the construction and D (or homeomorphic copy) has not been thrown away, then $g_N(D)$ intersects at most one of the planes P_m.

Now, some elements $g \in H_1'$ may be topological letter H's and involve arcs connecting dog bones in their definition (or "defining sequence"). Observe that the shrinking can be accomplished so that these arcs contribute no more than $\varepsilon/6$ to the diameter of $g_N(g)$. A defining sequence for g may involve a pair of dog bones and an arc at each stage. In any case, we can construct g_N so that $\operatorname{diam} g_N(g) < \varepsilon$. Consequently, H_1' is shrinkable.

It follows from [5] that $E^3/G_1' \simeq E^3$.

REMARKS. It was not necessary to connect the dog bones with arcs. We could have made the dog bones touch as in Figures 6 and 7 to produce a simple arc intersecting each element $g \in H$ in exactly one point. In this case, some elements of H_1 would be tame arcs and others would be topolocial letter X's .

In our example constructed to answer Daverman's question, we could have made these dog bones touch in vertical intervals. Compare the corresponding usc decompositions to those in Section IV.

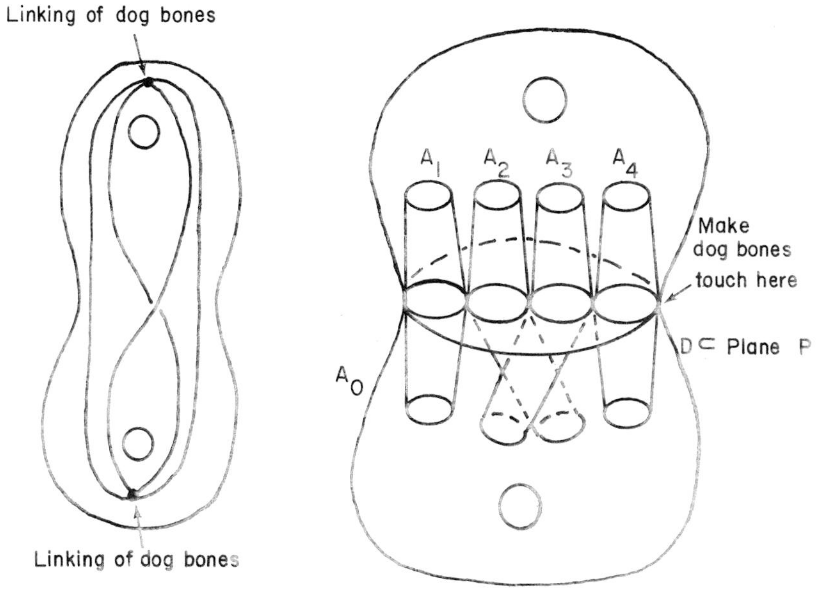

Fig. 6.

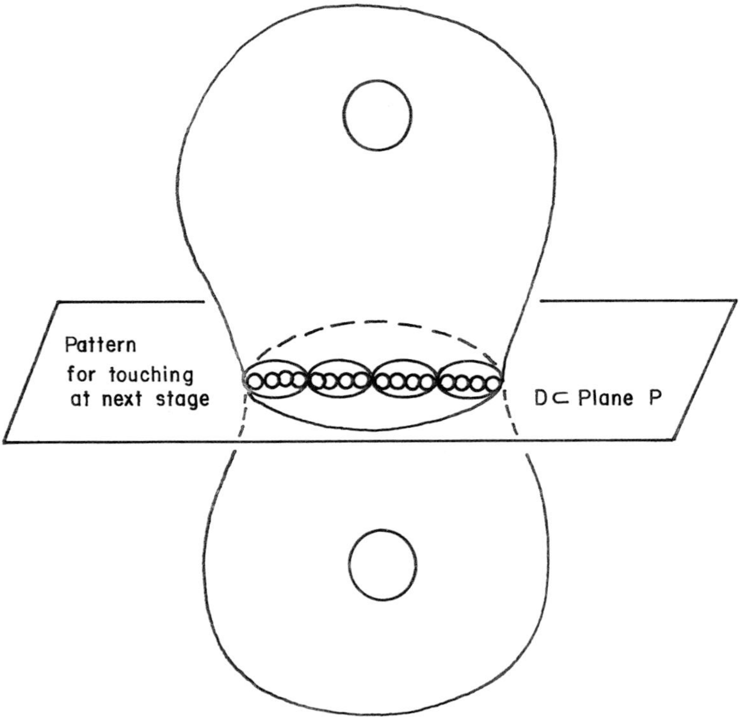

Fig. 7.

IV. DAVERMAN'S QUESTION

Daverman's version of the question answered above is as follows: *Does there exist an usc decomposition* G *of* E^3 *such that (a) each element of* H *is a simple arc, (b)* H^* *is a 2-cell, (c)* H *is a continuous collection, (d)* $E^3/G \neq E^3$ *, (e)* P(H) *has dimension* 1 *, and (f) if* $H' \subset H$ *and* P(H') *has dimension* 0 *, then* $E^3/G' \simeq E^3$ *where* H' *is the collection of nondegenerate elements of the usc decomposition* G' ?

We provide an affirmative answer to Daverman's question. Instead of constructing a simple arc A intersecting each element $g \in H$ in exactly one point, we construct a disk D which intersects each $g \in H$ in a simple arc. In fact, we can thicken the arc A into a ribbon. We could have used cylinders and rectangular disks for our chains C_i in the previous section. See Figure 8.

It should be clear that we can obtain a 2-cell $D \simeq [0,1]\times[0,1]$ (a unit disk) such that (1) $g \cap D \simeq ([0,1],x)$ for some $x \in [0,1]$ (indeed, this can be a straight line interval perpendicular to P) and (2) for the arc A of the previous section, we can now have $A \simeq (0,[0,1])$. Denote the top edge of D by B_1 and the bottom edge by B_2 . See Figure 9. Note that $B_2 = A$.

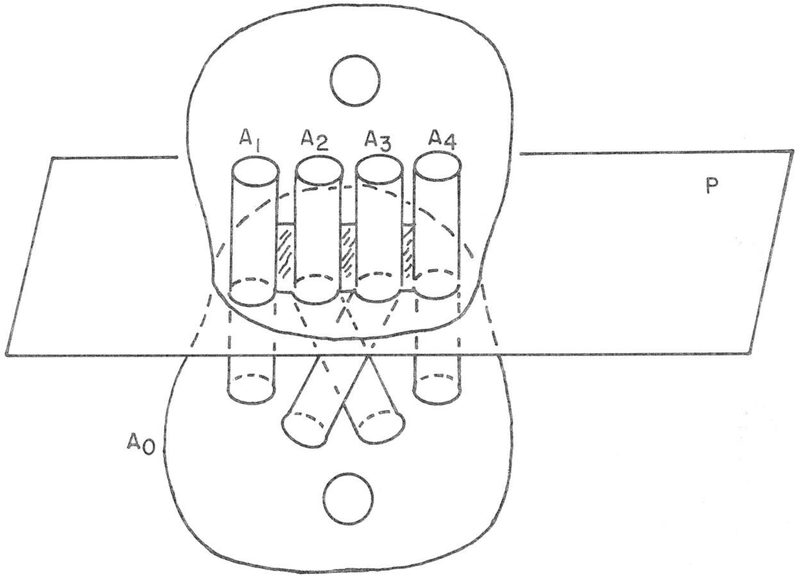

Fig. 8.

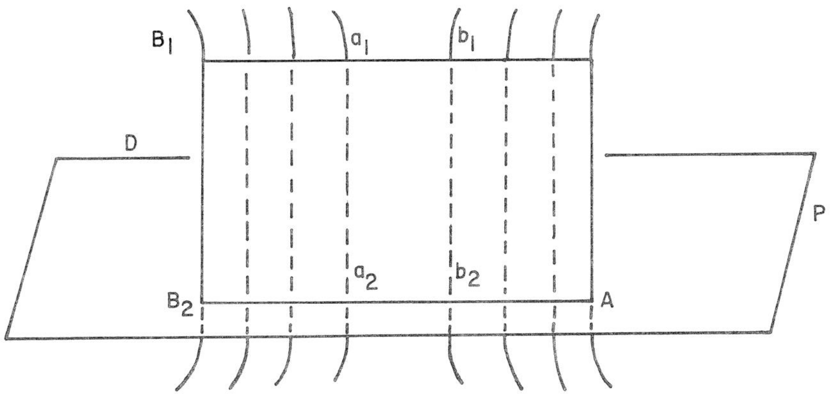

Fig. 9.

For each $g \in H$, $g-D$ consists of two (topological) half open intervals g_1 with an endpoint in B_1 and g_2 with an endpoint in B_2 . Let H_i, i=1,2 , denote the collection of all $\bar{g}_i$ for $g \in H$. Let G_i, i=1,2 denote the usc decomposition of E^3 whose collection of nondegenerate elements is H_i . Now, each G_i (in particular, H_i) is a shrinkable collection. It follows from either [5] or [6] that $E^3/G_i \simeq E^3$.

Let H_j, j=3,4 , denote the collection of all closed intervals $[a_i,b_i]$ such that (1) (a_i,b_i) is an open interval in B_i, i=1,2 , respectively, (2) for $g \in H$, $g \cap (a_i,b_i) = \emptyset$, and (3) there exists $g,h \in H$ such that $g \cap B_i = \{a_i\}$ and $h \cap B_i = \{b_i\}$. Also, let G_j, j=3,4 , denote the usc decomposition of E^3 such that H_j is the collection of nondegenerate elements of G_j .

We shall indicate how to shrink G_i, i=1,2 , so that H_j, j=3,4 , respectively, remains a null sequence of tame arcs. That is, we shall describe a pseudo-isotopy which shrinks each element g of H_1 (similarly for H_2) inside itself to an endpoint which moves each element of H_3 (similarly for H_4) to a polygonal arc.

First, each element of H_1 can be considered as a polygonal arc. In fact, if we use M. K. Fort's construction [4] of Bing's Dog Bone decomposition, then we can consider each element of H_1 as a straight line interval. Recall that Fort's construction has the property that each element of H is a polygonal arc with just two bends each in one of two horizontal planes. Our ribbon (2-cell) can be placed so as to intersect each element of H in the interval between those two planes. Without loss of generality, we can assume that the open neighborhood in the shrinkability definition contains A_0 .

First, let f_1 be a homeomorphism of E^3 onto itself which is the identity outside of A_0 and below the plane P . Now, f_1 shrinks each of the four top halves of dog bones (A_i, i=1,2,3,4) at the first stage *in themselves* such that if $x \in g \in H_1$, then $f_1(x) \in g$. The shrinking is upwards towards the endpoints of the elements of H_1. If e is an upper endpoint of $g \in H_1$, then $f_1(e) = e$. Actually, f_1 is the identity outside of a small neighborhood of the four dog bones A_i and below P. Also, f_1 takes each element of H_3 to a horizontal interval halfway towards the set E of all top endpoints of the elements of H_1.

Next, construct a homeomorphism f_2 of E^3 onto itself such that $f_2 = f_1$ outside of a small neighborhood of the top halves of the sixteen dog bones at the second stage of the construction and below a plane halfway towards E. Again, f_2 shrinks the top halves of dog bones in themselves upwards such that if $x \in g \in H_1$, then $f_2(x) \in g$. Again, f_2 is fixed on E. Now, f_2 moves certain elements of H_3 (all except three) upwards to a horizontal plane within $N_{1/4}(E)$. The three exceptions contain the three intervals on the top edge of D which connect the four dog bones A_i (see Figure 8). These become horseshoe-shaped with horizontal intervals within $N_{1/2}(E)$ determined by f_1 but with endpoints moved upwards inside of elements of H_1.

We continue the process to obtain a sequence of homeomorphisms $\{f_i\}$ which converges uniformly to a continuous mapping f whose point inverses are the elements of H_1. Clearly, this sequence $\{f_i\}$ is part of a pseudo-isotopy which shrinks the elements of H_1 in themselves to their upper endpoints. At the same time, the elements of H_3 are moved to a null sequence of polygonal arcs. Now, $E^3/G_1 \simeq E^3$.

By a theorem of Bing [2], an usc decomposition of E^3 whose collection of nondegenerate elements is a null sequence of tame arcs has a decomposition space which is homeomorphic to E^3. Thus, $E^3/G_3 \simeq E^3$.

In fact, we consider (1) $P_1:E^3 \Rightarrow E^3/G_1 \simeq E^3$, (2) $P_2:E^3/G_1 \Rightarrow E^3/P_1(G_2) \simeq E^3$ where $P_1(G_2)$ means the usc decomposition of E^3/G_1 whose collection of nondegenerate elements is the collection $\{P_1(g) | g \in H_2\}$, (3)

$P_3 : E^3/P_1(G_2) \Rightarrow E^3/P_2P_1(G_2) \simeq E^3$, and (4) $P_4 : E^3/P_2P_1(G_2) \Rightarrow E^3/P_3P_2P_1(G_2) \simeq E^3$.

The 2-cell D is "transformed" into a 2-cell $P_4P_3P_2P_1(D)$. Consider the closed 2-cells $P_4P_3P_2P_1(\sigma)$ where $\sigma \subset D$ such that (1) σ has edge arcs $[a_i,b_i] \subset B_i$ where $[a_i,b_i] \in H_j$, $i=1,2;j=3,4$, respectively, and (2) the other two edges are contained in elements $g,h \in H$.

We can fill σ with a nice continuous collection of straight line intervals $[p,q]$ each parallel to B_i with one endpoint $p \in g$ and the other endpoint $q \in h$. See Figure 10.

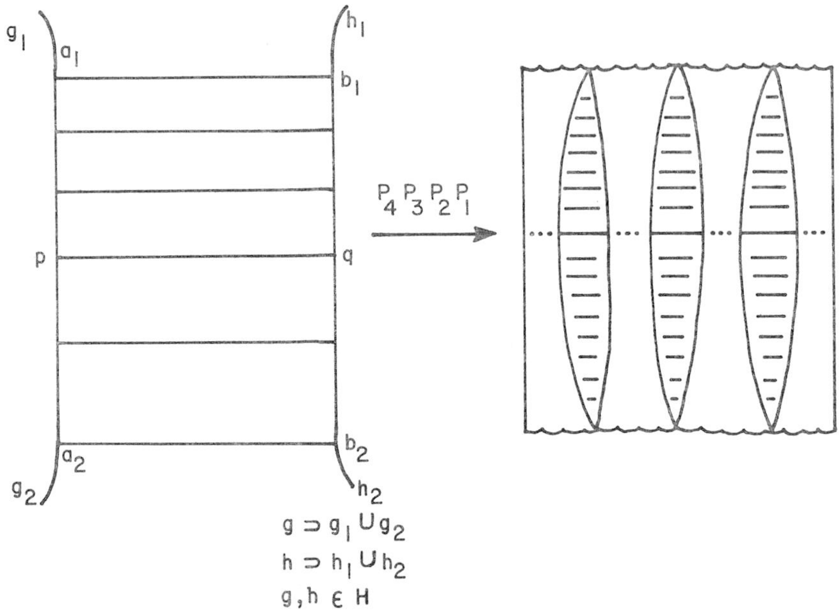

Fig. 10.

Let G_5 be the usc decomposition of E^3 (actually, $(((E^3/G_1)/G_2)/G_3)/G_4 = E^3$) such that the collection H_5 of nondegenerate elements of G_5 is the collection of the various arcs $P_4P_3P_2P_1[p,q]$ for the closed intervals $[p,q]$ filling the countably infinite number of 2-cells σ described above. The intervals $[a_i,b_i]$ are not in H_5 since they are mapped to points (not arcs) by P_3 and P_4 .

It is not too difficult to see that G_5 (in particular, H_5) is shrinkable. Let an open neighborhood U containing the elements of H_5, and $\varepsilon > 0$ be given. There is a 2-cell $\hat{D} \subset D$ such that $P_4P_3P_2P_1(\hat{D})$ contains the elements in the collection $\hat{H}_5 = \{g \in H_5 \mid \text{diam } P_4P_3P_2P_1(g) \geq \frac{\varepsilon}{2}\}$ and D does not intersect any element in H_1, H_2, H_3, or H_4. In fact, for an appropriate $\delta > 0$, and for $D \simeq [0,1]\times[0,1]$, we can consider $\hat{D}$ to correspond to $[\delta,1-\delta]\times[0,1]$. There is a homeomorphism ϕ of $\hat{D}$ which is the identity on $\partial\hat{D}$ and the elements of $H_5-\hat{H}_5$, and which takes each $g \in \hat{H}_5$ to sufficiently small size that diam $P_4P_3P_2P_1(\phi(g)) < \varepsilon$. We can also choose ϕ so that it is the identity on $P_4P_3P_2P_1(\hat{D}-U)$. Since $P_4P_3P_2P_1$ is a homeomorphism on $\hat{D}$, the image $P_4P_3P_2P_1(\hat{D})$ is tame. Hence, $P_4P_3P_2P_1\phi$ can be extended to a homeomorphism ξ of E^3 and ξ can be the identity on E^3-U, and on the nondegenerate elements in $H_5-\hat{H}_5$ and in $P_4P_3P_2P_1(H_1 \cup H_2 \cup H_3 \cup H_4)$. Hence, H_5 is shrinkable and $E^3/G_5 \simeq E^3$.

The 2-cells $P_4P_3P_2P_1(\sigma)$ are mapped by P_5 to simple arcs.

We are ready to describe an example which answers Daverman's question.

Let G_6 be the usc decomposition of E^3 such that the collection H_6 of nondegenerate elements are exactly (1) those sets $P_5P_4P_3P_2P_1(\sigma)$ where σ is a 2-cell in D described above and (2) the remaining arcs $P_5P_4P_3P_2P_1(\beta)$ where $\beta = q \cap D$ for some $g \in H$ and not included in an element described in (1). Note that H_6 contains the collection $\{P_5P_4P_3P_2P_1(g) \mid g \in H\}$. It follows that $E^3/G_6 \not\simeq E^3$, that is,

$$(((((E^3/G_1)/G_2)/G_3)/G_4)/G_5)/G_6 \not\simeq E^3 .$$

The decomposition G_6 can be viewed as an alteration G_6' of our first modification of Bing's Dog Bone decomposition (used to answer Starbird's question) where the interval threaded through H has been widened into a thin 2-cell. That is, elements of H_6' are arcs and topological H's with a 2-cell for a horizontal bar. It is Bing's example with certain 2-cells (a countable number) sewed to certain pairs of elements of H. It should be clear that H_6' is not shrinkable and consequently, H_6 is not shrinkable. Thus, H_6 fills up a topological 2-cell (as a continuous collection) in a homeomorphic copy of E^3. The decomposition space E^3/G_6 is not homeomorphic to E^3 since H_6 is not shrinkable.

As in the example which answered Starbird's question, it can be shown that if $H' \subset H_6$ is such that $P_6(H')$ has dimension 0, then $E^3/G' \simeq E^3$ where G' is the usc decomposition whose collection of nondegenerate elements is H'. Thus, the answer to Daverman's question is affirmative.

V. OTHER EXAMPLES

Our technique can be applied to other examples. In the case of the example of Armentrout and Bing [1], the construction of the arc A through H is replaced by putting in various open intervals which approach elements of H as a $\sin 1/x$ curve. However, $P(H_1)$ *is an arc* where H_1 represents the modified collection of nondegenerate elements.

It does not appear that we can modify this example [1] using our techniques to answer Daverman's question.

REFERENCES

1. Armentrout, S. and R. H. Bing, "A toroidal decomposition of E^3," *Fund. Math.* *60*(1967), 81-87.

2. Bing, R. H., "Upper semicontinuous decompositions of E^3," *Annals Math.* *65*(1957), 363-374.

3. ______, "A decomposition of E^3 into points and tame arcs such that the decomposition space is topologically different from E^3," *Annals Math.* *65*(1957), 484-500.

4. Fort, M. K., Jr., "A note concerning a decomposition space defined by Bing," *Annals Math.* *65*(1957), 501-504.

5. McAuley, Louis F., "Some upper semicontinuous decompositions of E^3 into E^3," *Annals Math.* *73*(1961), 437-457.

6. Voxman, Wm. L., "On the shrinkability of decompositions of 3-manifolds," *Trans. Amer. Math. Soc.* *150*(1970), 27-39.

7. Whyburn, G. T., *Analytic Topology*, Amer. Math. Soc. Colloquium Publications 28, 1942.

WHICH DISPERSED DIAFACTORIZATION STRUCTURES ON *Top* ARE HEREDITARY?

Austin Melton

Kansas State University, Manhattan

Notation 1. *Top* denotes the category of topological spaces and continuous functions.

Definition 2. Let K be a category. A *source* in K is a pair $(X,(f_i:X \to Y_i)_I)$ where X is a K-object and each $f_i:X \to Y_i$ is a K-morphism. X is said to be the *domain of the source*. The indexing collection I is assumed to be a class. Alternate notations for a source include (X,f_i), $(f_i)_I$, and (X,F).

Definition 3. If K is a category, if E is a subclass of the morphisms of K, and if M is a collection of sources of K, then (E,M) is called a *diafactorization structure* on K if

i) whenever $(X,(f_i:X \to Y_i)_I)$ is a source in K, then there exists a morphism $e:X \to W \in E$ and a source $(W,(m_i:W \to Y_i)_I) \in M$ such that $f_i = m_i \circ e$ for each $i \in I$;

ii) whenever for each $j \in J$, $k_j \circ a = b_j \circ h$ with $a:W \to X \in E$, $(X,(k_j:X \to Z_j)_J)$ a K-source, $h:W \to Y$ a K-morphism, and $(Y,(b_j:Y \to Z_j)_J) \in M$, then there exists a K-morphism $d:X \to Y$ such that $d \circ a = h$ and $b_j \circ d = k_j$ for each $j \in J$; and

iii) E and M are closed under composition with isomorphisms. (If k is a K-isomorphism and if $(m_i)_I \in M$, then whenever $(m_i \circ k)_I$ is defined, it is in M.)

ISBN 0-12-481820-X

Remarks 4. a) (E,M) of Definition 3 has been called a factorization structure; however, since (E,M) has the diagonalization property ii) as well as the factorization property i), we call such structures diafactorization structures.

b) If (E,M) is a diafactorization structure on K, then K is said to be an (E,M)-category.

c) Some known results about diafactorization structures are given in the next theorem; a more complete list is found in [HSV].

THEOREM 5. *If* K *is an* (E,M)*-category, then*

i) E *is contained in the class of* K*-epimorphisms;*

ii) E *and* M *determine each other by the diagonalization property;*

iii) if $(X,(f_i)_I)$ *is a* K*-source and if there exists* $J \subseteq I$ *with* $(X,(f_j)_J) \in M$, *then* $(X,(f_i)_I) \in M$ *also.*

Definition 6. Let K be a category; let $f:X \to Y$ be a K-morphism; and let W be a K-object. f is said to be W-*extendible* if whenever $g:X \to W$ is a K-morphism, then there exists a K-morphism $g':Y \to W$ such that $g' \circ f = g$. If A is a subcategory of K, then f is said to be A-*extendible* if f is W-extendible for each A-object W.

Definition 7. a) Let K be an (E,M)-category and let (C,D) be a diafactorization structure on K. (C,D) is said to be an (E,M)-*dispersed diafactorization structure* on K or simply a *dispersed diafactorization structure* on K if there exists a full, isomorphism-closed subcategory A of K with $C = \{f \in E : f$ is A-extendible$\}$.

b) The morphisms in $\{f \in E : f$ is A-extendible$\}$ are called A-*concentrated* E-*morphisms*.

Definition 8. Let (C,D) be a diafactorization structure on *Top*.

(C,D) is said to be a *hereditary diafactorization structure* if there is a diafactorization structure (C',D') on *Top* such that $C = \{f:X \to Y$ a continuous function: for every open subset U of Y, $f\big|^{U}_{f^{-1}U} : f^{-1}U \to U \in C'\}$.

Remarks 9. a) In [HSV] it is shown that whenever K is an (E,M)-category and A is a full, isomorphism-closed subcategory of K, then $\{f \in E$: f is A-extendible$\}$ is the left factor of a diafactorization structure on K. In [HSV] it is further shown that the right factor of this dispersed diafactorization structure is $\{(X,F)$ a K-source: $(X,F \dot{\cup} S(X,A)) \in M\}$ where $S(X,A)$ is the collection of all K-morphisms with domain X and codomain in A. ($\dot{\cup}$ represents disjoint union.) The members of this right factor are called *A-dispersed sources.*

b) In [CD] it is shown that whenever (C',D') is a diafactorization structure on *Top*, then $\{f:X \to Y$ a continuous function: for every open subset U of Y, $f\big|^{U}_{f^{-1}U} \in C'\}$ is the left factor of a diafactorization structure on *Top*. In general, no explicit formula for determining the right factor of this hereditary diafactorization structure is known, but since the left factor is known, then by ii) of Theorem 5 the diafactorization structure is determined. (Actually the work in [CD] is only concerned with diafactorization structures for single functions with the added stipulation that the left factor be a subclass of the epimorphisms (= surjective continuous functions) of *Top*. However, in Remarks 1.3(2) and (3) of [HSV] a method for obtaining a diafactorization structure (C^*,D^{**}) on *Top* from a diafactorization structure (C^*,D^*) for singletons with $C^* \subseteq$ epimorphisms is given. Thus, the results in [CD] hold for diafactorization structures.)

c) If (C',D') is a diafactorization structure on *Top*, then the class $\{f:X \to Y$ a continuous function: for every open subset U of Y, $f\big|^{U}_{f^{-1}U} \in C'\}$ is denoted by hereditarily-C' and abbreviated by her-C'. Also, $f\big|^{U}_{f^{-1}U}$ is abbreviated by f^U .

d) Two well known diafactorization structures on *Top* are (epimorphisms, extremal monosources) and (quotients, monosources); we abbreviate these two diafactorization structures by (E,EMS) and (Q,MS), respectively. (A source in *Top* is a monosource if and only if it is a point-separating source; a source in *Top* is an extremal monosource if and only if it is a point-separating initial source.)

Question 10. In [CD] P. J. Collins and R. Dyckhoff begin with one (E,EMS)-dispersed diafactorization structure on *Top* and two (Q,MS)-dispersed diafactorization structures on *Top*, and they form the corresponding hereditary diafactorization structures. Then they want to determine if the three new diafactorization structures differ from the beginning three. They show that the six diafactorization structures are distinct; to do this Collins and Dyckhoff construct continuous---and some rather complicated---functions which are used to show that the six left factors are distinct. From this work in [CD] we are led to ask: if we begin with all (E,EMS)-dispersed diafactorization structures and all (Q,MS)-dispersed diafactorization structures and if for each of these we form the corresponding hereditary diafactorization structure, then when do we obtain new diafactorization structures? To answer this question we determine the intersections of the following four collections: the collection of (E,EMS)-dispersed diafactorization structures, the collection of (Q,MS)-dispersed diafactorization structures, the collection of hereditarily (E,EMS)-dispersed diafactorization structures, and the collection of hereditarily (Q,MS)-dispersed diafactorization structures.

Notations and Definitions 11. Let R be the collection of full, homeomorphism-closed subcategories of *Top*; let T be the collection of diafactorization structures on *Top*; let T_Q be the collection of diafactorization structures (C,D) on *Top* with $C \subseteq Q$; let T^* be the collection of hereditary diafactorization structures on *Top*; and let T^*_Q be the collection of hereditary diafactorization structures (C^*, D^*) on *Top* with $C^* \subseteq Q$. We define the following functions:

$F: R \to T$ by $FA = (cA, -)$

where $cA = \{f \in E : f$ is A-extendible$\}$ and where $(cA, -)$ is the diafactorization structure with left factor cA;

$F_Q: R \to T_Q$ by $F_Q A = (c_Q A, -)$

where $c_Q A = \{f \in Q : f$ is A-extendible$\}$;

$F^*: R \to T^*$ by $F^* A = (\text{her-}cA, -)$; and

$F^*_Q: R \to T^*_Q$ by $F^*_Q A = (\text{her-}c_Q A, -)$.

Definition 12. Let $(X, (f_i : X \to Y_i)_I)$ be a source in *Top*. A subset X' of X is called a *fiber* of $(X, (f_i)_I)$ if there exist $y_i \in Y_i$ such that $X' = \bigcap_I f_i^{-1}(y_i)$. Each nonempty fiber of (X, f_i) is of the form $\bigcap_I f_i^{-1} f_i(x)$ with $x \in X$.

Remarks 13. a) From [HSV] we have that (homeomorphisms, sources) and (quotients with indiscrete fibers, sources with T_0-fibers) are in $F[R] \cap F_Q[R]$; (homeomorphisms, sources) = $F(Top) = F_Q(Top)$ and (quotients with indiscrete fibers, sources with T_0-fibers) = $F(T_0) = F_Q(T_0)$ where T_0 is the category of T_0-spaces and continuous functions. Since her-homeomorphisms = homeomorphisms and her-(quotients with indiscrete fibers) = (quotients with indiscrete fibers), then $F(Top) = F_Q(Top) = F^*(Top) = F^*_Q(Top)$ and $F(T_0) = F_Q(T_0) = F^*(T_0) = F^*_Q(T_0)$.

b) In [HSV] it is shown that if A is a subcategory of *Top* and if B is the epi-reflective hull of A [respectively, the quotient-reflective hull of A], then the (E,EMS)- [respectively, the (Q,MS)-] dispersed diafactorization structures determined by A and B are equal. From [HSV] we also have that there exists a bijection between the epi-reflective subcategories of *Top* and the diafactorization structures in $F[R]$ and a bijection between the quotient-reflective subcategories of *Top* and the diafactorization structures in $F_Q[R]$. Thus, if $(C,D) \in F[R] \cap F_Q[R]$, then there is a quotient-reflective subcategory B of *Top* such that $(C,D) = FB = F_Q B$. Likewise, if $(C^*,D^*) \in F^*[R] \cap F_Q^*[R]$, then there is a quotient-reflective subcategory B^* such that $(C^*,D^*) = F^*B^* = F_Q B^*$. (Actually, the first results mentioned in this Remark b) are obtained for an arbitrary category and associated diafactorization structure in [HSV].)

PROPOSITION 14. {*(homeomorphisms, sources), (quotients with indiscrete fibers, sources with* T_0*-fibers)*} $= F[R] \cap F_Q[R] = F[R] \cap F_Q^*[R] = F^*[R] \cap F_Q[R] = F^*[R] \cap F_Q^*[R]$.

Proof. Case i). Let $(C,D) \in F[R] \cap F_Q[R]$; thus, there is a quotient-reflective subcategory B such that $(C,D) = FB = F_Q B$. If we suppose that B is neither *Top* nor T_0, then B is a subcategory of T_1, the category of T_1-spaces and continuous functions. Consider the continuous function $f_1 : Z \to I_2$ where $Z = \{a,b\}$ with topology $\{Z,\{a\},\emptyset\}$ and $I_2 = \{r,s\}$ with topology $\{I_2,\emptyset\}$ and where $f_1(a) = r$ and $f_1(b) = s$. Since the only open set containing b is Z itself, then f_1 is T_1-extendible, and thus, it is B-extendable. Therefore, since f_1 is an epimorphism, then $f_1 \in cB$. However, since f_1 is not a quotient, then $f_1 \notin c_Q B$. Hence, B is *Top* or T_0.

Case ii). If we assume that $(C,D) \in F[R] \cap F_Q^*[R]$, then again the corresponding quotient-reflective subcategory B is *Top* or T_0 because her-$c_Q B$ being a subclass of $c_Q B$ implies that $f_1 \notin$ her-$c_Q B$ if B is a subcategory of T_1.

Cases iii) and iv) follow by similar arguments.

Remark 15. To answer the question in Question 10 we now only need to determine $F[R] \cap F^*[R]$ and $F_Q[R] \cap F_Q^*[R]$.

PROPOSITION 16. *{(homeomorphisms, sources), (quotients with indiscrete fibers, sources with* T_0*-fibers),* $(Q,MS)\} = F_Q[R] \cap F_Q^*[R]$. *And* $(Q,MS) = F_Q S$ *where* S *is the full, homeomorphism-closed subcategory of Top with object class* $\{W$ *a topological space:* $|W| \leq 1\}$.

Proof. Since each quotient is S-extendible, then $F_Q S = (Q,MS)$, and since her-quotients = quotients, then $F_Q^* S = (Q,MS)$.

Let $(C,D) \in F_Q[R] \cap F_Q^*[R]$, and let B be the quotient-reflective subcategory of *Top* such that $(C,D) = F_Q B = F_Q^* B$. If we assume that B is neither *Top* nor T_0, then B is a subcategory of T_1, and if we assume that $B \neq S$, then B contains a two-point discrete space, call it $D_2 = \{d_1, d_2\}$. Consider the function $f_2: M \to Z$ where $M = \{m_1, m_2, m_3\}$ with topology $\{M, \{m_1, m_2\}, \{m_1\}, \{m_2\}, \emptyset\}$ and $f_2(m_1) = f_2(m_2) = a$ and $f_2(m_3) = b$. (Z is defined in Proposition 14.) Since the only open set containing m_3 is M, then f_2 is T_1-extendible, and thus, it is B-extendible. Therefore, since f_2 is a quotient, then $f_2 \in c_Q B$. However, $\{a\}$ is an open subset of Z ; and $f_2^{\{a\}}: \{m_1, m_2\} \to \{a\}$ is not D_2-extendible; and thus, $f_2^{\{a\}} \notin c_Q B$. Hence, $f_2 \notin$ her-$c_Q B$. It follows that B is either *Top*, T_0, or S, and therefore, (C,D) is (homeomorphisms, sources), (quotients with indiscrete fibers, sources with T_0-fibers), or (Q,MS).

LEMMA 17. *If* I *is the category of indiscrete spaces and continuous functions, then* $FI = F^*I =$ *(bijections, initial sources).*

Proof. Let $f: X \to Y$ be a bijection. Suppose that W is an indiscrete space and that $g: X \to W$ is a continuous function. $g \circ f^{-1}: Y \to W$ is a function such that $(g \circ f^{-1}) \circ f = g$, and $g \circ f^{-1}$ is continuous because W is an indiscrete space. Thus, $f \in cI$. Let $h: A \to B \in cI$, and let A' be the indiscrete space with underlying set equal to the underlying set of A, and let $k: A \to A'$ be the identity function on the underlying sets. There exists a continuous function $k':B \to A'$ such that $k' \circ h = k$, and since k is injective, then h is also injective. Hence, h is a bijection. Therefore, $F\mathcal{I}$ = (bijections,-) , and since her-bijections = bijections, then $F^*\mathcal{I}$ = (bijections,-) also.

Denote the right factor of $F\mathcal{I}$ by $d\mathcal{I}$. As stated in Remark 9a) $d\mathcal{I} = \{(X,F)$ a source in $\mathit{Top}:(X,F \dot{\cup} S(X,\mathcal{I})) \in \mathrm{EMS}\}$. Suppose that $(X,(f_i)_I) \in d\mathcal{I}$ and further suppose that $(W,(h_i)_I)$ is a source in Top and that $s:W \to X$ is a function such that $h_i = f_i \circ s$ for each $i \in I$. For each $f:X \to Y \in S(X,\mathcal{I})$ we have the continuous function $f \circ s:W \to Y$, and thus, we have $(X,(f_i)_I \dot{\cup} S(X,\mathcal{I})) \in \mathrm{EMS}$, $(W,(h_i)_I \dot{\cup} (f \circ s)_{f \in S(X,\mathcal{I})})$ a source in Top and $s:W \to Y$ a function such that $h_i = f_i \circ s$ for each $i \in I$ and $(f \circ s) = f \circ s$ for each $f \in S(X,\mathcal{I})$. Therefore, since $(X,(f_i)_I \dot{\cup} S(X,\mathcal{I}))$ is initial, then s is a continuous function; hence, (X,f_i) is also initial. Suppose that $(A,(t_j)_J)$ is an initial source. If A' is the indiscrete space whose underlying set is the underlying set of A and if $k:A \to A'$ is the identity function on the underlying sets, then $k \in S(X,\mathcal{I})$, and consequently, $(A,(t_j)_J \dot{\cup} S(A,\mathcal{I}))$ is a point-separating initial source, i.e., $(A,(t_j)_J \dot{\cup} S(A,\mathcal{I})) \in \mathrm{EMS}$. Hence, $(A,t_j) \in d\mathcal{I}$.

Remark 18. Let $R_0 = \{\mathcal{B} \in R:\mathcal{B}$ is a subcategory of $\mathcal{T}_0\}$, and let $R^0 = \{\mathcal{B}' \in R:\mathcal{B}'$ is not a subcategory of $\mathcal{T}_0\}$. From [M] we have that there exists an order preserving bijection $L:R_0 \to R^0$ such that $L\mathcal{B} = [\mathcal{B} \cup I_2]$ where $[\mathcal{B} \cup I_2]$ is the epi-reflective hull of I_2 and the objects of $\mathcal{B}$. In particular, $L\mathcal{S} = \mathcal{I}$ where $\mathcal{S}$ is defined as in Proposition 16 and $\mathcal{I}$ is defined as in Lemma 17, and $L\mathcal{T}_0 = \mathit{Top}$. Since the epi-reflective hull of $\mathcal{B} \cup \{I_2\}$ is $L\mathcal{B}$, then to show that an epimorphism f is $L\mathcal{B}$-extendible, it need only be shown that f is $(\mathcal{B} \cup \{I_2\})$-extendible.

PROPOSITION 19. *{(homeomorphisms, sources), (quotients with indiscrete fibers, sources with* T_0*-fibers), (bijections, initial sources),* (E,EMS)} = $F[R] \cap F^*[R]$.

Proof. Since each epimorphism is S-extendible, then $FS = (E,EMS)$, and since her-epimorphisms = epimorphisms, then $F^*S = (E,EMS)$ also.

Let $(C,D) \in F[R] \cap F^*[R]$, and let B be the epi-reflective subcategory of *Top* such that $(C,D) = FB = F^*B$.

Case i). $B \in R_0$. If $B \neq T_0$, then B is a subcategory of T_1, and if $B \neq S$, then the discrete space $D_2 = \{d_1,d_2\}$ is in B. Consider the continuous function $f_3 : M \to N$ where M is defined in the proof of Proposition 16 and $N = \{n_1,n_2,n_3\}$ with topology $\{N,\{n_1,n_2\},\emptyset\}$ and $f_3(m_i) = n_i$ for $i=1,2,3$. Since the only open set containing m_3 is M, then f_3 is T_1-extendible, and thus, f_3 is B-extendible. Hence, $f_3 \in cB$. But since $f_3^{\{n_1,n_2\}}$ is not D_2-extendible, then $f_3 \notin$ her-cB.

Case ii). $B \in R^0$; thus, there exists $D \in R_0$ such that $B = [D \cup I_2]$. If B is neither *Top* nor I, then D is a subcategory of T_1 and D contains D_2. Since f_3 is a bijection, then f_3 is I_2-extendible, and by case i) f_3 is D-extendible. Hence, f_3 is B-extendible, and consequently, $f_3 \in cB$. However, since D_2 is in B, then $f_3 \notin$ her-cB.

It follows that (C,D) is one of the four diafactorization structures listed in the statement of this proposition.

Answer 20. The diafactorization structures (homeomorphisms, sources) and (quotients with indiscrete fibers, sources with T_0-fibers) are (E,EMS)- and (Q,MS)-dispersed and hereditary. (Q,MS) is (Q,MS)-dispersed and hereditary, and (E,EMS) and (bijections, initial sources) are (E,EMS)-dispersed and hereditary. If (C,D) is any other (E,EMS)- or (Q,MS)-dispersed diafactorization structure, then (her-C,-) is neither (E,EMS)- nor (Q,MS)-dispersed. The simplicity of this answer is striking because according to [HSV] there is a proper class of (E,EMS)-dispersed diafactorization structures on *Top* and there is a proper class of (Q,MS)-dispersed diafactorization structures on *Top*.

REFERENCES

[CD] Collins, P. J. and R. Dyckhoff, "Connexion properties and factorization theorems," *Quaestiones Math.* *2*(1977), 103-112.

[HSV] Herrlich, H., G. Salicrup and R. Vázquez, "Dispersed factorization structures," *Can. J. Math* *31*(1979), 1059-1071.

[M] Marny, Th., "On epireflective subcategories of topological categories," *Gen. Topology Appl.* *10*(1979), 175-181.

SOME HYPERSPACES HOMEOMORPHIC TO SEPARABLE HILBERT SPACE

Mark Michael

Louisiana State University, Baton Rouge
and
Southeast Missouri State University, Cape Girardeau

Throughout this announcement, X is a non-degenerate Peano continuum, 2^X denotes the hyperspace consisting of all closed subsets of X, and $C(X)$ denotes the hyperspace consisting of all subcontinua of X. We represent the Hilbert cube Q as $\Pi_{i=1}^{\infty} I_i$, where $I_i = [-1,1]$ for all i, and let $\Sigma = \{x \in Q\colon \sup|x_i| < 1\}$ and $\sigma = \{x \in \Sigma\colon x_i = 0$ for almost all $i\}$.

It is well-known that 2^X is homeomorphic to Q, and if X contains no free arc, $C(X)$ is also homeomorphic to Q. However, neither the original proof by Curtis and Schori [5] nor the more recent proof by Torunczyk [8] provides an explicit homeomorphism. Thus we are left with a very broad question: if H is a hyperspace of X (i.e., a subspace of 2^X defined in terms of X), how is H positioned in 2^X? This report states certain conditions under which $(2^X,H)$ is homeomorphic to (Q,Σ) or to (Q,σ).

A subset B of a metric space M has the *compact absorption property* and is said to be a *capset* in M if B can be expressed as $\cup_{i=1}^{\infty} B_i$, where

(1) for each i, $B_i \subset B_{i+1}$ and B_i is a compact Z-set in M,

(2) for each $\epsilon > 0$, positive integer j, and compact subset K of M, there exists a positive integer k and an embedding $f\colon K \to B_k$ such that $f|K \cap B_j = id$ and $d(f,id) < \epsilon$.

ISBN 0-12-481820-X

The definition of *finite-dimensional compact absorption property* (and *f-d capset*) is identical except that K and the B_i are taken to be finite-dimensional.

The set Σ is a capset in Q while σ is an f-d capset in Q. The (finite-dimensional) compact absorption property is preserved by homeomorphisms of pairs. Furthermore, if B and B′ are both capsets or both f-d capsets in Q, then $(Q,B) \approx (Q,B')$. Thus Σ and σ are topologically characterized as subsets of Q.

The identification of (f-d) capsets in Q is aided by the following two lemmas. The first is in the folklore (see [2], pp. 402-403). The second is a correction by Curtis [3] of earlier work by Kroonenberg [6] and by the author [7].

Lemma 1. *If* B *is a* σ-Z-*set (i.e., the union of countably many Z-sets) in* Q *and contains a capset in* Q, *then* B *is itself a capset in* Q. *If* B *is the union of countably many finite-dimensional Z-sets in* Q *and contains an f-d capset in* Q, *then* B *is also an f-d capset.*

Lemma 2. *Let* $B = \bigcup_{i=1}^{\infty} B_i$, *where* $B_1 \subset B_2 \subset B_3 \subset \ldots$ *is a tower of Z-sets in* Q. *Suppose there exists a deformation* $g: Q \times [0,1] \to Q$ *(with* $g_0 = \mathrm{id}$*) such that, for each* $\delta > 0$, *there is an integer* n *so that* $g[Q \times [\delta,1]] \subset B_n$.

(1) B *is a capset in* Q *if, for each* i, $B_i \approx Q$ *and* B_i *is a Z-set in* B_{i+1}.

(2) B *contains an f-d capset in* Q *if, for each* i, *there is an integer* j *and an embedding* $h: B_i \times [0,1] \to B_j$ *such that* $h(x,0) = x$, *for all* $x \in B_i$.

Let H be a subspace of 2^X. We say H has the *inclusion property* if $F \in H$ whenever $F \in 2^X$ and $F \supset E$, for some $E \in H$. More generally, H has the *growth property* if $F \in H$ whenever $F \in 2^X$ and, for some $E \in H$, $F \supset E$ and each component of F meets E. We say

H is a *finitely-generated inclusion* (resp., *growth*) *hyperspace* provided there exists $\alpha \subset H$ such that α is finite and H is the smallest hyperspace containing α and having the inclusion (resp., growth) property.

Theorem. *Suppose* $\mathcal{Q}$ *is a copy of* Q *in* 2^X *and* H *is a dense* σ-Z-*set in* $\mathcal{Q}$ *with the growth property.*

(1) H *is a capset in* $\mathcal{Q}$ *if either* X *contains no free arc or* H *has the inclusion property.*

(2) H *contains an f-d capset in* $\mathcal{Q}$ *if* X *is a finite graph.*

Comments on proof. A tower of finitely-generated growth or inclusion hyperspaces is constructed so as to satisfy the hypotheses of Lemma 2. By Lemma 1, the tower need only be contained in H. The elements of the tower are copies of Q for part (1), collapsible polyhedra for part (2).

Corollary 1. *Let* $H = \{F \in 2^X: \text{int } F \neq \emptyset\}$. *Then* H *is a capset in* 2^X, *and if* X *contains no free arc,* $H \cap C(X)$ *is a capset in* C(X) .

Corollary 2. *Let* C *be the Cantor set. If* H *is either* $\{F \in 2^X: \dim F > 0\}$ *or* $\{F \in 2^X: F \not\approx C\}$, *then* H *is a capset in* 2^X .

Corollary 3. *Let* $H = \{F \in 2^X$: F *has finitely many components*$\}$. *If* X *contains no free arc, then* H *is a capset in* 2^X . *If* X *is a finite graph, then* H *is an f-d capset in* 2^X .

Earlier results by Kroonenberg [6] and by Curtis [2] can also be obtained from the theorem.

It is known that $Q\setminus\Sigma$ and $Q\setminus\sigma$ are each homeomorphic to the separable Hilbert space ℓ^2 . Consequently, each of the corollaries above gives rise to a corresponding corollary pertaining to hyperspaces homeomorphic to ℓ^2. The following is an example.

Corollary 3′. *If either* X *contains no free arc or* X *is a finite graph, then* $\{F \in 2^X$: F *has infinitely many components*$\}$ *is homeomorphic to* ℓ^2.

Since the two alternative hypotheses in Corollary 3′ are at opposite ends of the spectrum with regard to the existence of free arcs, it is reasonable to suspect the hypotheses are unnecessary. This conjecture led Curtis to define a *boundary set* in Q as a dense σ-Z-set in Q whose complement in Q is homeomorphic to ℓ^2. He subsequently confirmed the conjecture and also proved that $\{F \in 2^X : |F| < \aleph_0\}$ is a boundary set in 2^X [4].

Incidental to the study of boundary sets was the revised statement of Lemma 2 and the discovery of a pathological space having many of the nice topological properties of ℓ^2 [1].

The proofs of the theorem and its corollaries will be published elsewhere.

REFERENCES

1. Anderson, R.D., D. W. Curtis, and Jan van Mill, "A fake topological Hilbert space," *Trans. Amer. Math. Soc.*, submitted.

2. Curtis, D.W., "Hyperspaces of non-compact metric spaces," *Comp. Math. 40*(1980), 139-152.

3. ____________, *Boundary sets in the Hilbert cube*, in preparation.

4. ____________, *Boundary sets in hyperspaces*, in preparation.

5. Curtis, D.W. and R. M. Schori, "Hyperspaces of Peano continua are Hilbert cubes," *Fund. Math. 101*(1978), 19-38.

6. Kroonenberg, Nelly, "Pseudo-interiors of hyperspaces," *Comp. Math. 32* (1976), 113-131.

7. Michael, Mark, *Sigma-compact subsets of hyperspaces*, Ph.D. dissertation, Louisiana State University, 1979.

8. Torunczyk, Henryk, "On CE-images of the Hilbert cube and characterizations of Q-manifolds," *Fund. Math.*, to appear.

DIHEDRAL GROUP ACTIONS I

Deane Montgomery

Institute for Advanced Study
Princeton, New Jersey

C. T. Yang

University of Pennsylvania, Philadelphia

I. INTRODUCTION

Milnor [3] has proved that the dihedral group D_q , q odd, cannot act freely on a mod 2 n-sphere. In this paper it is shown that this theorem remains true for a more general class of manifolds under certain conditions. For example, it will follow that D_q cannot act freely on RP^{4n+1} , although it can be seen that D_q acts freely on RP^{4n-1} . As Petrie observed, Theorem 1 has some relation to Lee, "Semicharacteristic classes," *Topology 12* (1973), 183-199.

II. DEFINITIONS AND KNOWN RESULTS

It will be assumed that all manifolds and actions are smooth, and that all homology is taken mod 2 . The symbol Z_2 is used for the coefficient field and as a transformation group. In the first case, the unit of Z_2 is denoted by 1 , and in the second case, the generator of Z_2 is denoted by α .

Let X be a manifold on which Z_2 acts with fixed point set $F(Z_2) = F$. Then [1,2] there is the exact sequence

$$\cdots \xrightarrow{j} H^{\sigma}_{k+1}(X) \xrightarrow{\partial} H^{\sigma}_{k}(X) \oplus H_k(F) \xrightarrow{i} H_k(X) \xrightarrow{j} H^{\sigma}_{k}(X) \xrightarrow{\partial} \cdots$$

ISBN 0-12-481820-X

to be denoted by $seq(Z_2,X,F)$ and called the Smith homology sequence for the action. If X' is another manifold on which Z_2 acts with fixed point set F' and $f:X\to X'$ is equivariant, then f induces a homomorphism

$$f:seq(Z_2,X,F) \to seq(Z_2,X',F') .$$

This means that the following commutes:

$$\begin{array}{ccccccccc}
\cdots\to H^{\sigma}_{k+1}(X') & \xrightarrow{\partial} & H^{\sigma}_k(X') \oplus H_k(F') & \xrightarrow{i} & H_k(X') & \xrightarrow{j} & H^{\sigma}_k(X') & \to\cdots \\
\uparrow f & & \uparrow f & & \uparrow f & & \uparrow f & \\
\cdots\to H^{\sigma}_{k+1}(X) & \xrightarrow{\partial} & H^{\sigma}_k(X) \oplus H_k(F) & \xrightarrow{i} & H_k(X) & \xrightarrow{j} & H^{\sigma}_k(X) & \to\cdots
\end{array} .$$

(2.1) *Smith's Theorem*. Let X be a manifold with the mod 2 homology of S^n on which Z_2 acts with $F = F(Z_2)$. Then for some integer r, $-1\le r\le n$, F has the mod 2 homology of S^r. Further $H^{\sigma}_k(X)$ is isomorphic to Z_2 or 0 according as $r<k\le n$ or not, and

$$\partial^{n-r}i^{-1}:H_n(X) \to H_r(F)$$

is an isomorphism where reduced groups are used.

For the group D_q, q odd, we take β to be a generator of the invariant subgroup Z_q and α an element of order 2 such that

$$\alpha\beta\alpha = \beta^{-1} .$$

(2.2) *Poincare duality*. Let M be a closed connected manifold and A a closed subset with finite homology. Then

$$H_k(M-A) \simeq H_{n-k}(M,A) .$$

For any basis $\{b_1,\dots,b_m\}$ of $H_k(M-A)$ there is a dual basis $\{b_1^*,\dots,b_m^*\}$ of $H_{n-k}(M,A)$ such that

$$b_i \cap b_j^* = \delta_{ij}$$

where $b_i \cap b_j^*$ denotes the intersection number mod 2 .

Later the D_q manifold M will be required to be connected and closed and to satisfy one or both of the following conditions:

(i) $n = 2m+1$ and the trace of

$$\beta^{-2}: \bigoplus_{k=0}^{m} H_k(M) \to \bigoplus_{k=0}^{m} H_k(M)$$

is equal to 1.

(ii) Let $\dim F = r$, and in $\mathrm{seq}(Z_2, M, F)$, concerning

$$\partial^{n-r}: H_n^{\sigma}(M) \to H_r^{\sigma}(M) \oplus H_r(F) ,$$

it is assumed that $H_n^{\sigma}(M)$ is mapped to 0 in the first term and isomorphically onto in the second.

Results to be proved are:

THEOREM 1. *Let* M^n *be a smooth connected closed* D_q *manifold,* $n = 2m+1$. *If (i) holds then* D_q *cannot act freely.*

THEOREM 2. *Let* M^n *be a smooth connected closed* D_q *manifold,* $n = 2m+1$. *If (i) and (ii) hold and* $F(Z_q) = \emptyset$ *then* $\dim F \geq (n-1)/2$. *If in addition* F *is on r-sphere* mod 2, *then* $\dim F = (n-1)/2$.

If M^n is a mod 2-sphere, $n = 2m+1$, all of the assumptions of Theorem 2 are true, so $r = (n-1)/2$. Bredon also noticed this case.

III. A LEMMA

Let M^n be a closed D_q manifold with $F(Z_q) = \emptyset$ and $n = 2m+1$. Define an action of Z_2 on $M \times M$ as

$$\alpha(x,y) = (y,x) .$$

As in [3] let

$$f,h: M \to M \times M$$

be the Z_2 equivariant imbeddings defined by

$$f(x) = (x, \alpha x) , \quad h(x) = (\beta x, \beta\alpha x) .$$

Then $F(Z_q) = \emptyset$ implies that for any $x \in M$, $\beta\alpha x \neq \alpha\beta x$ so that

$$h(M) \subset M \times M - f(M) .$$

Hence h induces a homomorphism

$$h: \mathrm{seq}(Z_2, M, F) \to \mathrm{seq}(Z_2, M \times M - f(M), \Delta - f(M)) .$$

Since for any $x \in M$, $\beta\alpha x \neq \alpha\beta x$, F is a proper subset of M, so that $H_n(F) = 0$ and

$$i: H_n^\sigma(M) \to H_n(M)$$

is an isomorphism. Let u be the generator of $H_n(M)$ and let

$$u^\sigma = i^{-1} u ,$$

that is, u^σ is the generator of $H_n^\sigma(M)$.

LEMMA. *Let* M^n, $n = 2m+1$, *be a closed* D_q *manifold with* $F(Z_q) = \phi$ *and satisfying (i). Then* hu^σ *is not contained in*

$$\text{image}[H_n(M\times M - f(M)) \xrightarrow{j} H_n^\sigma(M\times M - f(M))] .$$

Hence $\partial hu^\sigma \neq 0$.

For any $k = 0,\ldots,m$, let B_k be a basis of $H_k(M)$ and let B_k^* be the basis of $H_{n-k}(M)$ dual to B_k. Then for any $b \in B_k$ there is a unique $b^* \in B_k^*$ such that for any $b' \in B_k^*$, $b \cap b' = 1$ or 0 according as $b' = b^*$ or not. Then

$$A = \bigcup_{k=0}^{m} (B_k \times B_k^* \cup B_k^* \times B_k)$$

is a basis of $H_n(M\times M)$. Since the homomorphism $f: H_*(M) \to H_*(M\times M)$ is injective, it follows that the homomorphism $H_*(M\times M) \to H_*(M\times M, f(M))$ induced by inclusion is surjective. Hence by Poincare duality, the homomorphism

$$H_n(M\times M - f(M)) \to H_n(M\times M)$$

induced by inclusion is injective. Because of this fact, any element of $H_n(M\times M - f(M))$ may be written as a linear combination of elements of A as if it is an element of $H_n(M\times M)$.

Let $\Delta: M \to M\times M$ be the diagonal imbedding. Then for any $b\times b' \in A$, $(b\times b') \cap \Delta u = 1$ if and only if $b \cap b' = 1$. Therefore

$$\Delta u = \sum_{k=0}^{m} \sum_{b\in B_k} (b\times b^* + b^*\times b) ,$$

and

$$fu = \sum_{k=0}^{m} \sum_{b \in B_k} (b \times \alpha b^* + b^* \times \alpha b) ,$$

and

$$hu = \sum_{k=0}^{m} \sum_{b \in B_k} (\beta b \times \beta\alpha b^* + \beta b^* \times \beta\alpha b) .$$

Suppose $hu^\sigma = jv$ for some $v \in H_n(M\times M - f(M))$. Then

$$ijv = ihu^\sigma = hiu^\sigma = hu .$$

Let $B_k = \{b_1^k, \ldots, b_{t_k}^k\}$ and let

$$\alpha b_\lambda^k = \sum_{\mu=1}^{t_k} \alpha_{\lambda\mu}^k b_\mu^k \qquad \lambda = 1, \ldots, t_k$$

where $\alpha_{\lambda\mu}^k \in Z_2$. Then for any $\lambda , \mu = 1, \ldots, t_k$

$$\alpha_{\lambda\mu}^k = \alpha b_\lambda^k \cap b_\mu^{k*} = b_\lambda^k \cap \alpha b_\mu^{k*} .$$

Therefore

$$\alpha b_\mu^{k*} = \sum_{\lambda=1}^{t_k} \alpha_{\lambda\mu}^k b_\lambda^{k*} .$$

Using this gives:

$$\begin{aligned} ijv &= \sum_{k=0}^{m} \sum_{\mu=1}^{t_k} \beta b_\mu^k \times \beta\alpha b_\mu^{k*} + \sum_{k=0}^{m} \sum_{\lambda=1}^{t_k} \beta b_\lambda^{k*} \times \beta\alpha b_\lambda^k \\ &= \sum_{k=0}^{m} \sum_{\lambda,\mu=1}^{t_k} \alpha_{\lambda\mu}^k \beta b_\mu^k \times \beta b_\lambda^{k*} + \sum_{k=0}^{m} \sum_{\lambda,\mu=1}^{t_k} \alpha_{\lambda\mu}^k \beta b_\lambda^{k*} \times \beta b_\mu^k \\ &= \sum_{k=0}^{m} \sum_{\lambda,\mu=1}^{t_k} \alpha_{\lambda\mu}^k (\beta b_\mu^k \times \beta b_\lambda^{k*} + \beta b_\lambda^{k*} \times \beta b_\mu^k) . \end{aligned}$$

Consider now the following commutative diagram

$$\begin{array}{ccccccc} \cdots \longrightarrow H_n^\sigma(M \times M) \oplus H_n(\Delta) & \xrightarrow{i} & H_n(M \times M) & \xrightarrow{j} & H_n^\sigma(M \times M) & \xrightarrow{\partial} \cdots \\ \uparrow & & \uparrow & & \uparrow & \\ \cdots \longrightarrow H_n^\sigma(M\times M-f(M) \oplus H_n(\Delta - f(M)) & \xrightarrow{i} & H_n(M\times M-f(M)) & \xrightarrow{j} & H_n^\sigma(M\times M-f(M)) & \xrightarrow{\partial} \cdots \end{array}$$

where the vertical arrows are from inclusion. Let v in $H_n(M\times M-f(M))$

have image v'' in $H_n(M\times M)$. Then in $H_n(M\times M)$ we define v' as follows:

$$\text{(A)}\quad v'' = v' + \sum_{k=0}^{m} \sum_{\lambda,\mu=1}^{t_k} \alpha^k_{\lambda\mu} (\beta b^{k*}_\lambda \times \beta b^k_\mu) \ .$$

Of course v' , as well as the other terms above, have carriers in $M\times M$, although conceivably this may not be true in $M\times M - f(M)$. Since the terms in ijv (they are the same in ijv'') are symmetric, it follows that

$$ijv' = 0 \qquad \text{in} \quad M\times M \ .$$

In $M\times M$ the set of all $\alpha b^k_\lambda \times b^{k*}_\mu, b^{k*}_\mu \times \alpha b^k_\lambda$ ($\lambda,\mu=1,\dots,t_k; k=0,\dots,m$) also form a base of $H_n(M\times M)$. As a consequence v' may be written as follows:

$$v' = \sum_{k=0}^{m} \sum_{\lambda,\mu=1}^{t_k} (\gamma^k_{\lambda\mu} \alpha b^k_\lambda \times b^{k*}_\mu + \delta^k_{\lambda\mu} b^{k*}_\mu \times \alpha b^k_\lambda) \ .$$

Hence

$$ijv' = 0 = \sum_{k=0}^{m} \sum_{\lambda,\mu=1}^{t_k} (\gamma^k_{\lambda\mu} + \delta^k_{\lambda\mu}) (\alpha b^k_\lambda \times b^{k*}_\mu + b^{k*}_\mu \times \alpha b^k_\lambda)$$

and

$$\gamma^k_{\lambda\mu} = \delta^k_{\lambda\mu}$$

We wish to compute $v' \cap f(u)$:

$$v' \cap f(u) = \sum_{k=0}^{m} \sum_{\lambda,\mu=1}^{t_k} \gamma^k_{\lambda\mu} (\alpha b^k_\lambda \times b^{k*}_\mu + b^{k*}_\mu \times \alpha b^k_\lambda) \cap \sum_{k=0}^{m} \sum_{\nu=1}^{t_k} (b^k_\nu \times \alpha b^{k*}_\nu + b^{k*}_\nu \times \alpha b^k_\nu) \ .$$

Notice that

$$\alpha b^k_\lambda \times b^{k*}_\mu \cap b^k_\nu \times \alpha b^{k*}_\nu = 0$$

$$\alpha b^k_\lambda \times b^{k*}_\mu \cap \alpha b^{k*}_\nu \times b^k_\nu = 1 \qquad \text{only when} \quad \lambda = \mu = \nu \ .$$

Similarly,

$$b^{k*}_\mu \times \alpha b^k_\lambda \cap b^k_\nu \times \alpha b^{k*}_\lambda = 1 \qquad \text{only when} \quad \lambda = \mu = \nu$$

$$b^{k*}_\mu \times \alpha b^k_\lambda \cap \alpha b^k_\nu \times b^k_\nu = 0 \ .$$

Therefore

$$v' \cap f(u) = \sum_{k=0}^{m} \sum_{\nu=1}^{t_k} (\gamma_{\nu\nu}^{k} + \gamma_{\nu\nu}^{k}) = 0 .$$

This implies that v' has a carrier in $M \times M - f(M)$. To see this, proceed as follows. The relation $v' \cap f(u) = 0$ means that a singular cycle v', in general position, must intersect $f(M)$ in an even number of points. At each of these points we may assume that v' is a cell locally and that this cell is orthogonal to $f(M)$ at the point of intersection. Let $[a,b]$ be a smooth arc of $f(M)$ joining two points of the intersection. Let N be

$$[a,b] \times \sigma^n , \qquad \sigma^n \text{ a closed n-cell,}$$

and assume N imbedded in $M \times M$ so that

$$N \cap f(M) = [a,b]$$

and so that the ends $a \times \sigma^n$ and $b \times \sigma^n$ are in v'. Now, adding mod 2, replace v' by

$$v' + B \qquad (\text{where } B = \partial N),$$

which is homologous to v' and is a carrier of v' without a and b as points of intersection with $f(M)$. Proceeding in this way gives a carrier of v' in $M \times M - f(M)$ as asserted.

Now consider (A). Since v' and v'' have carriers in $M \times M - f(M)$, the same must be true for the double sum in (A). We may then consider (A) as an equation in $M \times M - f(M)$, and we may of course write the double sum in its equal original form. This gives

$$v - v' = \sum_{k=0}^{m} \sum_{\lambda=1}^{t_k} \beta b^{k*} \times \beta\alpha b_{\lambda}^{k} .$$

Since $v - v' \in H_n(M \times M - f(M))$,

$$0 = (v-v') \cap f(u) = \sum_{k=0}^{m} \sum_{\lambda=1}^{t_k} (\beta b_{\lambda}^{k*} \times \beta\alpha b_{\lambda}^{k}) \cap f(u)$$

$$= \sum_{k=0}^{m} \sum_{\lambda=1}^{t_k} (\beta b_{\lambda}^{k*} \times \alpha\beta\alpha b_{\lambda}^{k}) \cap \Delta u$$

$$= \sum_{k=0}^{m} \sum_{\lambda=1}^{t_k} (b_\lambda^{k*} \times \beta^{-2} b_\lambda^{k}) \cap \Delta u$$

$$= \sum_{k=0}^{m} \text{trace } \beta^{-2} | H_k(M) .$$

This contradicts our assumption and proves the lemma.

It follows that $\partial h u^\sigma \neq 0$.

IV. PROOFS OF THEOREMS

Assume k, $1<k\leq n$, is such that $\partial^k h i^{-1} u = 0$, but $\partial^{k-1} h i^{-1} u \neq 0$. The relevant part of the commutative diagram of Smith sequences is as follows, where for brevity $X = M\times M - f(M)$ and $F = F(Z_2)$.

$$\begin{array}{ccccccc} \cdots \to H^\sigma_{n-k+1}(M) \oplus H_{n-k+1}(F) & \xrightarrow{i} & H_{n-k+1}(M) & \xrightarrow{j} & H^\sigma_{n-k+1}(M) & \xrightarrow{\partial} & H^\sigma_{n-k}(M) \oplus H_{n-k}(F) \to \cdots \\ \downarrow h & & \downarrow h & & \downarrow h & & \downarrow h \\ \to H^\sigma_{n-k+1}(X) \oplus H_{n-k+1}(\Delta - f(M)) & \xrightarrow{i} & H_{n-k+1}(X) & \xrightarrow{j} & H^\sigma_{n-k+1}(X) & \xrightarrow{\partial} & H^\sigma_{n-k}(X) \oplus H_{n-k}(\Delta - f(M)) \to \end{array} \quad (*)$$

This has been written in general but to prove Theorem 1, we assume $F = \emptyset$, so $\Delta - f(M) = \Delta$ and the homology groups of F are 0 . We are not using reduced groups here. Since the diagram commutes and since $H_{n-k}(F) = 0$, $\partial^{k-1} h i^{-1} u$ is entirely in $H^\sigma_{n-k+1}(X)$ and its component in $H_{n-k+1}(\Delta - f(M))$ is 0 .

Therefore

$$\partial^{k-1} h i^{-1} u = jv$$

for some v in $H_{n-k+1}(X)$ and since $k > 1$,

$$ijv = i\partial^{k-1} h i^{-1} u = 0 .$$

But $H_{n-k+1}(X) \simeq H_{n-k+1}(M\times M)$ which has a basis B consisting of elements of the form $b\times b'$ such that $b\times b' \in B$ implies $b'\times b \in B$ and

$$ij(b\times b') = b\times b' + b'\times b .$$

But $ijv = 0$, so v must be a linear combination of elements $b\times b$ and hence

$$jv = 0 .$$

This contradiction shows that such a k, contrary to assumption, cannot exist.

However, Z_2 acts freely on M so $H_0^\sigma(M) = Z_2$, and not freely in X, which includes Δ, so $H_0^\sigma(X) = 0$. This shows that such a k must exist and this proves Theorem 1.

The proof of Theorem 2 is closely related to the above. To prove the first part of this theorem, assume it is false, that is, assume $\dim F < (n-1)/2$.

By hypothesis property (ii) is satisfied, so there is an integer $r = \dim F$ and in the map

$$\partial^{n-r}: H_n^\sigma(M) \to H_r^\sigma(M) \oplus H_r(F) ,$$

∂^{n-r} takes $H_n^\sigma(M)$ to zero in the first term and takes it isomorphically onto the second term. Now

$$h: H_r(F) \to H_r(\Delta - f(M))$$

and since F bounds in M, $h(F)$ bounds in $\Delta - f(M)$, because $h(F)$ is a translate of $f(F)$ and because $r < (n-1)/2$.

Referring to *, it is seen that, since $H_j(F) = 0$, $j > r$, the situation is similar to that in the proof of Theorem 1. There exists a k, $1 < k \leq n-r$, as described in the proof of Theorem 1, and on the other hand such a k cannot exist as before. This completes the proof of the first part of Theorem 2.

To prove the second part of Theorem 2, the same methods apply. It is now true by hypothesis that F is a mod 2 homology S^r. If $r \neq (n-1)/2$ then $h(F)$ bounds in $\Delta - f(M)$.

Remarks. Assume M^{2m+1} is a mod 2 sphere and that $F(Z_q) = \emptyset$ so that $\dim F = m$. *Then* $F(Z_2)$ *and* $\beta F(Z_2)$ *are linked.*

It can be seen that $F(Z_2) \cap \beta F(Z_2) = \emptyset$. The following diagram is commutative:

$$\begin{array}{ccc} F(Z_2) & \xrightarrow{\ h\ } & \Delta\text{-}F(Z_2) \\ & \searrow & \big\downarrow \text{projection} \\ & & M\text{-}F(Z_2) \end{array}$$

The horizontal and vertical arrows induce isomorphisms of homology. Therefore the same is true for the slanted arrow and this proves linking.

REFERENCES

1. Borel, A., et al, *Seminar on Transformation Groups,* Annals of Mathematics Studies no. 46, Princeton Univ. Press, New Jersey, 1960.

2. Floyd, E. E., "On periodic maps and the associated characteristic of associated spaces," *Trans. Amer. Math. Soc.* *72*(1952), 138-147.

3. Milnor, J. W., "Groups which act on S^n without fixed points," *Amer. J. Math.* *79*(1957), 623-630.

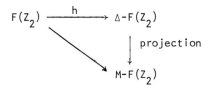

DIRECTED SETS WHICH CONVERGE

Mary Ellen Rudin

University of Wisconsin, Madison

The problem which we solve here had its origins in category theory and was suggested to me by A. Stralka through F. B. Jones by letter.

Let $\langle P, \leqq \rangle$ be a partially ordered set. For $a \in P$ and $A \subset P$, use $\uparrow a$ to denote $\{b \in P \mid a \leqq b\}$ and use $\uparrow A$ to denote $\cup\{\uparrow a \mid a \in A\}$. The set A is said to be *directed* if for every a and b in A, there is c in A with $c \in (\uparrow a) \cap (\uparrow b)$. If $1 \in P$, A is said to *converge to* 1, if $\cap\{\uparrow a \mid a \in A\} = 1$. If $\mathfrak{F}$ is a family of subsets of P, $\mathfrak{F}$ is said to be *directed* if for every A and B in $\mathfrak{F}$, there is C in $\mathfrak{F}$ with $C \subset (\uparrow A) \cap (\uparrow B)$. Similarly also $\mathfrak{F}$ *converges to* 1, if $\cap\{\uparrow A \mid A \in \mathfrak{F}\} = 1$.

THEOREM: *If* $\mathfrak{F}$ *is a collection of finite subsets of* P *which is directed and converges to* 1, *then there is a subset of* $\cup\mathfrak{F}$ *which is directed and converges to* 1.

Proof. Assume the hypothesis of the theorem.

Let $B = \{a \in \cup\mathfrak{F} \mid 1 \notin \uparrow a\}$ and let $\mathfrak{G} = \{F - B \mid F \in \mathfrak{F}\}$. Since $1 = \cap\{\uparrow F \mid F \in \mathfrak{F}\}$, for every $F \in \mathfrak{F}$, $(F - B)$ is nonempty and $1 = \cap\{\uparrow(F - B) \mid F \in \mathfrak{F}\}$; thus $\mathfrak{G}$ converges to 1. If $(F - B)$ and $(G - B)$ belong to $\mathfrak{G}$ for some F and G in $\mathfrak{F}$ there is an $H \subset (\uparrow F) \cap (\uparrow G)$ in $\mathfrak{F}$, and $(H - B) \subset \uparrow(F - B) \cap \uparrow(G - B)$; so $\mathfrak{G}$ is directed. Thus $\mathfrak{G}$ is a family of finite subsets of $\cup\mathfrak{F}$, $\mathfrak{G}$ is directed and converges to 1, and $1 = \cap\{\uparrow a \mid a \in \mathfrak{G}\}$.

ISBN 0-12-481820-X

Now, assuming that there is *no* directed subset of $\cup\mathcal{G}$ converging to 1, by transfinite induction we select $p_\alpha \in P$ for each ordinal $\alpha < 2^{|P|}$.

If $G \in \mathcal{G}$ definie $G_\alpha = \{x \in G | p_\beta \not< x$ for any $\beta < \alpha\}$; our induction hypothesis on α will be that $G_\alpha \neq \phi$ for all $G \in \mathcal{G}$.

If $G \in \mathcal{G}$ and $\alpha = 0$, then $G_\alpha = G$ and, since $1 \in (\uparrow G)$, $G_\alpha \neq \phi$.

If $G \in \mathcal{G}$ and α is a limit ordinal, then, since $G = G_0$ is finite and $\beta < \gamma < \alpha$ implies $G_\beta \supset G_\gamma$, $G_\alpha = G_\beta$ for some $\beta < \alpha$; since $G_\beta \neq \phi$ by assumption, $G_\alpha \neq \phi$.

Assume then that $\alpha < 2^{|P|}$ and $G_\alpha \neq \phi$ for $G \in \mathcal{G}$; we need to define p_α in such a way that $G_{\alpha+1} \neq \phi$

Consider $A = \cup\{G_\alpha | G \in \mathcal{G}\}$.

Since $A \subset \cup\mathcal{G}$ and $1 = \cap\{\uparrow a | a \in \cup\mathcal{G}\}$, $1 \in \cap\{\uparrow a | a \in A\}$. Since $G_\alpha = G \cap A \neq \phi$ for all $G \in \mathcal{G}$ and $\cap\{\uparrow G | G \in \mathcal{G}\} = 1$, $\{1\} \supset \cap\{\uparrow a | a \in A\}$. Thus $1 = \cap\{\uparrow a | a \in A\}$; i.e. A converges to 1. Since we are assuming that no directed subset of $\cup\mathcal{G}$ converges to 1, A cannot be directed. Hence there are p and q in A with $A \cap (\uparrow p) \cap (\uparrow q) = \phi$.

For $G \in \mathcal{G}$, define $G_p = G_\alpha - (\uparrow p)$ and $G_q = G_a - (\uparrow q)$. I claim that either no $G_p = \phi$ or no $G_q = \phi$. For suppose there were F and G in $\mathcal{G}$ with $F_p = \phi$ and $G_q = \phi$. There is an $H \in \mathcal{G}$ with $H \in (\uparrow F) \cap (\uparrow G)$; choose $x \in H_\alpha$ (which is nonempty by hypothesis). Choose $y \in F$ and $z \in G$ with $y < x$ and $z < x$. Observe that $y \in F_\alpha$ since, if $y \in \uparrow p_\beta$ for some $\beta < \alpha$, $x \in \uparrow p_\beta$ and hence $x \notin H_\alpha$. Since $F_p = F_\alpha - (\uparrow p)$, $y \in (\uparrow p)$; hence $x \in \uparrow p$. Similarly $z \in G_\alpha$ and $x \in \uparrow q$. But $H_\alpha \subset A$. Thus we have a contradiction since $x \in A \cap (\uparrow p) \cap (\uparrow q) = \phi$.

Hence one of $\{G_p | G \in \mathcal{G}\}$ and $\{G_q | G \in \mathcal{G}\}$ is a family of nonempty sets; say $\{G_p | G \in \mathcal{G}\}$ is a family of nonempty sets. Then definte $p_\alpha = p$; since $G_{\alpha+1} = G_p$, no $G_{\alpha+1} = \phi$.

Our choice of $p_\alpha \in \cup\{G_\alpha | G \in \mathcal{G}\}$ quarantees that $p_\alpha \neq p_\beta$ for any

$\beta < \alpha$. But there cannot be $2^{|P|}$ many distinct members of P so we have a directed subset of $\cup\mathcal{G} \subset \cup\mathfrak{F}$ which converges to 1 as desired.

WITNESSING NORMALITY

Franklin D. Tall

University of Toronto

I was very pleased to speak at a conference in honour of Burton Jones, since it was a theorem of his that set me on the mathematical path I have pursued for the past 13 years. I'd like to trace the developments proceeding from that theorem, and then prove some new results in the same spirit.

In preparation for my talk I consulted the Moore archives in Austin to see what R.L. Moore had to say about Burton. In 1971 Riverside was considering whether to give Burton a merit raise and they asked Moore for an appraisal of his work. I found the first and last pages of Moore's letter, which were highly favourable. Moore noted with approval that Jones had not written too many joint papers, and that he had introduced many new concepts. What I had really wanted to see, however, were Moore's thoughts on the normal Moore space roblem, especially concerning Jones' work. When I got to Riverside, John de Pillis was kind enough to find Moore's complete letter in the files. To my astonishment, Moore in his discussion of 20-odd papers failed to mention Jones' work in that area. (Burton later told me Moore was not fond of the normal Moore space problem.) I am sure, however, that I can on behalf of all those researchers who have seriously worked on the problem vouch for the great importance of Jones' results.

Briefly, for anyone unfamiliar with the normal Moore space problem, a *Moore space* is a regular space X possessing a sequence of open covers

ISBN 0-12-481820-X

$\{G_n\}_{n<\omega}$, called a *development*, such that for each point p in X and each open U containing p , for some n , $\bigcup\{g \in G_n : p \in g\} \subseteq U$. Moore spaces are somewhere in between first countable and metrizable spaces. In the course of attempts to discover topological characterizations of metrizability, the question arose of "what do you have to add to a Moore space to make it metrizable?". In the early 1930's Jones conjectured "normality", and that became known as the *normal Moore space conjecture*. While attempting to prove that conjecture, Jones came up with the amazing

THEOREM [J_1]. $2^{\aleph_0} < 2^{\aleph_1}$ *implies every* separable *normal Moore space is metrizable.*

Nowadays such a result would be unremarkable, but when it was pointed out to me in 1967 I was dumbfounded. What on earth did the hypothesis have to do with the conclusion? This result was surprising enough in 1967, but how could Burton come up with it 30 years earlier? Burton spent a lot of time trying to prove the hypothesis, which can't of course be done. Moore was amused by Jones' attempts - I guess he knew better.

Jones' other important contribution to the normal Moore space problem was *Jones' road space* (see, e.g. [J_2], a Moore space which Jones knew wasn't metrizable, but which he couldn't decide whether or not it was normal. It turned out no one else could either. Basically what Jones' space was is what is now called a *special Aronszajn tree:* a tree of height ω_1 with countable levels, no uncountable branches, and which is the union of countably many antichains. Such trees exist. The Moore space topology is generated by the intervals in the order. It's ironic that a generalization of Jones' theorem provided half of the definitive solution to the question of when his space is normal. Let me sketch how this happened. Jones' theorem was actually a corollary to a more general result he established:

THEOREM. $2^{\aleph_0} < 2^{\aleph_1}$ *implies no separable normal space has an uncountable closed discrete subspace.*

Heath [H] in 1964 proved the converse. In 1968 [T_1] I proved the consistency of restricted forms of the normal Moore space conjecture and Mary Ellen Rudin pointed out that these results implied the consistency of Jones' space being not normal. Around 1971, Šapirovskiĭ [Š] generalized Jones' results to get

THEOREM. $2^{\aleph_0} < 2^{\aleph_1}$ *implies no normal space of character* $\leq 2^{\aleph_0}$ *satisfying the countable chain condition has an uncountable closed discrete subspace.*

A couple of years later, Fleissner [F] proved

THEOREM. *Martin's Axiom plus the negation of the continuum hypothesis implies Jones' space is normal;* $\diamondsuit$ *implies it is not.*

In 1976 [T_2] I generalized Šapirovskiĭ to obtain

THEOREM. $2^{\aleph_0} < 2^{\aleph_1}$ *implies that in a normal space of character* $\leq 2^{\aleph_0}$, *for each uncountable closed discrete subspace* $\{x_\alpha\}_{\alpha<\omega_1}$, *there is an uncountable* $S \subseteq \omega_1$ *such that there exist pairwise disjoint open sets about the points* $\{x_\alpha\}_{\alpha \in S}$.

In 1978 Devlin and Shelah [DS] proved

THEOREM. $2^{\aleph_0} < 2^{\aleph_1}$ *implies Jones' space is not normal.*

Finally, Taylor [Ta] last year improved my theorem to get

THEOREM. $2^{\aleph_0} < 2^{\aleph_1}$ *implies that in a normal space of character* $\leq 2^{\aleph_0}$, *for each uncountable closed discrete subspace* $\{x_\alpha\}_{\alpha<\omega_1}$, *there is a* stationary $S \subseteq \omega_1$ *such that there exist pairwise disjoint open sets about the points* $\{x_\alpha\}_{\alpha \in S}$.

This can be used to get a trivial proof via the Pressing Down Lemma of the Devlin-Shelah result. Summing things up then, we see that if Burton had worked a bit harder, he could have generalized his own theorem in order to get his space to be not normal!

Let me move on now to some new results. Think of a normal space X and a closed discrete subspace Y. Take a local basis for each $y \in Y$ whose elements contain no other point of Y. Then there is a correspondence between functions which assign basic neighbourhoods to each point in Y, and open sets about Y. Given such a basic function f, and a $Z \subseteq Y$, the open set about Z determined by f (more precisely $f|Z$) may or may not be disjoint from the one it $(f|(Y - Z))$ determines about $Y - Z$. Since X is normal, there will be some f for which these open sets are disjoint. In general, one would expect $2^{|Y|}$ functions would be needed to *witness the normality of* (the $2^{|Y|}$ subsets of) *Y*. We consider the implications of assuming fewer than that many suffice. For simplicity we deal with the case $|Y| = \aleph_1$.

It is trivial to observe that if one function witnesses the normality of Y, then Y is *separated*, i.e. there exist pairwise disjoint open sets about the members of Y. It is less obvious that

THEOREM 1. *If* $\leq \aleph_1$ *functions witness the normality of* Y, Y *is separated.*

The idea of the proof goes back to [T_1]. We cast it in a somewhat peculiar form in order to get another result from it.

THEOREM 2. *BACH implies that if* $< 2^{\aleph_1}$ *functions witness the normality of* Y, Y *is separated.*

Recall [T_3] BACH is the conjunction of the continuum hypothesis with Baumgartner's version of generalized Martin's Axiom. Both theorems follow from an easy lemma:

LEMMA. *Let* P *be the extension ordering on countable partial functions from* Y *into* 2. *If* Y *is not separated but countable subsets of* Y *are, then for each basic function* f,

$$D_f = \{p \in P : p \text{ splits } f\}$$

is dense, where to say p splits f *means there are* $y, z \in \text{dom } p$ *such that* $p(y) \neq p(z)$ *and the neighbourhoods* $N_f(y)$, $N_f(z)$ *determined by* f *intersect.*

We leave the proof of the lemma to the reader. Since P is countably closed, for each collection of $\aleph_1$ dense sets there is a generic set meeting all of them. Since X is normal, Y is countably separated and so a generic G can be chosen to meet each D_f for the witnessing f's. G yields a partition of Y which - by density - splits each such f, contradicting normality. Theorem 2 also follows since P is well-known to satisfy the conditions for BACH.

Under a weaker hypothesis than Theorem 2, we get a weaker conclusion:

THEOREM 3. *Suppose* $2^{\aleph_0} < 2^{\aleph_1}$ *and* $<2^{\aleph_1}$ *functions witness the normality of* Y. *Then* Y *is* weakly separated, *i.e. there is an uncountable separated subset of* Y.

Proof. Since Y has $2^{\aleph_1}$ subsets and fewer than $2^{\aleph_1}$ functions witness normality, there must be one function f which witnesses the normality of a family $\mathcal{S}$ of $2^{\aleph_1}$ subsets of Y. Let us define an equivalence relation on the members of Y by

> $y \sim z$ if for each $S \in \mathcal{S}$ either y and z are both in S or neither is.

Y is thus decomposed into disjoint equivalence classes. f determines pairwise disjoint open sets about the equivalence classes; to see this, note that if $y \not\sim z$, there is an $S \in \mathcal{S}$ such that $y \in S$ and $z \in Y - S$. Also observe that each $S \in \mathcal{S}$ is a union of equivalence classes; for let E be an equivalence class intersecting S. If $E - S \neq \emptyset$ we obtain a contradiction. Since $2^{\aleph_0} < 2^{\aleph_1} = |\mathcal{S}|$, it follows that there must be $\aleph_1$ equivalence classes. Picking a point from each yields a weak separation.

CH settles the question of whether $2^{\aleph_0}$ functions witnessing normality yield a separation. On the other hand, if $2^{\aleph_0} = 2^{\aleph_1}$, there is a well-known example (see [H] or [CHT]) of a separable normal space including an uncountable closed discrete subspace.

Several problems remain. In order to save this note from utter triviality,

PROBLEM 1. Find a model of set theory in which there is a space X and a closed discrete Y such that fewer than $2^{\aleph_1}$ (better, $<2^{\aleph_0}$) functions witness the normality of Y, but Y is not (weakly?) separated.

PROBLEM 2. Prove that CH is not implied by the assertion that whenever $\leqq 2^{\aleph_0}$ functions witness the normality of a Y, Y is separated.

PROBLEM 3. Prove $2^{\aleph_0} < 2^{\aleph_1}$ does not imply that assertion.

Problem 1 is more difficult than it looks since it is tied up with the well-known hard problem of finding a model in which there is a family of fewer than $2^{\aleph_1}$ functions dominating all functions from ω_1 to ω.

REFERENCES

[CHT] Charlesworth, A., R. Hodel, F. D. Tall, "On a theorem of Jones and Heath concerning separable normal spaces," *Coll. Math. 34*(1975), 33-37.

[DS] Devlin, K. J., and S. Shelah, "A note on the normal Moore space conjecture," *Canad. J. Math. 31*(1979), 241-251.

[F] Fleissner, W. G., "When is Jones' space normal?," *Proc. Amer. Math. Soc. 50*(1975), 375-378.

[H] Heath, R. W., "Separability and $\aleph_1$-compactness," *Coll. Math. 12* (1964), 11-14.

[J_1] Jones, F. B., "Concerning normal and completely normal spaces," *Bull. Amer. Math. Soc. 43*(1937), 671-677.

[J_2] ____________, "Remarks on the normal Moore space metrization problem," *Ann. Math. Stud. 60*(1966), 115-120.

[S] Šapirovskiĭ, B., "On separability and metrizability of spaces with Souslin's condition," *Soviet Math. Dokl. 13*(1972), 1633-1638.

[T_1] Tall, F. D., "Set-theoretic consistency results and topological theorems concerning the normal Moore space conjecture and related problems," *University of Wisconsin, Madison, 1969; Dissert. Math 148*(1977), 1-53.

[T_2] ____________, "Weakly collectionwise Hausdorff spaces," *Top. Proc. 1* (1976), 295-304.

[T_3] ____________, "Some applications of a generalized Martin's Axiom,"

Trans. Amer. Math. Soc., to appear.

[Ta] Taylor, A. D, "Diamond principles, ideals and the normal Moore space problem," *Canad. J. Math.*, to appear.

CELL-LIKE MAPS WHICH DO NOT RAISE DIMENSION

John J. Walsh

University of Tennessee, Knoxville

This paper surveys the current status of the problem: Is there a cell-like dimension raising map? Our specific focus is on several results which show that certain cell-like maps do not raise dimension.

It has been know for some time that this problem is related to an old problem from dimension theory: Is there an infinite dimensional compactum which has finite cohomological dimension? The exact relationship was established by R.D. Edwards [Ed_1] who announced that a compactum has integral cohomological dimension $\leq n$ if and only if it is the image of a cell-like map defined on a compactum which has dimension $\leq n$. He also claimed that such examples exist if there is a family of maps $\{f_i: S^{n+p(i)} \longrightarrow S^n\}$, $p(i) > 0$, such that every finite composition in the following sequence is essential:

$$S^n \xleftarrow{f_1} S^{n+p(1)} \xleftarrow{\Sigma^{p(1)} f_2} S^{n+p(1)+p(2)} \xleftarrow{\Sigma^{p(1)+p(2)} f_3} \cdots .$$

Is there a precise homotopy theoretic restatement of the cell-like map problem (or the equivalent dimension theoretic question)?

It seems appropriate to point out that cell-like maps have played a central role in the study of both finite dimensional manifolds and Hilbert cube manifolds; the articles of J.W. Cannon [Ca], R.J. Daverman [Da], R.D. Edwards [Ed_2], and T.A. Chapman [Ch] contain a wealth of information about developments in these areas.

ISBN 0-12-481820-X

A non-empty compactum A is *cell-like* provided, for any embedding of A into an ANR, A is contractible in each of its neighborhoods. A map $f: X \to Y$ is *cell-like* provided it is a proper map with each of the sets $f^{-1}(y)$ being a cell-like set.

I. APPROXIMATE RIGHT INVERSES

For a finite open cover $\mathcal{U}$ of a compactum Y, let $N(\mathcal{U})$ denote the nerve of the cover, let $N(\mathcal{U})^{(n)}$ denote the n-skeleton of the nerve, and let $c: Y \to N(\mathcal{U})$ denote a canonical map. For a compact ANR X, a cell-like map $f: X \to Y$ satisfies the following lifting property:

> given an integer n and $\varepsilon > 0$, there is a $\delta > 0$ so that for any finite open cover $\mathcal{U}$ of Y with mesh $< \delta$, there is a map $r: N(\mathcal{U})^{(n)} \to Y$ such that $d(foroc(y), y) < \varepsilon$ for $y \in c^{-1}(N(\mathcal{U})^{(n)})$.

The map r is a "partial" approximate right inverse for f. If the map r can be defined on the entire nerve, then $r \circ c$ is a 2ε-map (that is, $\operatorname{diam}(roc)^{-1}(x) < 2\varepsilon$ for $x \in X$). In particular, if for each $\varepsilon > 0$ there is an open cover $\mathcal{U}$ and such a map r defined on the entire nerve, then $\dim Y \leq \dim X$. For example, if Y is finite dimensional, then we can choose $n = \dim Y$ and conclude that $\dim Y \leq \dim X$. There are situations where this strategy has been successful.

(i) *Completely Regular Cell-Like Maps.* A map $f: X \to Y$ is *completely regular* (in the sense of Dyer and Hamstrom [DH]) provided, for $\varepsilon > 0$ and $y \in Y$, there is a $\delta > 0$ such that, for each $y' \in N_\delta(y)$, there is a homeomorphism $h: f^{-1}(y) \to f^{-1}(y')$ with $d(x, h(x)) < \varepsilon$ for $x \in f^{-1}(y)$. As might be expected, such mappings possess a rich structure *provided Y is finite dimensional*: if the space of homeomorphisms of the fiber $f^{-1}(y)$ is locally contratible and Y is finite dimensional, then locally X is a product and f is the projection [DH]. There are examples with X and Y infinite dimensional and the fiber a Hilbert cube which are not locally trivial [KMW].

Question 1. If $f: X \to Y$ is a completely regular cell-like map and X is finite dimensional, then is Y finite dimensional?

If the fiber of f is a 1-cell (or finite tree), then locally f has a right inverse determined by endpoints and, hence, $\dim Y \leq \dim X$. If the fiber of f is a 2-cell, then locally f has a right inverse: the restriction of f to the union of the boundary circles is completely regular and results from [Wh] can be used to obtain a right inverse locally for the restriction.

This author does not know that the question has an affirmative answer for the case with the fiber a 3-cell.

(ii) *Convex Maps.* Suppose that $f: I^n \to Y$ is a cell-like map such that each set $f^{-1}(y)$ is convex; then the finite dimensionality of Y is established as follows: obtain a partial approximate right inverse $r: N(\mathcal{U})^{(0)} \to I^n$ and extend r linearly on each simplex to the entire nerve; the fact that each set $f^{-1}(y)$ is convex insures that the extension is an approximate right inverse.

(iii) *Three Dimensional Manifolds.* A space X is *aspherical* provided the homotopy groups of each component of X are trivial in dimensions greater than 1. A consequence of the Sphere Theorem [Pa] is that 3-manifolds contain an abundance of aspherical open sets. The pertinent facts are that cell-like subsets of 3-manifolds have arbitrarily small aspherical open neighborhoods and that the intersections of two aspherical open subsets of a non-compact 3-manifold is aspherical.

THEOREM. (Kozlowski-Walsh [KW_2]). *If* X *is a subset of a 3-manifold* M^3 *and* $f: X \to Y$ *is a cell-like map, then* $\dim Y \leq 3$.

A map $r: N(\mathcal{U}) \to M^3$ is constructed in two stages: first, r is defined on $N(\mathcal{U})^{(2)}$ for an appropriately chosen open cover $\mathcal{U}$ using the fact that f is cell-like and, second, r is extended with control using the "aspherical structure". (This argument is a variation due to R. Ancel of the argument in [KW_2] and it appears in detail in a set of seminar notes by R. Ancel).

Question 2. If $f: M^3 \to Y$ is a $\mathbb{Z}$-acyclic map, then is Y finite dimensional?

The above arguement breaks down not because of any loss of "aspherical structure" but rather because it is not clear that r can be defined on $N(\mathcal{U})^{(2)}$.

(iv) *Three Dimensional Polyhedra.* Although three dimensional polyhedra are substantially more complicated than 3-manifolds, it is not unreasonable to expect the preceding result to extend to this larger class of spaces. Unfortunately, the loss of the Sphere Theorem is quickly felt. The proof of the next theorem is similar to that for 3-manifolds except that a substantially more delicate analysis is required in order to establish the necessary "aspherical structure".

THEOREM (Kozlowski-Row-Walsh). *If* X *is a subset of* a *three dimensional polyhedron and* $f: X \to Y$ *is a cell-like map with* $\dim f^{-1}(y) \leq 1$ *for each* $y \in Y$, *then* $\dim Y \leq 3$.

An affirmative answer to the next question might allow the removal of the hypothesis "$\dim f^{-1}(y) \leq 1$".

Question 3. Does a 2-dimensional cell-like subset of three dimensional polyhedron have an aspherical neighborhood?

II. POLYHEDRAL MAPS

Let X be a subset of a polyhedron P; a map $f: X \to Y$ is *polyhedral* provided f is proper and each set $f^{-1}(y)$ is a non-empty subpolyhedron of P. Piecewise linear maps are polyhedral but the class of polyhedral maps is substantially larger. For example, piecewise linear maps do not raise dimension while every finite dimensional compactum is the image of a polyhedral map defined on a 1-dimensional compact subset of E^3. (Let $C * C \subset E^3$ be the join of a pair of Cantor sets contained in non-adjacent edges of a tetrahedron. For a finite dimensional compactum Y, there is a finite-to-one map α from C onto Y [Na]. Let $X = \cup\{\alpha^{-1}(y) * \alpha^{-1}(y) : y \in Y\} \subset C * C$ and define $f: X \to Y$ by requiring that $f(\alpha^{-1}(y) * \alpha^{-1}(y)) = y$; the map f is polyhedral and $\dim X = 1$).

A space X is *countable dimensional* provided that X is the countable union of finite dimensional subspaces. The following arguments resulted from a conversation with R.J. Daverman; it is entirely possible that the theorem is already known.

THEOREM. *If* X *is a subset of a polyhdron* P *and* $f: X \to Y$ *is a polyhedral map, then* Y *is countable dimensional.*

Proof. (by induction on the dimension of P). Let $f = \ell \circ m$ be the monotone-light factorization of f, say $m: X \to Z$ and $\ell: Z \to Y$. The map m is polyhedral, the map ℓ finite-to-one, and both maps are proper. Since the image of a countable-to-one proper map defined on a countable dimensional space is countable dimensional [Na; p. 85], it remains to show that Z is countable dimensional and, for this, it suffices to show that points of Z have arbitrarily small neighborhoods whose frontiers are countable dimensional.

Let $z \in Z$ and let N be a "small" closed polyhedral neighborhood of $m^{-1}(z)$; note that $\dim \mathrm{Fr}N < \dim P$ and that $m(N)$ is a neighborhood of z . Since the map m is monotone, $\mathrm{Fr}\, M(N) \quad m(\mathrm{Fr}N)$. The restriction of m to $\mathrm{Fr}N$ is polyhdral and, therefore, we can conclude inductively that $m(\mathrm{Fr}N)$ is countable dimensional.

COROLLARY. *If* X *is a complete subset of a polyhdron* P *and* $f: X \to Y$ *is a cell-like polyhdral map, then* Y *is finite dimensional.*

Proof. The space Y is complete since it is the image of a proper map defined on a complete space [Va]. The previous theorem yields that Y is countable dimensional. Complete countable dimensional spaces have a transfinite dimension [HW] and integral cohomological dimension and dimension agree for such spaces [Ku]. Since the integral cohomological dimension of Y is finite ($\leq \dim P$) , the dimension of Y is finite.

The assumption that X is complete is unnecessary provided the next question has an affirmative answer.

Question 4. Do integral cohomological dimension and dimension agree for countable dimensional spaces?

III. COHOMOLOGICAL COMPUTATIONS

The following characterization represents a particularly useful point of view when comparing the concepts of dimension (denoted dim) and integral cohomological dimension (denoted c-dim).

THEOREM ([HW]). *A space* X *has dimension* $\leq n$ *if and only if, for each closed subset* $A \subset X$ *and each map* $f: A \to S^n$ *, there is an extension* $\tilde{f}: X \to S^n$.

Let K_n denote a CW-complex which is an Eilenberg-MacLane space with $\pi_i(K_n) = \pi_i(S^n)$ for $i \leq n$ and $\pi_i(K_n) = 0$ for $i \geq n + 1$.

Definition. A space X has *integral cohomological dimension* $\leq n$ if, for each closed subset $A \subseteq X$ and each map $f: A \to K_n$, there is an extension $\tilde{f}: X \to K_n$.

A classical result is that, since $K_1 = S^1$, $\dim X \leq 1$ if and only if $\text{c-dim}\, X \leq 1$ and the corollaries which follow exploit this fact. The reader is referred to [Ku] for a further treatment of cohomological dimension.

The following result is implicitly contained in [Dy_1].

THEOREM. (Dyer) *If* $f: X \to Y$ *is a cell-like open map with each set* $f^{-1}(y)$ *being a* 1-dimensional AR *(absolute retract), then* $\text{c-dim}\, Y <$ $\text{c-dim}\, X$.

COROLLARY. *If* $\dim X \leq 2$ *and* $f: X \to Y$ *is a cell-like open map with each set* $f^{-1}(y)$ *a point or* 1-dimensional AR, *then* Y *is finite dimensional.*

The above theorem is false without the assumption that f is open. For example: Let $X = I \times \{0\} \cup C \times I$ for a Cantor set $C \subset I$. Let G be the decomposition of X whose elements are $\{t\} \times I$ for t a "non-endpoint" of C and $(\{t\} \times I) \cup ([t,s] \times \{0\}) \cup (\{s\} \times I)$ for (s,t) a complementary domain of C. The quotient map $\pi: X \to X/G$ has only arcs as point inverses but $\dim X = \dim X/G = 1$ (notice that $X/G \cong I$).

Assuming that Y is infinite dimensional, there is a variation of the above theorem a statement and proof of which will appear in a joint paper with D. Wilson currently being prepared. It has the follow consequence.

COROLLARY (Walsh-Wilson). *If* $\dim X \leq 2$ *and* $f: X \to Y$ *is a cell-like map with each set* $f^{-1}(y)$ *a point or* 1-dimensional AR, *then* Y *is finite dimensional.*

Although its proof is entirely different, this corollary is not unrelated to results in Section 1 since a consequence of a reduction discussed in [KW_1] is the following.

COROLLARY. *If* X *is a* 3-dimensional ANR *and* $f: X \to Y$ *is cell-like with each set* $f^{-1}(y)$ *a point or* 1-dimensional AR, *then* Y *is finite dimensional.*

REFERENCES

[Ca] Cannon, J. W., "The recognition problem: what is a topological manifold?," *Bull. Amer. Math. Soc.* *84*(1978),832-866.

[Ch] Chapman, T. A., *Lectures on Hilbert Cube Manifolds*, Regional conference series in mathematics no. 28, Amer. Math. Soc., Providence, R.I., 1976.

[Da] Daverman, R. J., "Products of cell-like decompositions," *Topology and Appl.* *11*(1980),121-139.

[Dy_1] Dyer, E., "Certain transformations which lower dimension," *Ann. of Math.* *63*(1956),15-19.

[Dy_2] _______, "Regular mappings and dimension," *Ann. of Math.* *67*(1958), 119-149.

[DH] Dyer, E., and M. E. Hamstrom, "Completely regular mappings," *Fund. Math.* *45*(1958), 103-118. MR 19, 1187.

[Ed_1] Edwards, R.D., "A theorem and a question related to cohomological-dimension and cell-like maps," *Notices Amer. Math. Soc.* *25*(1978), A-259.

[Ed_2] _______, "The topology of manifolds and cell-like maps," *Proceedings of the International Congress of Mathematicians*, Helsinki, 1978.

[HW] Hurewicz, W., and H. Wallman, *Dimension Theory*, Princeton University Press, Princeton, N. J., 1941.

[KW_1] Kozlowski, G., and J. Walsh, "The cell-like mapping problem," *Bull. Amer. Math. Soc.* *2*(1980)

[KW_2] _______, "Cell-like mappings on 3-manifolds," preprint.

[KMW] Kozlowski, G., J. van Mill, and J. Walsh, "AR-maps obtained from cell-like maps," to appear in *Proc. Amer. Math. Soc.*

[KRW] Kozlowski, G., H. Row, and J. Walsh, "Cell-like mappings with 1-dimensional fibers on 3-dimensional polyhedra," preprint.

[Ku] Kuźminov, V. J., "Homological dimension theory," *Russian Math. Surveys* *23*(1968), 1-45.

[Na] Nagami, K., *Dimension Theory*, Academic Press, New York and London, 1970.

[Pa] Papakyriakopoulos, C. D., "On Dehn's Lemma and the asphericity of knots," *Ann. of Math.* *66*(1957), 1-26.

[Sch] Schori, R. M., "The cell-like mapping problem and hereditarily infinite-dimensional compacta," preprint.

[Va] Vaǐnšteǐn, I. A., "On closed mappings of metric spaces," *Dokl. Akad. Nauk. SSSR* *57*(1947) 319-321. MR 9, 153.

[Wh] Whitney, H., "Regular families of curves," *Ann of Math.* *34*(1933), 244-270.

AXIOMS FOR CUTPOINTS

L. E. Ward, Jr.

University of Oregon, Eugene

I. INTRODUCTION

Suppose (X,T) is a connected topological space and x,y and z are elements of X. If $X-\{y\}$ is the union of separated sets A and B with $x \in A$ and $z \in B$, we say that y *separates* x and z, and we introduce the notation $Sxyz$ to denote this. We list a few of the properties of the ternary relation S, all of which are trivial or can be proved easily.

A1. *If* $x,y \in X$ *then* $Sxyx$ *cannot occur.*

A2. *If* $Sxyz$ *then* $Szyx$.

A3. *If* $Sxyz$ *then* $Sxzy$ *cannot occur.*

A4. *If* $Sxyz$ *and if* $w \in X-\{y\}$ *then* $Swyz$ *or* $Sxyw$.

What has transpired here is that one structure on X (the topology T) has been used to define a new structure (the *cutpoint structure* S) on X. The cutpoint and cyclic element theory of G. T. Whyburn [12] may be regarded as the theory of the triple (X,T,S). Moreover, a substantial part of that theory is concerned primarily with the properties of S and only incidentally with the topology T. Therefore it is natural to consider the problem of developing cyclic element theory solely from the structure (X,S), and that is the general problem to which this paper is addressed. The relation S describes the *global* structure of a connected space quite accurately. (By the term *global* we refer here to those

ISBN 0-12-481820-X

properties of the space which are external to the cyclic elements themselves.) The *internal* properties of the individual cyclic elements are, of course, lost irretrievably. Consequently, a space which has few cutpoints cannot be studied effectively by means of the relation S, but nearly all information remains available for spaces which are sufficiently rich in cutpoints.

The literature includes only a few investigations in this spirit. An early paper of C. J. Harry [2] is closest to ours, although his axioms are somewhat different, and the applications to Peano spaces are limited in scope. A more modern and comprehensive theory is that of Rado and Reichelderfer [6] who were also interested in applications to the theory of area [7]. Their approach was based on axioms for the binary relation of conjugacy between points of X. While this approach is more general than ours, it does not discriminate between continua as effectively. Another study due to Wallace [8] was based on set separation, but it is more distantly related to our concerns.

In this paper we will present only an outline of the theory which can be developed. Many results and all of the proofs will appear in a later, more comprehensive paper.

II. THE AXIOMS AND THEIR CONSEQUENCES

Let X be a nonempty set and suppose S is a ternary relation on X which satisfies the axioms A1-A4 above. The pair (X,S) is called a *cutpoint structure*, and if x, y and z are elements of X such that $Sxyz$ then y is a *cutpoint* of X. Simple examples show that A1-A4 are independent.

2.1 THEOREM. *If* (X,S) *is a cutpoint structure, then the following statements hold.*

A5. *If* $x,y \in X$ *then* $Sxxy$ *cannot occur.*

A6. *If* Saxb *and* Sayb *then* Saxy *or* Sayx *or* x = y .

A7. *If* Saxy *and* Sayz *then* Saxz *and* Sxyz .

If (X,S) is a cutpoint structure, then we may define the *closure* $\bar{S}$ of S as follows:

$\bar{S}xyz$ *if and only if* Sxyz *or* x = y *or* y = z .

The analogs of A1-A7 for $\bar{S}$ are succinct and easily established.

2.2 THEOREM. *If* (X,S) *is a cutpoint structure, then the following statements hold.*

B1. *If* $\bar{S}xyx$ *then* x = y .

B2. *If* $\bar{S}xyz$ *then* $\bar{S}zyx$.

B3. *If* $\bar{S}xyz$ *and* $\bar{S}xzy$ *then* y = z .

B4. *If* $\bar{S}xyz$ *and* $w \in X$ *then* $\bar{S}wyz$ *or* $\bar{S}xyw$.

B5. *If* $x,y \in X$ *then* $\bar{S}xxy$.

B6. *If* $\bar{S}axb$ *and* $\bar{S}ayb$ *then* $\bar{S}axy$ *or* $\bar{S}ayx$.

B7. *If* $\bar{S}axy$ *and* $\bar{S}ayz$ *then* $\bar{S}xyz$ *and* $\bar{S}axz$.

It will surprise no one that B1-B4 can be employed to characterize cutpoint structures.

2.3 THEOREM. *Let* X *be a set and suppose* $\hat{S}$ *is a ternary relation on* X *which satisfies B1-B4. If* $S = \hat{S} - \{(x,y,z) \in X^3:$ *any two of* x, y *and* z *are equal*$\}$ *, then* (X,S) *is a cutpoint structure and* $\hat{S} \subset \bar{S}$. *In addition, if* $\hat{S}$ *satisfies B5 then* $\hat{S} = \bar{S}$.

Still another formulation of cutpoint structures can be made in terms of partial order relations. The author and others (see, for example, the bibliography in [11]) have utilized the so-called cutpoint partial orders of connected spaces to study the structure of continua. The connection between these partial orders and the relation S is clear.

Given a cutpoint structure (X,S) and an element $e \in X$, we define

$\Gamma_e = \{(x,y) \in X^2 : \bar{S}exy\}$.

Further convenient notation is the following. If $x,y \in X$ then

$$x\Gamma_e = \{y \in X : (x,y) \in \Gamma_e\} ,$$

$$\Gamma_e y = \{x \in X : (x,y) \in \Gamma_e\} ,$$

and it is also helpful to write $x \leq_e y$ (or $x \leq y$) for $(x,y) \in \Gamma_e$. As usual, $x <_e y$ (or $x < y$) means $x \leq_e y$ (or $x \leq y$) and $x \neq y$.

2.4 THEOREM. *If* (X,S) *is a cutpoint structure and* $e \in X$ *, then* Γ_e *is a partial order. Moreover,*

P1. *If* $x \in X$ *then* $e \leq_e x$ *,*

P2. *If* $x \in X$ *then* $\Gamma_e x$ *is simply ordered with respect to* Γ_e .

2.5 THEOREM. *Suppose* X *is a set,* $e \in X$ *and* Γ *is a partial order on* X *which satisfies P1 and P2. If* $\hat{S}_e = \{(x,y,z) \in X^3 : y \leq z$ *and* $y \not\leq x$ *or* $y \leq x$ *and* $y \not\leq z\}$ *, then* $\hat{S}_e$ *satisfies B1-B5. If* $S = \hat{S}_e - \{(x,y,z) \in X^3 :$ *any two of* x, y *and* z *are equal}* , *then* $\Gamma = \Gamma_e$. *Moreover,* $\bar{S} = \bigcup\{\hat{S}_e : e \in X\}$.

III. CONJUGATE POINTS

Suppose (X,S) is a cutpoint structure and x and y are elements of X . If there is no $z \in X$ such that $Sxzy$, we say that x and y are *conjugate*, and the symbol $x \sim y$ is used to denote this relation.

3.1 THEOREM. *If* (X,S) *is a cutpoint structure, then the following statements hold.*

C1. *If* $x \in X$ *then* $x \sim x$.

C2. *If* $x \sim y$ *then* $y \sim x$.

C3. *If* $x_1 \sim x_2 \sim \cdots \sim x_n \sim x_1$ *then* $x_i \sim x_j$ *for all* $i,j = 1,\ldots,n$.

C4. *If* $Spxq$ *and* $p \sim z \sim q$ *then* $x = z$.

C5. *If* $Sabc$ *and* $c \sim x$ *then* $b = x$ *or* $Sabx$.

Rado and Reichelderfer [6] used C1, C2 and C3 as a set of axioms for the conjugacy relation. (See the remarks in Section I.) The crucial axiom is C3, which they called *cyclic transitivity*.

Following the classical usage we define $E(a,b) = \{x \in X : Saxb\}$, and we note that $E(a,b) \cup \{a,b\} = \{x \in X : \bar{S}axb\} = \Gamma_a b$, and hence this set is simply ordered relative to Γ_a.

We are led next to the concept of endpoint. An element e of a connected space X is an *endpoint* if e has arbitrarily small neighborhoods with one point boundary. We write $E(X)$ to denote the set of endpoints of X. Although this definition does not translate directly to cutpoint structures, there are two analogs which suffice nicely. If (X,S) is a cutpoint structure, then an element e of X is said to be *terminal* if $E(e,x) \cap E(e,y) \neq \emptyset$ whenever $x \neq e \neq y$. The set of terminal points is denoted $T(X)$. The element e is said to be *peripheral* if e is not a cutpoint and if $E(e,x) \neq \emptyset$ for all $x \neq e$. The set of peripheral points is denoted $P(X)$.

3.2 THEOREM. *No terminal point is a cutpoint and hence* $T(X) \subset P(X)$.

3.3 THEOREM. *If* X *is a connected topological space, then* $T(X) = P(X)$. *If* X *is a* T_1*-space, then* $E(X) \subset T(X)$.

IV. COMPLETE CUTPOINT STRUCTURES

Many of the more interesting results in the classical theory demand a compactness hypothesis of some kind. A typical theorem is the assertion that an element of a continuum is either a cutpoint, and endpoint, or is conjugate to some other point. This theorem is due originally to Kuratowski and Whyburn [4] for metric continua, and more general versions have been established by Wallace [9] and Lehman [5]. For this and other results in the setting of cutpoint structures, the following completeness hypothesis suffices.

A cutpoint structure (X,S) is *complete* provided $\Gamma_a b = E(a,b) \cup \{a,b\}$ is complete as an ordered set for each $a,b \in X$.

4.1 THEOREM. *If* (X,S) *is a complete cutpoint structure, if* $x \in X-T(X)$ *, and if* x *is not a cutpoint, then there exists* $y \in X-\{x\}$ *such that* $x \sim y$.

4.2 COROLLARY. *If* (X,S) *is a complete cutpoint structure, then* $P(X) = T(X)$.

We apply Theorem 4.1 in a topological setting somewhat more general than has been published heretofore. A space is *rim compact* if each of its elements has arbitrarily small neighborhoods with compact boundary.

4.3 THEOREM. *If* X *is a rim compact, connected Hausdorff space, then* $T(X) = E(X)$.

4.4 COROLLARY. *If* X *is a connected Hausdorff space which is either compact or rim compact and locally connected, and if* $x \in X$ *, then* x *is an endpoint, a cutpoint or* $x \sim y$ *for some* $y \in X-\{x\}$.

V. CYCLIC ELEMENTS

The definition of a simple link can be taken directly from the classical usage. Suppose (X,S) is a cutpoint structure and x is an element of X which is neither a cutpoint nor a terminal point. Then the set $\{y \in X : x \sim y\}$ is a *simple link* of X . Theorem 4.1 can now be paraphrased as follows: *in a complete cutpoint structure, the simple links are nondegenerate.*

The appropriate definition of E_o-set is not so readily come by since we lack a class of subsets to be termed connected. However, Whyburn [13] has observed that if X is a connected, locally connected Hausdorff space then the E_o-sets are characterized as those subsets of the form

$\{x \in X : a \sim x \sim b\}$ where a and b are distinct conjugate elements of X. We take this as the definition of E_o-set.

5.1 THEOREM. *In a cutpoint structure each nondegenerate simple link is an E_o-set and each E_o-set which contains a noncutpoint is a simple link.*

A subset of a cutpoint structure which is both a simple link and an E_o-set is called a *true cyclic element*. We defer to another paper most of the details of the development of the cyclic element theory for cutpoint structures, except to remark that there is nothing unexpected in that development. A subset N of a cutpoint structure (X,S) is a *nodal set* if N is nondegenerate and if there exists $y \in N$ such that $Sxyz$ for all $x \in N-\{y\}$ and all $z \in X-N$. This element y, which is called the *boundary* of N, is seen to be unique. A nodal set N is *prime* if it is minimal with respect to a certain boundary y, and in this case $N-\{y\}$ is an *open prime nodal set.* An *A-set* is a set whose complement is the union of open prime nodal sets. If a and b are distinct elements of X then

$$C(a,b) = \bigcap\{A : A \text{ is an } A\text{-set and } a,b \in A\}$$

is the *cyclic chain from* a to b. The following result is nontrivial.

5.2 THEOREM. *If* (X,S) *is a complete cutpoint structure and if* a *and* b *are distinct elements of* X, *then* $C(a,b) = \Gamma_a b \cup \{E : E$ *is an* E_o-*set meeting* $\Gamma_a b$ *in precisely two points*$\}$.

VI. INTRINSIC TOPOLOGIES FOR CUTPOINT STRUCTURES.

There are several natural ways to impose a topology in a cutpoint structure. To a degree these are reminiscent of the various intrinsic topologies of a lattice [1] or a partially ordered set [10]. For cutpoint structures we expect, of course, that in the case where S arises from a connected topology, these natural topologies would have much in common with the given topology; always excepting that the intrinsic topologies cannot reveal any of the internal structure of the true cyclic elements. It turns out that, to effect this, the interval topologies of the partial orders Γ_e are not appropriate. The following variations better suit our purposes.

If (X,S) is a cutpoint structure, then the collection of all sets $x\Gamma_e$, where $e \in X$ and x is a cutpoint, is taken as a subbase for the closed sets. The resulting topology is called the *cutpoint topology* and is denoted T_c. If e and x are both arbitrary elements of X, we obtain a finer topology, denoted T'_c and called the *augmented cutpoint topology*. One sees readily that T'_c is a T_1-topology and that, relativized to the true cyclic elements, T_c is indiscrete and T'_c is the cofinite topology.

6.1 THEOREM. *If* (X,T) *is a semi-locally connected continuum then* $T'_c \subset T$.

The topology T_c admits a useful description in terms of convergence. If (X,S) is a cutpoint structure, we define a convergence structure on X by letting $x_\alpha \to x$ mean that if $Sabx$ then eventually $Sabx_\alpha$. The topology induced by this convergence structure is T_c.

Another intrinsic topology arises by taking the open prime nodal sets as a subbase for the open sets. This topology, denoted T_n and called the *nodal topology*, can also be augmented to yield a finer T_1 topology T'_n, and we have

$$\begin{array}{ccccc} & & T'_c & & \\ & \subset & & \subset & \\ T_c & & & & T'_n \\ & \subset & & \subset & \\ & & T_n & & \end{array}$$

Let us call a cutpoint structure (X,S) *dendritic* if, whenever x and y are distinct elements of X, there exists $p \in X$ such that $Sxpy$.

The following result is not difficult to prove.

6.2 THEOREM. *If* (X,S) *is a cutpoint structure, then the following statements are equivalent.*

(1) (X,S) *is dendritic.*

(2) T_c *is Hausdorff.*

(3) T_n *is Hausdorff.*

A continuum X is a *tree* if each two distinct elements of X are separated by some third element.

6.3 THEOREM. *If* (X,T) *is a tree then* $T = T_c$.

6.4 THEOREM. *If* (X,S) *is a complete cutpoint structure, then* T'_n *is a compact topology. Moreover, relative to* T_n *each set* $C(a,b)$ *is connected and the cutpoints of* (X,S) *are precisely the* T_n- *and* T_c-*cutpoints.*

Up to this final paragraph we have avoided any discussion of functions, continuity, monotonicity, etc., but it is clear how such an inquiry should proceed. If (X,S) and (Y,T) are cutpoint structures, then a function $f:(X,S) \to (Y,T)$ is continuous if it is continuous relative to the topologies T_c on X and Y. In the detailed sequel to this paper, this and related topics will be explored in detail.

REFERENCES

1. Birkhoff, G., *Lattice Theory*, rev. ed., New York, 1948.

2. Harry, C. J., "Concering the geometry of acyclic sets," *Amer. J. Math.* *56*(1934), 233-253.

3. Jones, F. B., "Almost cyclic elements and simple links of a continuous curve," *Bull. Amer. Math. Soc.* *46*(1940), 775-783.

4. Kuratowski, K. and G. T. Whyburn, "Sur les elements cycliques et leurs applications," *Fund. Math.* *16*(1930), 305-331.

5. Lehman, B., "Cyclic element theory in connected and locally connected Hausdorff spaces," *Canadian J. Math.* *28*(1976), 1032-1050.

6. Rado, T. and P. Reichelderfer, "Cyclic transitivity," *Duke Math. J.* *6* (1940), 474-485.

7. Rado, T., *Length and Area*, New York, 1948.

8. Wallace, A. D., "Separation spaces," *Ann. Math.* *42*(1941), 687-697.

9. ______, "Monotone transformations," *Duke Math. J.* *9*(1942), 487-506.

10. Ward, A. J., "On relations between certain intrinsic topologies in partially ordered sets," *Proc. Cambridge Ph. Soc.* *51*(1955), 254-261.

11. Ward, L. E., Jr., "Partially ordered spaces and the structure of continua," *Proc. Conf. on Metric Spaces, Gen. Metric Spaces and Continua*, Guilford College, Greensboro, N. Carolina, 1980.

12. Whyburn, G. T., *Analytic Topology*, New York, 1942.

13. ______, "Cutpoints in general topological spaces," *Proc. Nat'l. Acad. Sci.* *61*(1968), 380-387.

RECENT ADVANCES IN CONTINUA THEORY

David C. Wilson

University of Florida, Gainesville

This note is a summary of results from two papers [5] and [6] written together with Alice Mason and John Walsh.

The first paper gives sufficient conditions under which a space cannot be hereditarily indecomposable.

Definition. A continuum is *decomposable* if it can be written as the union of two proper subcontinua. A continuum is *indecomposable* if it is not decomposable. A continuum is *hereditarily indecomposable* if every subcontinuum is indecomposable.

The pseudo-arc is the standard example of a space which is hereditarily indecomposable. In 1951 Bing [1] constructed higher dimensional examples. However, in 1960 M. Brown [2] proved the following curious theorem.

THEOREM (Brown). *If* $X = \varprojlim_i \{S_i, f_i\}$ *, where each* S_i *is the Euclidean n-sphere,* $n \geqq 2$ *, and each* $f_i : S_i \rightarrow S_{i-1}$ *has non-zero degree, then* X *cannot be hereditarily indecomposable.*

In [5] this result is generalized to the following theorem.

THEOREM A. *If* $X = \varprojlim_i \{M_i, f_i\}$ *, where each* M_i *is a closed orientable PL n-manifold,* $n \geqq 2$ *, where* $\sup_i\{\mathrm{rk}\, H_1(M_i)\} < +\infty$ *, and where each* $f_i : M_i \rightarrow M_{i-1}$ *has non-zero degree, then* X *cannot be hereditarily indecomposable.*

ISBN 0-12-481820-X

The examples of Bing [1] can be modified slightly to show that the hypothesis $\sup_i\{rk\, H_1(M_i)\} < +\infty$ is necessary. An immediate consequence of Theorem A is that no closed orientable PL n-manifold, $n \geq 2$, can have the shape of an hereditarily indecomposable continuum. Similar results hold in the nonorientable case.

An important result in continua theory is Jones' Aposyndetic Decomposition Theorem [4]. For the purposes of this note it is sufficient to note that this theorem gives rise to a continuous decomposition G of a certain type of 1-dimensional continuum M such that the quotient map $\pi : M \to M/G$ is monotone. Rogers [7] showed that π is completely regular and acyclic.

THEOREM B. *If* f *is a monotone completely regular map of a 1-dimensional continuum* X *onto a nondegenerate continuum* Y *, then* f *is cell-like. (i.e. Each* $f^{-1}(y)$ *is a dendrite.)*

An immediate consequence of Theorem B is that the map π mentioned above must be cell-like.

An example can be constructed to show that the hypothesis of complete regularity is necessary. In particular, there is a 1-dimensional space X and a monotone map $f : X \to I = [0,1]$ such that each $f^{-1}(y)$ is acyclic but not cell-like. The point-inverse sets can be taken to be the Case-Chamberlain example [3] "wedged" with itself an infinite number of times.

REFERENCES

1. Bing, R. H. "Higher dimensional hereditarily indecomposable continua," *Trans. Amer. Math. Soc. 71* (1951), 267-273.

2. Brown, M., "On the inverse limit of Euclidean n-spheres," *Trans. Amer. Math. Soc. 96* (1960), 129-134.

3. Case, J.H. and R. E. Chamberlain, "Characterizations of tree-like continua," *Pacific J. Math* (1960), 73-84.

4. Jones, F. B., "On a certain type of homogeneous plane continuum," *Proc. Amer. Math Soc.* 6 (1955), 735-740.

5. Mason, A., and J.J. Walsh, D.C. Wilson, "Inverse limits which are not hereditarily indecomposable, preprint.

6. Mason, A. and D.C. Wilson, "Monotone Mappings on n-dimensional continua, preprint.

7. Rogers, J.J.,Jr., "Completely regular mappings and homogeneous, aposyndetic continua," preprint.

MAPPINGS WITH 1-DIMENSIONAL ABSOLUTE NEIGHBORHOOD RETRACT FIBERS

David C. Wilson

University of Florida, Gainesville

This paper is a summary of results from a collaboration with John J. Walsh. The completed manuscript will have the same title and will appear elsewhere.

All spaces in this paper are metric. They symbol $\dim X$ denotes the covering dimension of X. The cohomology theory we use is Cech with integer coefficients. The notation $c\text{-}\dim X$ denotes the integral cohomological dimension of X and is defined as follows: (i) $c\text{-}\dim X \leqq n$ if for each closed subset $A \subseteq X$ the inclusion induced homomorphism $i^* : H^n(X) \to H^n(A)$ is surjective. (ii) $c\text{-}\dim X = n$ if $c\text{-}\dim X \leqq n$ but $c\text{-}\dim X \not\leqq n-1$. Two facts of note are that $\dim X \geqq c\text{-}\dim X$ and $\dim X = c\text{-}\dim X$ for finite dimensional spaces.

The 1956 paper of E. Dyer [1] implicitly contains the following result: If $f:X \to Y$ is an open map between compacta and there is an $\varepsilon > 0$ such that each $f^{-1}(y)$ is a 1-dimensional ANR (absolute neighborhood retract) containing no simple closed curve of diameter less than ε, then the cohomological dimension of Y is one less than the cohomological dimension of X. The fact that cohomological dimension and covering dimension are known to agree when the former is $\leqq 1$ yields: If $f: X \to Y$ is an open cell-like map between compacta with each $f^{-1}(y)$ a 1-dimensional AR (absolute retract) or a point and $\dim X \leqq 2$ then $\dim Y \leqq 2$. In turn a reduction described in [4] based on results from

ISBN 0-12-481820-X

[5] yields: If $f: X \to Y$ is an open cell-like map with each $f^{-1}(y)$ a 1-dimensional AR or a point and X is an ANR with $\dim X \leq 3$, then $\dim Y \leq 3$ or equivalently Y is an ANR.

Definition. A space is *countable dimensional* it can be written as the countable union of zero dimensional spaces.

MAIN THEOREM. *If* $f: X \to Y$ *is a proper surjection between metric spaces where* $\dim X \leq 2$ *and where each* $f^{-1}(y)$ *is an* ANR *having dimension* ≤ 1, *then* Y *is countable dimensional.*

COROLLARY A. *If* $f: X \to Y$ *is a proper sujection between metric spaces where* X *is complete and* $\dim X \leq 2$ *and where each* $f^{-1}(y)$ *is an* AR *having dimension* ≤ 1, *then* $\dim Y \leq 2$.

COROLLARY B. *If* $f: X \to Y$ *is a proper surjection where* X *is a locally compact* ANR *and* $\dim X \leq 3$ *and where each* $f^{-1}(y)$ *is an* AR *having dimension* ≤ 1, *then* $\dim Y \leq 3$ *or equivalently,* Y *is an* ANR.

COROLLARY C. *If* $f: E^3 \to Y$ *is a proper surjection where each* $f^{-1}(y)$ *is an* ANR *having dimension* ≤ 1, *then* Y *is countable dimensional.*

The first two corollaries add to the current knowledge of cell-like maps which do not raise dimension. Corollary B can be viewed as an extension of the result in [3] that cell-like maps defined on 3-manifolds do not raise dimension and of the result in [2] that cell-like maps defined on 3-dimensional poly-hedra with 1-dimensional point-inverses do not raise dimension. The techniques used in this paper are radically different from those used in the latter two papers.

At the heart of the paper is the following version of the result of Dyer mentioned above which does not assume openness. Let $f: X \to Y$ be a proper surjection between metric spaces where each $f^{-1}(y)$ is an ANR having dimension ≤ 1 and where $\dim X \leq n$. If Y is not countable dimensional, then there is a subset $Z \subseteq Y$ and subset $B \subseteq f^{-1}(Z)$ where

$\dim Z = +\infty$, c-dim $Z \leq n-1$, $f(B) = Z$, and $f|B$ is a cell-like map.

Even if X is locally compact, the set Z is not necessarily locally compact. It is this circumstance and not simply an attempt to merely generalize which forces us to consider proper maps defined on metric spaces.

REFERENCES

1. Dyer, E., "Certain transformations which lower dimension," *Ann. of Math 63* (1956) 15-19.

2. Kozlowski, G., W. H. Row and J. Walsh, "Cell-like mappings with 1-dimensional fibers on 3-dimensional polyhedra," to appear in *Houston J. of Math.*

3. Kozlowski G., and J. Walsh, "Cell-like mappings on 3-manifolds," preprint.

4. ______________, "The cell-like mapping problem," *Bull. Amer. Math. Soc. 2* (1980).

5. Sieklucki, K., "A generalization of a theorem of K. Borsuk concerning the dimension of ANR-sets," *Bull. Acad. Polon. Sci. Ser. Sci., Math Astonom. Phys.10* (1962) 433-436. (Also see correction 12 (1964).)

SECTION IV

MODERN ANALYSIS AND SET THEORY

STRUCTURE OF BANACH SPACES: RADON-NIKODÝM AND OTHER PROPERTIES

Robert C. James*

Claremont Graduate School, Claremont, California

The purpose of this paper is to trace a path through the development of certain related types of structural properties of general Banach spaces. Although the order of introduction of concepts will largely be the historical order, the methods of proof will be simplified drastically in some cases because of the advantage of hindsight. Most proofs suggested are elementary and direct. In particular, a direct elementary proof will be given for the equivalence of the Radon-Nikodým property and non-containment of bushes (Theorem 7). From this, it follows trivially that spaces isomorphic to uniformly convex spaces have the Radon-Nikodým property, including all L^p-spaces with $1 < p < \infty$ and all finite-dimensional spaces.

The properties along our path will interlace with other concepts at various stages. No attempt will be made to survey the complete structural theory of Banach spaces, since this would be far too ambitious. However, the present discussion will involve many important Banach-space concepts.

Condition (f) of Theorem 1 was introduced originally as an aid to proving that a Banach space is reflexive if each continuous linear functional attains its supremum on the unit ball (see [21, pg. 206], [22, pp. 114-115], and [24]). We will not discuss that theorem, but will concentrate on other concepts that were suggested originally by conditions (e)

*Supported in part by NSF-MCS 78-01890.

ISBN 0-12-481820-X

and (f). The proof of Theorem 1 also gives an easy proof of the Eberlein-Smulian theorem [13], i.e., that (a) and (c) are equivalent, or with slight modifications that (b) and (c) are equivalent if U_X is replaced by any weakly closed bounded subset [22, Theorem 6]. With Theorem 1, as well as with many later theorems, basic ideas of proofs will be suggested, with references given for complete arguments if possible.

The following notation will be used: U_X denotes the unit ball of X, $\hat{X}$ denotes the natural image of X in its second dual X^{**}, ANP denotes *asymptotic-norming property*, KMP denotes *Krein-Milman property*, and RNP denotes *Radon-Nikodým property*.

THEOREM 1. *For a Banach space* X, *the following are equivalent.*

(a) X *is reflexive.*

(b) U_X *is weakly compact.*

(c) U_X *is weakly sequentially compact.*

(d) $\cap K_n \neq \emptyset$, *whenever* $\{K_n\}$ *is a decreasing sequence of bounded closed convex nonempty subsets of* X.

(e) *There is no sequence* $\{z_n\}$ *in* U_X *and positive number* ε *for which it is true that, for each* $n \geqq 1$,

$$\mathit{dist}(\mathit{conv}\{z_1,\ldots,z_n\}, \mathit{conv}\{z_{n+1},z_{n+2},\ldots\}) > \varepsilon .$$

(f) X *does not have* one structure, *i.e., it is* not *true that, for some* $\theta > 0$ *(for any* θ *with* $0 < \theta < 1$*), there is a sequence* $\{(z_n,g_n)\}$ *with* $z_n \in U_X$, $g_n \in U_{X^*}$, *and*

$$g_i(z_j) = \theta \quad \mathit{if} \quad i \leqq j, \quad g_i(z_j) = 0 \quad \mathit{if} \quad i > j .$$

Proof. If X is reflexive, then (b) follows from the facts: $\hat{X} = X^{**}$, $U_{X^{**}}$ is w*-compact, and the w* and w topologies agree on X^{**}. To see that (b) $\Rightarrow$ (c), let $\{x_i\}$ be an arbitary sequence in U_X and choose a sequence $\{f_n\}$ in X^* that is total over $\mathrm{cl}[\mathrm{lin}\{x_i\}]$. Choose a subsequence $\{y_i\}$ of $\{x_i\}$ such that $\lim_{i\to\infty} f_n(y_i)$ exists for each n, which implies $\{y_i\}$ has a unique accumulation point $\bar{y}$. Suppose there is an f in X^* such that $\lim_{i\to\infty} f(y_i)$ either does not exist or is not $f(\bar{y})$. Then there is a subsequence $\{z_i\}$ of $\{y_i\}$ such that $\lim_{i\to\infty} f(z_i)$ exists and is not $f(\bar{y})$, but $\bar{y}$ is a w-accumulation point of $\{z_i\}$. The proof of (c) $\Rightarrow$ (d) is very easy. Merely let $x_n \in K_n$ for each n; then any limit of a subsequence of $\{x_n\}$ belongs to $\cap K_n$. For (d) $\Rightarrow$ (e), assume (e) is false for X and let $K_n = \mathrm{cl}[\mathrm{conv}\{x_i: i > n\}]$; then $\cap K_n = \emptyset$. If (f) is false when the quantifier "for some $\theta > 0$" is used, then (e) is false, since if $\Sigma_{n+1}^{\infty}\mu_i = 1$, then

$$g_{n+1}(\Sigma_1^n \lambda_i z_i) = 0 \quad \text{and} \quad g_{n+1}(\Sigma_{n+1}^{\infty} \mu_i z_i) = \theta ,$$

so that $\|\Sigma_1^n \lambda_i z_i - \Sigma_{n+1}^{\infty} \mu_i z_i\| \geq \theta$. To show that (f), with the quantifier "for any θ with $0 < \theta < 1$", implies X is reflexive, suppose X is not reflexive and $0 < \theta < 1$. Choose $F \in X^{**}$ with $\mathrm{dist}(F,\hat{X}) > \theta$ and $\theta < \|F\| < 1$. Then choose g_1 with $\|g_1\| < 1$ and $F(g_1) = \theta$, and choose z_1 with $g_1(z_1) = \theta$ and $\|z_1\| < 1$. Now choose $\{(z_n, g_n)\}$ inductively, so that:

(i) $\|g_n\| < 1$,

(ii) $F(g_n) = \theta$ for each n,

(iii) $g_n(z_j) = 0$ if $j < n$,

(iv) $\|z_n\| < 1$,

(v) $g_i(z_n) = \theta$ if $i \leq n$.

If (z_i, g_i) has been chosen for each $i < n$, use Helly's Condition, first to choose g_n to satisfy (i),(ii), and (iii), and then to choose z_n to satisfy (iv) and (v) (see[22,pg. 115]).

The first use of (f) of Theorem 1 was for the next theorem. This theorem led to the conjecture that uniformly nonoctahedral (non-$\ell_1^{(3)}$) Banach spaces are reflexive, that a Banach space is reflexive if it is uniformly non-$\ell_1^{(n)}$ for some n ([15], pg. 144],[20,pg. 546]), and much later that spaces of type 2 are reflexive, all of which were proved to be false ([25],[26],[28]).

DEFINITION. A Banach space X is *uniformly convex* if for any

$\varepsilon > 0$ there is a $\delta > 0$ such that no x and y in U_X satisfy

$$\|1/2(x+y)\| > 1 - \delta \quad \text{and} \quad \|1/2(x-y)\| > \varepsilon;$$

X is *uniformly nonsquare* if there is a $\Delta > 0$ such that no x and y in U_X satisfy

$$\|1/2(x+y)\| > 1 - \Delta \quad \text{and} \quad \|1/2(x-y)\| > 1 - \Delta.$$

THEOREM 2. *A Banach space* X *is reflexive if* X *is uniformly convex or if* X *is uniformly nonsquare.*

Proof. Suppose X is not reflexive and $\varepsilon < 1/2$ and δ are positive numbers as described in the above definition of uniform convexity. Choose θ so that $2\varepsilon < \theta < 1$ and $\theta > 1 - \delta$, and let $\{(z_i, g_i)\}$ be as in (f) of Theorem 1. Then

$$g_1[1/2(z_1+z_2)] = \theta \quad \text{implies} \quad \|1/2(z_1+z_2)\| \geqq \theta > 1 - \delta,$$
$$g_2[1/2(z_2-z_1)] = \tfrac{1}{2}\theta \quad \text{implies} \quad \|1/2(z_2- z_1)\| \geqq \tfrac{1}{2}\theta > \varepsilon .$$

The first proofs that X is reflexive if X is uniformly convex are given in [33],[29], and [36]. The proof that a Banach space is reflexive if it is uniformly nonsquare also uses (f), but the argument is not as short [20].

DEFINITION. A Banach space X has the *finite-bush property* if there is a positive *separation constant* ε and, for each positive integer n, there is a partially ordered subset B_n of U_X for which each member has finitely many successors, B_n has a first member x_{11}, and:

(i) $\|x-y\| > \varepsilon$ if y is a successor of x in B_n.

(ii) Each member of B_n can be joined to x_{11} by a linearly ordered chain of successive members of B_n.

(iii) If the chain joining x_{11} to x has fewer than n members, then x has at least two successors and x is the average of its successors.

The definition of the *finite-tree property* is obtained by replacing "at least two successors" in (iii) by "exactly two successors". A *bush* (or *tree*) is a partially ordered set B in U_X for which each member has finitely many successors, B has a first member x_{11}, (i) and (ii) are satisfied, and each x in B has at least two successors (exactly two successors) and x is the average of its successors.

The finite-tree property was introduced as a sufficient condition for nonisomorphism with a uniformly convex space [23, Theorems 6 and 7], with the hope it would also be necessary--as is now known to be the case (see Theorem 4). The space ℓ_1 has the finite-tree property with separation constant 1: Given n, let x_{11} be the average of $\{e_1,\cdots,e_{2^n}\}$, let the two successors of x_{11} be the averages of $\{e_1,\cdots,e_{2^{n-1}}\}$ and $\{e_{2^{n-1}+1},\cdots,e_{2^n}\}$, etc. The spaces c_o and $L_1[0,1]$ have obvious trees.

THEOREM 3. *Each nonreflexive Banach space has the finite-tree property* [*23, Theorem 5*].

Proof. Suppose X is not reflexive. Use (e) of Theorem 1 to get a subset $\{z_n\}$ of U_X and a positive number ε such that, for each $n \geqq 1$,

$$\operatorname{dist}(\operatorname{conv}\{z_1,\cdots,z_n\},\ \operatorname{conv}\{z_{n+1},z_{n+2},\cdots\}) > \varepsilon.$$

Given n, let x_{11} be the average of $\{z_1,\cdots,z_{2^n}\}$, let the two successors of x_{11} be the averages of $\{z_1,\cdots z_{2^{n-1}}\}$ and $\{z_{2^{n-1}+1},\cdots z_{2^n}\}$, etc.

THEOREM 4. *For any Banach space X, the following are equivalent to X not having the finite-tree property.*

(a) X *is isomorphic to a uniformly nonsquare space.*

(b) X *is isomorphic to a uniformly convex space.*

(c) X *is isomorphic to a uniformly smooth space.*

(d) X *is isomorphic to a space that is both uniformly convex and uniformly smooth.*

A Banach space being *uniformly smooth* means its norm is uniformly Fréchet differentiable on the unit sphere. It would be too much of a diversion to discuss the proof of Theorem 4 now (see [14]). Instead, we go to the next theorem.

THEOREM 5. *A Banach space is not reflexive if it contains a bush.*

Proofs for this theorem have been known since 1967. The easiest proof uses the fact that if K is the closure of the convex span of a bush with separation constant ε in a reflexive Banach space, then K is weakly compact (Theorem 1) and there is a closed convex subset C of K such that the diameter of $K-C$ is less than ε [35,Lemma]. This is impossible, since if any bush element x belongs to $K-C$, then all its successors belong to C and convexity of C implies $x \in C$. Another proof is implicit in [40], which essentially contains a proof that weakly compact subsets cannot contain bushes.

Another proof of Theorem 5 was very difficult, but actually proved a stronger result: The dual X^* of a Banach space X is not separable if X^* contains a bush. However, this did not seem particularly interesting until the relation between RNP and bushes became known much later. Because of the easier proof above, this proof was forgotten and lost. The ideas have been resurrected and used to prove Theorem 9, to be discussed later.

The original purpose of Theorem 5 was to prove that if a Banach space X has the finite-tree property, then there is a nonreflexive Banach space Y that is *finitely representable* in X, meaning that each finite-dimensional subspace of Y is nearly isometric to a subspace of X [23,pg. 169]. This was done by showing that X having the finite-tree property implies there is a space Y that has a tree and is finitely representable in X [23, Lemma B, pg. 170]. This then gave the next theorem.

DEFINITION. A Banach space X is *super-reflexive* if no nonreflexive Banach space is finitely representable in X.

THEOREM 6. *A Banach space* X *is super-reflexive if and only if*

X *does not have the finite-tree property.*

It also is true that a Banach space is super-reflexive if and only if it does not have the finite-bush property. To show this, one can show that X is not isomorphic to a uniformly convex space if X has the finite-bush property.

If Y is a nonreflexive Banach space and $Y = cl(\cup_1^\infty Y_n)$, where each Y_n is finite-dimensional, then X is reflexive and not super-reflexive if $X = \Pi_1^\infty Y_n$, where the ℓ_2-norm is used for defining the product.

Although a reflexive Banach space cannot contain a bush, it is not true that every nonreflexive Banach space contains a bush. It has long been known that ℓ_1 does not contain a tree [23, Theorem 1, pg. 163]. We will see that ℓ_1 does not contain any bush, since ℓ_1 has RNP.

Let $(S, \mathcal{a}, \mu)$ be a measure space and X a Banach space. A *simple function* is a function $s = \Sigma_{i=1}^n x_i \chi(A_i)$, where $\chi(A_i)$ is the characteristic function of A_i, $A_i \in \mathcal{a}$ with $\mu(A_i) < \infty$, and each $x_i \in X$. We let

$$\int_S s\,d\mu = \Sigma_{i=1}^n x_i \mu(A_i).$$

A function $f\colon S \to X$ is *measurable* if there is a sequence of simple functions $\{s_n\}$ such that $\lim_{n\to\infty} \|f - s_n\| = 0$ a.e., or equivalently if f^{-1} maps open sets onto measurable sets and f is *almost separably valued*, in the sense that there is a subset S_o of S for which $\mu(S - S_o) = 0$ and $f(S_o)$ is separable. A measurable function f is *Bochner integrable* if, for each $\varepsilon > 0$, there is a simple function s such that $\int_S \|f - s\|\,d\mu < \varepsilon$. If f is integrable, then $\int_S f\,d\mu = \lim \int_S s\,d\mu$, where the limit is taken as $\int_S \|f - s\|\,d\mu$ goes to 0.

DEFINITION. A Banach space X has RNP if, for any finite-measure space $(S, \mathcal{a}, \mu)$ and any μ-continuous, X-valued measure λ with $\|\lambda\|$ finite, there is an X-valued Bochner-integrable function f such that, for each $A \in \mathcal{a}$,

$$\lambda(A) = \int_A f\,d\mu .$$

It was proved in 1936 that uniform convexity implies RNP [7, Theorem 6] and that spaces with a boundedly complete basis have RNP [11]. In 1940, it was proved that subspaces of separable duals have RNP [12], and in 1943 that reflexive spaces have RNP [38, Corollary 5.6]. All of these are special cases of the more general result that a Banach space has RNP if each separable subspace is ismorphic to a subspace of a separable dual (Theorem 10).

THEOREM 7. *For a Banach space* X , *the following are equivalent.*

(a) X *has RNP.*

(b) X *has RNP with respect to Lebesgue measure on* $[0,1)$.

(c) X *does not contain a bush.*

Proof. Suppose first that X contains a bush. Let $\{I_{ni}\}$ be the "bush" of subintervals of $[0,1)$ for which $I_{11} = [0,1)$; I_{11} has the same number of equal subintervals $\{I_{2i}\}$ as there are successors $\{x_{2i}\}$ of x_{11} in the given bush; for each i, I_{2i} has the same number of equal subintervals $\{I_{3j}\}$ as there are successors $\{x_{3j}\}$ of x_{2i}, and we let $x_{3j} \leftrightarrow I_{3j}$ for each j, etc. We let each subinterval used be left-closed and right-open. Now let $\lambda(I_{ni}) = x_{ni}|I_{ni}|$ for each I_{ni} and, for each Lebesgue-measurable subset E of $[0,1)$, let

$$\lambda(E) = \lim \Sigma_1^\infty \lambda(J_i),$$

where $E \subset \cup_1^\infty J_i$, $\{J_i\}$ is a pairwise disjoint collection chosen from $\{I_{ni}\}$, and the limit is taken as the Lebesgue measure of $\cup_1^\infty J_i - E$ goes to zero. It is easy to show that λ is a measure and that λ has no Radon-Nikodým derivative with respect to Lebesgue measure on $[0,1)$.

We know now that (a) $\Rightarrow$ (b) $\Rightarrow$ (c). We will show that (c) $\Rightarrow$ (a). Let $(S,\mathcal{a},\mu)$ be a finite-measure space and let λ be an X-valued measure with $\|\lambda\|$ finite that is μ-continous. It is easy to show that μ-continuity of λ implies there is a countable partition $\{S_i\}$ of S such that, for each i , $\|\lambda(E)/\mu(E)\|$ is bounded for

$E \subset S_i$ and $\mu(E) \neq 0$. If for each i the Radon-Nikodým derivative f_i of the restriction to S_i of a measure λ exists and $\|\lambda\| < \infty$, then $\Sigma_1^\infty f_i$ is a Radon-Nikodým derivative for λ. Thus we can assume without loss of generality that

$$\left\|\frac{\lambda(E)}{\mu(E)}\right\| < 1 \quad \text{if} \quad \mu(E) \neq 0 .$$

Define E to be ε-*pure* [39] if, whenever A and B are subsets of E with positive μ-measure, we have

$$\left\| \frac{\lambda(A)}{\mu(A)} - \frac{\lambda(B)}{\mu(B)} \right\| < \varepsilon .$$

Suppose that, for each $\varepsilon > 0$, S has a countable partition into ε-pure subsets. Then for each positive integer n there is a partition $\{E_i^n : i \geq 1\}$ such that each E_i^n is n^{-1}-pure, each partition is a refinement of the preceding one, and each $\mu(E_i^n) \neq 0$. For each n, let f_n be defined by

$$f_n(s) = \frac{\lambda(E_i^n)}{\mu(E_i^n)} \quad \text{if} \quad s \in E_i^n .$$

Now we let $f = \lim f_n$ and use $\|f(s) - f_n(s)\| \leq n^{-1}$ to obtain $\lambda(A) = \int_A f d\mu$, by noting that

$$\|\lambda(A) - \int_A f d\mu\| \leq \|\lambda(A) - \Sigma \frac{\lambda(A \cap E_i^n)}{\mu(A \cap E_i^n)} \mu(A \cap E_i^n)\| + n^{-1}\|\mu\| = n^{-1}\|\mu\| ,$$

where the sum is over all i with $\mu(A \cap E_i^n) \neq 0$.

Thus if X does not have RNP, then there is an $\varepsilon > 0$ such that S does not have a partition into ε-pure subsets. Let $A = \bigcup_1^\infty E_i$, where the sets $\{E_i\}$ are pairwise disjoint and for each n the value of $\mu(E_n)$ is at least half as large as possible with E_n being ε-pure and $E_n \subset A - \bigcup_{i=1}^{n-1} E_i$. Then A may be empty, but in any case $F = S - A$ has no ε-pure subsets and $\mu(F) \neq 0$. Since F is not ε-pure, there is a subset F_1 of F, with $\mu(F_1) \neq 0$, $\mu(F - F_1) \neq 0$, and

$$\left\| \frac{\lambda(F_1)}{\mu(F_1)} - \frac{\lambda(F)}{\mu(F)} \right\| \geq \tfrac{1}{2}\varepsilon > \tfrac{1}{3}\varepsilon .$$

Then $F - F_1$ is not ε-pure and there is a subset F_2 of $F - F_1$ with $\mu(F_2) \neq 0$ and

$$\left\| \frac{\lambda(F_2)}{\mu(F_2)} - \frac{\lambda(F)}{\mu(F)} \right\| > \tfrac{1}{3}\varepsilon .$$

If $\mu[F - (F_1 \cup F_2)] \neq 0$, choose $F_3 \subset F - (F_1 \cup F_2)$ similarly, etc. If at each choice, $\mu(F_i)$ is made at least half as large as possible, then

$$\frac{\lambda(F)}{\mu(F)} = \sum \beta_i \frac{\lambda(F_i)}{\mu(F_i)} ,$$

where $\beta_i = \mu(F_i)/\mu(F)$, $\Sigma_1^\infty \beta_i = 1$, and the sum has at least two terms. If repetitions are allowed, we can represent $\lambda(F)/\mu(F)$ as an average of $\lambda(F_i)/\mu(F_i)$ with an arbitarily small error in averaging. Since no F_i has an ε-pure subset, this process can be continued to give an *approximate bush* with separation constant $\frac{1}{3}\varepsilon$ and errors in averaging whose sum σ is less than $\frac{1}{6}\varepsilon$. It is easy to show that for each member of this approximate bush, the natural-weighted average of the followers of x that can be connected to x by a chain of k successive bush members has a limit $\bar{x}$ as $k \to \infty$. If each x is replaced by such a limit, we have a bush with separation constant $\frac{1}{3}\varepsilon - 2\sigma > 0$ [16].

It is easy to show that a uniformly convex space cannot contain a bush. Let $\|\bar{x}\|$ be nearly M, where $M = \sup\{\|x\|: x \text{ in the bush}\}$. Since the average of $\frac{1}{2}(\bar{x} + x_i)$ is $\bar{x}$, if the average is over all successors $\{x_i\}$ of $\bar{x}$, we have $\|\frac{1}{2}(\bar{x} + x_k)\| \geq \|\bar{x}\|$ for some k. Since $\|x_k\| \leq M$, the line segment $[\bar{x}, x_k]$ is nearly in the sphere $\{x: \|x\| = M\}$. Thus all super-reflexive spaces have RNP, which includes all L^p-spaces with $1 < p < \infty$ and all finite-dimensional spaces.

If a dual of a separable Banach space X is not separable, then X^*

contains a tree [41, pg. 222] and therefore fails RNP. Conversely, if X^* is separable, then X^* does not contain any bush and therefore has RNP. To show that separable duals do not contain bushes, we will introduce ANP.

DEFINITION. Let κ be I, II, or III. For a Banach space X to have ANP-κ means that X has an equivalent norm for which there is a norming set Φ which has the property that $\{w_i\}$ satisfies (κ) below if $\|w_i\| = 1$ for each i and $\{w_i\}$ is *asymptotically normed* by Φ, meaning that, for each positive ε, there exist $\phi \in \Phi$ and N such that $\phi(w_i) > 1 - \varepsilon$ if $i > N$.

(I) The sequence $\{w_i\}$ converges strongly.

(II) Some subsequence of $\{w_i\}$ converges strongly.

(III) $\cap_1^\infty K_n \neq \emptyset$, where $K_n = \mathrm{cl}[\mathrm{conv}\{w_i: \; i > n\}]$.

It is easy to see that ℓ_1 has ANP-II for its usual norm. We just let the norming set be the set of all members of ℓ_∞ with only finitely many nonzero components, each +1 or -1. However, the norming set cannot contain the sequence ϕ with all components +1, since ϕ asymptotically norms the natural basis of ℓ_1 very well. Since ℓ_1 has ANP-II, it has ANP-III and it will follow from Theorem 9 that ℓ_1 has RNP.

All reflexive Banach spaces have ANP-III, since any bounded sequence $\{w_i\}$ in a reflexive space has a w-accumulation point; any accumulation point w belongs to K_n of III for each n, since closed convex sets are w-closed. In fact, if $\{w_i\}$ is asymptotically normed and $\|w_i\| = 1$ for each i, then $\|w\| = 1$. It also is easy to show that uniformly convex Banach spaces have ANP-I. Suppose $\{w_i\}$ is asymptotically normed by Φ and $\|w_i\| = 1$ for each i. If $\phi(w_i) > 1 - \delta$ when $i > N$, then $\phi[\frac{1}{2}(w_i+w_j)] > 1 - \delta$ and $\|\frac{1}{2}(w_i+w_j)\| > 1 - \delta$ if $i > N$ and $j > N$. Thus uniform convexity implies $\{w_i\}$ is Cauchy.

If a Banach space X is separable, then the three ANP properties are equivalent and will be denoted simply by ANP [27, Theorem 1.2].

THEOREM 8. *If a Banach space is isomorphic to a subspace of a separable dual, then it has ANP.*

Proof. It is known that if X^* is separable, then X has an equivalent norm for which X^* is locally uniformly convex ([1] and [9, Theorem 2, pg. 118]). Also, all subspaces of a Banach space have ANP if the space has ANP. Thus it is sufficient to consider only separable duals X^* that are locally uniformly convex. Let the norming set be $U_{\hat{X}}$. Suppose $\{w_i\}$ is a sequence in X^* for which $\|w_i\| = 1$ for each i and $\{w_i\}$ is asymptotically normed by $U_{\hat{X}}$. Let w be a w*-accumulation point of $\{w_i\}$. If $\|\hat{x}\| \leqq 1$ and $\hat{x}(w_i) = w_i(x) > 1 - \varepsilon$ if $i > N$, then $w(X) \geqq 1 - \varepsilon$. Thus $\|w\| = 1$ and local uniform convexity implies $\{w_i\}$ converges to w.

THEOREM 9. *If a Banach space* X *has ANP-III, then* X *has RNP.*

The proof of this theorem involves proving that a separable Banach space with a bush cannot have ANP. The proof is not easy [27, Theorem 1.8]. Although it can be simplified somewhat and still yield a proof of the theorem that a separable dual cannot contain a bush, the extra effort to prove Theorem 9 seems worthwhile.

It follows from Theorem 7 that RNP is separably determined, so the next theorem follows from Theorems 8 and 9 (also see [12], [34, Theorem 3.5], and [41 , Theorem A]).

THEOREM 10. *If each separable subspace of a Banach space* X *is isomorphic to a subspace of a separable dual, then* X *has RNP.*

Many examples have been given of Banach spaces and classes of Banach spaces that have RNP [10, pp. 217-219]. At one time, all such examples were special cases of the preceding theorem, and it was conjectured that a separable Banach space with RNP is isomorphic to a subspace of a separable dual [42]. However, examples are known now of separable Banach spaces that have RNP and are not isomorphic to subspaces of separable duals ([5] and [32]). Some of these provide examples of separable spaces with ANP that are not isomorphic to a subspace of any separable dual [27, Theorem 2.1]. It remains an open question whether RNP implies ANP for separable Banach spaces. Even if this is true, it might be false for nonseparable spaces. Although RNP is separably determined, this is not known to be true for any ANP-κ. In fact, it is not known whether Theorem 10 is valid for ANP.

The following concept of dentability was introduced by Rieffel, who proved that (b) of Theorem 11 implies RNP [39]. Proofs that RNP implies (b) are given in [8] and [17]. Before RNP and (b) were known to be equivalent, the equivalence of RNP and a concept called σ-dentability was known [31].

DEFINITION. A set K in a Banach space *can be* ε*-dented* if it contains an x for which x does not belong to the closure of the convex span of $K-B(x,\varepsilon)$, where $B(x,\varepsilon)$ is the ε-ball with center at x. The set K is *dentable* if K can be ε-dented for each positive ε.

THEOREM 11. *For a Banach space* X, *the following are equivalent.*

(a) X *does not contain a bush.*

(b) *All bounded closed convex nonempty subsets of* X *are dentable.*

(c) *All bounded nonempty subsets of* X *are dentable.*

Proof. The implication (b) $\Rightarrow$ (c) is an easy exercise [39, Prop.2]. Also, (c) $\Rightarrow$ (a) follows from the fact that if x belongs to a bush B with separation constant ε, then x is the average of its successors, all of which belong to $B - B(x,\varepsilon)$. Let us now show that (a) $\Rightarrow$ (b). Suppose K is a bounded closed convex subset of X and K is not dentable. That is, there is an $\varepsilon > 0$ such that K cannot be ε-dented. Let x_{11} be any member of K. Then x_{11} can be approximated arbitrarily closely as a convex combination of members of $K - B(x_{11},\varepsilon)$. Since repetitions are allowed, this convex combination can be an average:

$$\|x_{11} - n^{-1}\Sigma_{i=1}^{n} x_{2i}\| < \varepsilon_1 ,$$

where ε_1 is the *error in averaging*. This can be repeated for each x_{2i} and a new error in averaging ε_2. This process gives an approximate bush. As in the proof of Theorem 7, it follows that X contains a bush if $\Sigma_1^{\infty} \varepsilon_i < \frac{1}{2}\varepsilon$.

We have discussed several properties equivalent to RNP, and one possibly stronger property (ANP). Let us now discuss briefly a possibly weaker property.

DEFINITION. A Banach space X has KMP if each bounded closed convex subset is the closed convex span of its extreme points.

THEOREM 12. *If a Banach space has RNP, then it has KMP* [*37, Theorem 2 (Lindenstrauss)*].

Although the proof of this thorem is not difficult, it is interesting that ℓ_1 was known to have RNP as early as 1936 [7, pg. 412], but was not proved to have KMP until 1966 [30]. To prove Theorem 12, one shows first that if each bounded closed convex subset of X has an extreme point, then X has KMP [30, Lemma 1]. To do this, let C be the closure of the convex span of the extreme points of K. If $K - C \neq \emptyset$, let $f = c$ be a hyperplane that separates some point of K from C and choose d so $f = d$ is a supporting hyperplane H of $K - C$. The Bishop-Phelps theorem [3, Corollary 4] gives a supporting hyperplane arbitrarily near H that contains points of $K - C$; by hypothesis, this supporting set has an extreme point, which is an extreme point of K not in C. Now we can complete the proof by showing that each bounded closed convex subset of K has an extreme point, using the fact that RNP implies K is dentable: Suppose each $\varepsilon_i > 0$ and $\varepsilon_i \to 0$. If x is a $\frac{1}{2}\varepsilon$-denting point, then x can be separated from $\mathrm{cl}[\mathrm{conv}\ K - B(x,\frac{1}{2}\varepsilon_1)]$ by a hyperplane $f = c$. We can repeat the preceding argument, using the Bishop-Phelps theorem to get a supporting set K_1 of K that has diameter less than ε_1 . Any extreme point of K_1 is an extreme point of K . Now repeat this with K_2 and ε_2 , etc. Then $\cap_1^\infty K_n$ is a single point, which is an extreme point of each K_n and of K.

The question of whether KMP implies RNP has produced a great deal of interest and work. It remains unresolved, although many related results contribute to understanding of the problem, particularly its difficulty. Some of the seemingly most relevant of these are the following sufficient conditions for X to have RNP (all but the last one also are necessary conditions):

1. Each bounded closed nonempty subset has an extreme point [19, Theorem 4].

2. Each bounded weakly-closed nonempty subset has an extreme point [4].

3. Each bounded closed convex subset of X is the closed convex span of its strongly exposed points [37, Theorem 9, pg. 85].

4. X contains no weighted tree [16, Theorem B, pg. 65] (although there is a Banach space with a bush that has no trees [6]).

5. X is a dual and X has KMP [18].

REFERENCES

1. Asplund, E.,"Averaged norms," *Israel J. Math 5*(1967), 227-233.

2. Beck, A.,"A convexity condition in Banach spaces and the strong law of large numbers," *Proc. Amer. Math. Soc.13*(1962), 329-334.

3. Bishop, E. and R. R. Phelps, "The support functionals of a convex set," *Proc. Symp. Pure Math., vol. 7(convexity), Amer. Math. Soc.* (1963), 27-35.

4. Bourgain, J.,"A geometric characterization of the Radon-Nikodým property in Banach spaces," *Compositio Math. 36*(1978), 3-6.

5. Bourgain,J. and F. Delbaen, "A special class of L^∞-spaces," to appear in *Acta Math.*

6. Bourgain,J. and H.P. Rosenthal, "Martingales valued in certain subspaces of L^1,"Univ. Texas Austin, *Tech. Report #7*(1979).

7. Clarkson, J.A., "Uniformly convex spaces," *Trans. Amer. Math. Soc. 40*(1936), 396-414.

8. Davis, W. J. and R. R. Phelps, "The Radon-Nikodým property and dentable sets in Banach spaces," *Proc. Amer. Math. Soc.* *45*(1974), 119-122.

9. Diestal, J., *Geometry of Banach Spaces--Selected Topics*, Springer-Verlag (Berlin and New York), 1975.

10. Diestal, J. and J. J. Uhl, Jr., "Vector Measures," *Math. Surveys No. 15*, *Amer. Math. Soc.* (Providence, R. I.), 1977.

11. Dunford, N. and A. P. Morse, "Remarks on the preceding paper of James A. Clarkson," *Trans. Amer. Math. Soc.* *40*(1936), 415-420.

12. Dunford, N. and B. J. Pettis, "Linear operations on summable functions," *Trans. Amer. Math. Soc.* *47*(1940), 323-392.

13. Eberlein, W. F., "Weak compactness in Banach spaces," *Proc. Nat. Acad. Sci. U.S.A.* *33*(1947), 51-53.

14. Enflo, P., "Banach spaces which can be given an equivalent uniformly convex norm," *Israel J. Math.* *13*(1972), 281-288.

15. Giesy, D. P., "On a convexity condition in normed linear spaces," *Trans. Amer. Math. Soc.* *125*(1966), 114-146.

16. Ho, A., "The Radon-Nikodým property and weighted trees in Banach spaces," *Israel J. Math.* *32*(1979), 59-66.

17. Huff, R. E., "Dentability and the Radon-Nikodým property," *Duke Math. J.* *41*(1974), 111-114.

18. Huff, R. E. and P. D. Morris, "Dual spaces with the Krein-Milman property have the Radon-Nikodým property," *Proc. Amer. Math. Soc.* *49* (1975), 104-108.

19. Huff, R. E. and P. D. Morris, "Geometric characterizations of the Radon-Nikodým property in Banach spaces," *Studia Math.* *56*(1976), 157-164.

20. James, R. C., "Uniformly nonsquare Banach spaces," *Ann. of Math.* *80* (1964), 542-550.

21. ________, "Characterizations of reflexivity," *Studia Math.* *23* (1964), 205-216.

22. ________, "Weak compactness and reflexivity," *Israel J. Math.* *2* (1964), 101-119.

23. ________, "Some self-dual properties of normed linear spaces," Symposium on infinite-dimensional topology (1967), *Ann. of Math. Studies* *69*(1972), 159-175.

24. ________, "Reflexivity and the sup of linear functionals," *Israel J. Math.* *13*(1972), 289-300.

25. ________, "A nonreflexive Banach space that is uniformly non-octahedral," *Israel J. Math.* *18*(1974), 145-155.

26. ________, "Nonreflexive spaces of type 2," *Israel J. Math.* *30*(1978), 1-13.

27. James, R. C. and A. Ho, "The asymptotic-norming and Radon-Nikodým properties for Banach spaces," to appear in *Arkiv för Matematik.*

28. James, R. C. and J. Lindenstrauss, *The octahedral problem for Banach spaces*, Proc. seminar on random series, convex sets, and geometry of Banach spaces, Aarhus (Denmark), 1974.

29. KaKutani, S., "Weak topology and regularity of Banach spaces," *Proc. Imp. Acad. Tokyo 15*(1939), 169-173.

30. Lindenstrauss, J., "On extreme points in ℓ_1," *Israel J. Math. 4*(1966), 59-61.

31. Maynard, H., "A geometric characterization of Banach spaces possessing the Radon-Nikodým property," *Trans. Amer. Math. Soc., 185*(1973), 493-500.

32. McCartney, P.W. and R. C. O'Brien, "A separable Banach space with the Radon-Nikodým property that is not isomorphic to a subspace of a separable dual," *Proc. Amer. Math. Soc. 78*(1980), 40-42.

33. Milman, D., "On some criteria for the regularity of spaces of the type (B)," *Dokl. Akad. Nauk SSSR 20*(1938), 243-246.

34. Namioka, I., "Neighborhoods of extreme points," *Israel J. Math. 5* (1967), 145-152.

35. Namioka, I. and E. Asplund, "A geometric proof of Ryll-Nardzewski's fixed-point theorem," *Bull. Amer. Math. Soc. 73*(1967), 443-445.

36. Pettis, B. J., "A proof that every uniformly convex space is reflexive," *Duke Math. J. 5*(1939), 249-253.

37. Phelps, R. R., "Dentability and extreme points in Banach spaces," *J. Funct. Anal. 17*(1974), 78-90.

38. Phillips, R. S., "On weakly compact subsets of a Banach space," *Amer. J. Math. 65*(1943), 108-136.

39. Rieffel, M. A., *Dentable subsets of Banach spaces with applications to a Radon-Nikodým theorem in functional analysis*, Functional Analysis (Proc. Conf., Irvine, Colif., 1966), Academic Press (London), Thompson (Washington), 1967.

40. Ryll-Nardzewski, C., "On fixed points of semigroups of endomorphisms of linear spaces," *Proc. Fifth Berkeley Symp. Math. Stat. and Prob., Vol. II*(1967), 55-61.

41. Stegall, C., "The Radon-Nikodým property in conjugate Banach spaces, *Trans. Amer. Math. Soc. 206*(1976), 213-223.

42. Uhl, J. J., Jr., "A note on the Radon-Nikodým property for Banach spaces," *Rev. Roumaine Math. Pures Appl. 17*(1972), 113-115.

SEPARATION AND OPTIMIZATION IN FUNCTION SPACES

L. Asimow

University of Wyoming, Laramie

The *decomposability* of a convex set by an affine variety is a geometrically formulated condition which, loosely speaking, describes a local polyhedral configuration. Its attraction is that it can be used to infer separation theorems somewhat sharper than the standard geometric Hahn-Banach statements. This phenomenon has been long recognized in the finite dimensional Euclidean space setting where polyhedral separation is a standard topic in convex analysis and an indispensable tool in linear/convex programming theory.

It is, however, in the Banach space duality framework that the concept enjoys its greatest utility. The idea here is that if S is convex and M affine -- both *weak** closed subsets of a Banach dual space E^* -- then the decomposability of S by M leads to existence theorems for elements in E (as distinguished from E^{**}) satisfying certain separation properties. The presence of decomposability allows for the inductive construction of a Cauchy sequence of approximations in E whose limit has the desired features.

In this note we indicate how the decomposability of sets in a dual Banach space can be incorporated into the basic results of conjugate duality for optimization problems in order to yield sufficient conditions for the existence of dual solution pairs. This is a quite general setting and we sketch the relationship of function space results involving notions

ISBN 0-12-481820-X

such as peak sets and exact interpolation sets to this broader optimization/approximation context. A general reference for the function space applications is [1]. Some of the applications to optimization -- especially Fenchel duality and convex programming -- are given in [2].

Let E be a Banach space and let S, M be closed subsets such that

(1) S is a convex cone

(2) M is a subspace.

We consider the dual (polar) sets S^o, $M^o (=M^\perp)$ in E^* consisting of the elements $x^* \in E^*$ bounded above (and hence bounded above by 0) on S, M respectively. Then S^o, M^o are *weak** closed (cone and subspace, respectively) in E^*.

We say S^o is *decomposable* by M^o if there is an $\alpha > 0$ such that each $x^* \in S^o$ can be written as the sum $y^* + z^*$ with $y^* \in S^o \cap M^o$, $z^* \in S^o$ and

$$\| z^* \| \leqq \alpha \| z^* + M^o \| ,$$

where $\|z^* + M^o\|$ means the distance from z^* to M^o. If there is a norm bounded linear functional f which is identically 0 on M^o with

$$\| z^* \| \leqq \alpha f(x^*)$$

then we say S^o is *positively decomposable by* M^o. In this case $S^o \cap M^o$ is a *face* of S^o.

This definition, for S^o a cone, forms the basis for a general description of the decomposability of an arbitrary closed convex A by a closed variety N. For economy of space we will forego the details (abundant quantities of which may be found in [2]) of this, except to note one important case:

If A *is weak** *closed and norm bounded (hence weak** *compact) and* N *is a weak** *closed variety then* A *is decomposable by* N *if and only if for some* $\alpha > 0$,

$$A = \operatorname{conv}(A \cap N, Z): \quad Z = \{z \in A: \|z - N\| \geqq \alpha\} .$$

This conveys the essential geometric idea -- each element of A can be resolved into a (generalized, in the unbounded case) convex combination of something in $A \cap N$ and something else in A "far" from N.

We state the main result for the case where S^o is a polar cone, as above.

THEOREM. *If* S^o *is decomposable by* M^o *then* $S + M$ *is (norm) closed in* E.

EXAMPLE. The subspace M of E is called *proximinal* if each $x \in E$; has a best approximation (closest point) $y \in M$;

$$\|x - y\| = \|x - M\| .$$

It is an easily verified fact that M is proximinal if and only if $M + E_1$ is closed (in general, E_k denotes the elements of E of norm no more than k). Hence

If E_1^* *is decomposable by* M^o *then* M *is proximinal in* E.

We will discuss this further after developing the general conjugate duality.

Consider now *weak** closed convex sets A and B in E^* and let N denote the *weak** closed affine span of B. We denote by *ri*B the interior of B relative to N. The "polyhedral" separation theorem goes like this:

THEOREM. *Let* A *and* B *(as above) be such that*

(1) *ri* $B \neq \emptyset$ and $(ri\, B) \cap A = \emptyset$,

(2) A *is decomposable by* N.

Then there exists $a \in E$ *and* $\beta \in R$ *such that*

$$a \geqq \beta \text{ on } A \text{ and } a < \beta \text{ on } ri\, B$$

If E is finite dimensional it turns out that ([2]),

A *is polyhedral iff* A *is decomposable by each affine variety.*

Of course in this case, *ri* B is always non-empty (assuming $B \neq \emptyset$). To see what goes wrong if A is *not* polyhedral, let A be a disk in R^2 and B a line segment tangent to A at one of its endpoints. Then any linear f with greatest lower bound β on A and $f \leqq \beta$ on B must be *identically* β on B.

We proceed now with the conjugate duality theory in the general setting, using the notation, terminology and basic results in [4] . A function f on a Banach space X, with values in $(-\infty,+\infty]$ is *closed* and *convex* if its *epigraph*, *epi* f, is closed and convex in $X \times R$. If f is merely convex, we denote by $\bar{f}$ the *closure* of f, obtained as the function whose epigraph is $(epi\ f)^-$. Analogous notions for concave g and its *hypograph* pertain.

The *primal* problem is that of minimizing f (closed and convex) on X:

$$\text{(P)} \qquad \inf f(x);\ x \in X .$$

We consider this problem as embedded in a family of *perturbed* problems, indexed by elements in a Banach space U. This means there is a closed convex function F on $X \times U$ such that

$$f(x) = F(x,0) .$$

The perturbed problems are

$$(\text{P})_u \qquad \inf f(x,u);\ x \in X .$$

The optimal value function is given by

$$\phi(u) = \inf\{F(x,u) : x \in X\} ,$$

the solution value to $(\text{P})_u$.

The problem is dualized by computing conjugate functions on the duals V and Y of X and U respectively. Thus, we have the closed concave G on $Y \times V$ given by

$$G(y,v) = -F^*(v,-y) .$$

Then

$$\gamma(v) = \sup\{G(y,v) : y \in Y\} \quad \text{and} \quad g(y) = G(y,0)$$

leads to the dual problem for the concave function g on Y,

$$\text{(D)} \qquad \sup g(y) ; \; y \in Y .$$

We denote the solution values of (P) and (D) by α and β. Always $\beta \leqq \alpha$. The primal problem is *stable* if there exists $y_o \in Y$ such that

$$-\infty < \beta = g(y_o) = \alpha < \infty ,$$

that is, (P) and (D) have the same value and (D) has a solution vector, y_o. This solution, y_o, is called a *Kuhn-Tucker vector* for (P). Since (D) is in a Banach dual space with the *weak** topology it is generally more tractable than (P), yielding to the generous supply of compactness assured by the Alaoglu theorem. A standard hypothesis for stability is the continuity of ϕ at 0. This implies that the superlevel sets of g are norm bounded and hence compact.

The *a priori* assumption of continuity for g is rather restrictive and, following [2, 3], we prefer to give a geometric condition in the dual setting. One sees that for a finite β stability is equivalent to

$$\gamma(0) = \bar{\gamma}(0) .$$

Since *hypo* γ is obtained from the projection, P, of *hypo* G from $Y \times V \times R$ to $V \times R$, what is required is some reasonable principle assuring this projected image remains closed.

$$(*) \qquad \begin{array}{l} \text{For each } k > 0 \text{ there exists } \alpha(k) > 0 \text{ such that} \\ (hypo\ \gamma) \cap (V_k \times [-k,k]) \subset P[(hypo\ G) \cap (Y_{\alpha(k)} \times V \times R)]. \end{array}$$

The (*) condition, based on the Krein-Smulŷan theorem, guarantees that *hypo* γ is closed. Various versions, including local (*) conditions which yield $\gamma(0) = \bar{\gamma}(0)$, have been advanced in [3]. The set-up for Fenchel duality is considered in [2] where special formulations, suitable for that case, are given.

If (P) is stable and, in addition, has a solution $x_o \in X$ then x_o together with the Kuhn-Tucker vector y_o are said to form a *solution pair*.

The basic result is given next.

THEOREM. *If* $\gamma(0)$ *is finite and the* (*) *condition holds then* (P) *is stable. If, in addition, hypo* γ *is decomposable by the one-dimensional space* $\{0\} \times R$ *then there exists a solution* x_o *to* (P) *and hence, a solution pair* x_o, y_o *to* (P) *and* (D).

In the Fenchel duality set-up the function γ is an *inf-convolution* and it is possible to formulate the decomposability in terms related to the two separate functions convolved to form γ. For example let C be a closed convex set of X and suppose the problem is to determine if C is *proximinal* in X. The problem then is to solve, for each $z \in X$

$$\inf \|z - x\| + I_C(x); \quad x \in X$$

where I_C is the *indicator* of C: 0 on C and $+\infty$ elsewhere. The problem is stable for all z. Let C^* denote the set of elements in V bounded above on C with N the *weak** closed affine span. If C is a *subspace* then $N = C^\perp$ and $ri\, C^* = N \neq \emptyset$.

If E_1^* is decomposable by N and $ri\, C \neq \emptyset$ then C is proximinal.

Consider now a compact Hausdorff space K and a closed subset H. Let A be a closed subspace of C(K) (continuous complex functions with the uniform norm) such that A contains constants and separates points. To avoid technicalities we assume $H = (H_A^\perp)_\perp$ with respect to A, and that H is a G_δ. The former means

$$H = \{t \in K: a(t) = 0 \ \text{ for all } \ a \in H_A^\perp\},$$

where $H_A^\perp = \{a \in A;\ a \equiv 0 \ \text{ on } \ H\}$.

The decomposability of appropriate sets in the dual setting can be used to attack a number of questions involving A and H.

First, we consider conditions for H to be a *peak set* for A -- a set of the form $a^{-1}(1)$ for a function $a \in A$ satisfying $|a| < 1$ on $K \backslash H$. The issue here involves the state space S_A in A^* given by

$$S_A = \{x^* \in A^* : (1,x^*) = 1 = \|x^*\| \} .$$

Then S_A is a *weak** closed convex subset of A_1^* which contains a homeomorphic embedding of K under *evaluation*. For control over real *and* imaginary parts of functions we need to consider the *complex state space*

$$Z_A = \mathrm{conv}(S_A \cup -iS_A) .$$

Let $N = \{x^* \in A^* : (a,x^*) = 0 \text{ for all } a \in H_A^{\perp}\} = H_A^{\perp\perp}$.

If Z_A is positively decomposable by N then H is a peak set.

The analytic precursor for results of this type is the Bishop peak point theorem for function algebras. The decomposability becomes a geometric surrogate for the "1/4 - 3/4" property and resulting iteration, associated with this now classical theorem.

We consider next conditions for H to be an *exact norm interpolation set* for A, meaning that to each $a \in A$ there is a $b \in A$ with

$$b|_H = a|_H \quad \text{and} \quad \|b\|_K = \|a\|_H .$$

This amounts to $H^{\perp}$ being proximinal in A. The standard results are formulated in terms of annihilating measures, $A^{\perp}$, in $C(K)^*$, where the latter is regarded as identical to the space of regular Borel measures on K with total variation as norm. The measures $(A|_H)^{\perp}$ which annihilate A and have their support on H can be considered a subspace of $C(K)^*$.

The generalized version of the classical Rudin-Carleson theorem may be formulated this way:

If there exists k, $0 \leqq k < 1$, such that for each $\mu \in A^{\perp}$

$$\| \mu + (A|_H)^{\perp} \| \leq k \| \mu|_{K \setminus H} \|$$

then H is an exact norm interpolation set.

In the context of the conjugate duality theorem, one easily computes the dual problem and finds that the measure condition yields the stability of the primal problem (the (*)-condition is pertinent here). Next, it can be shown that with $k < 1$ the measure condition will, with somewhat more persuasion, yield the decomposability of A_1^* by $H_A^{\perp}$.

It is fashionable now to express these results in sharper terms using boundary measures and Choquet theory, as well as more refined concepts of distance in $C(K)$ than that measured by the uniform norm. The interested reader (the existence of which is, as far as we know, still an open question) will find these matters detailed in [1].

REFERENCES

1. L. Asimow and A. J. Ellis, *Convexity Theory and Its Applications in Functional Analysis*, LMS Mono. Ser. No. 16, Academic Press, London, 1980 (to appear).

2. L. Asimow and A. Simoson, "Decomposability and dual optimization in Banach spaces," *J. Math. Anal. and Appl.*, to appear.

3. P. Levine and J.C. Pomerol, "Sufficient conditions for Kuhn-Tucker vectors in convex programming," *SIAM J. Control and Opt. 17(6)*, (1979), 689-699.

4. R. T. Rockagellar, *Conjugate Duality and Optimization*, Regional Conference Series No. 16, SIAM Publications, Philadelphia, 1974.

FORMULAS FOR MINIMAL PROJECTIONS

Bruce L. Chalmers

University of California, Riverside

I. INTRODUCTION

Let X denote an arbitrary Banach space (with norm $\| \;\|$) and let $V = [v_1,\ldots,v_n] = [\vec{v}]$ denote an n-dimensional subspace of X. A projection P from X onto V is a linear map such that $Py = y$, for all $y \in V$. For $x \in X$ consider Px as an approximation to x from V. The error $\|x-Px\|$ is estimated as follows: for any $y \in V$,

$$\|x-Px\| = \|(I-P)(x-y)\| \le (1+\|P\|)\mathrm{dist}(x,V) .$$

Thus a "minimal projection" P_{min} yielding $\min\|P\|$ is sought.

Note that a bounded projection P can be identified with an n-dimensional subspace $P = [g_1,\ldots,g_n] = [\vec{g}]$ of the dual space X^*; i.e., $Px = \langle x,\vec{g}\rangle\cdot\vec{v} = \sum_{i=1}^{n} \langle x,g_i\rangle v_i$, where $\vec{v}$ is chosen dual to $\vec{g}$ (i.e., $\langle v_i,g_j\rangle = \delta_{ij}$).

For compactness of notation we will in the following write $X \subset L^\infty$ for $X \subset C$, and $X^* \subset L^1$ for $X^* \subset C^*$, where C denotes the continuous functions. As usual let $\frac{1}{q} + \frac{1}{p} = 1$ where $1 \le p \le \infty$; and if $X \subset L^p(Q)$, identify X^* with a subspace of $L^q(Q)$. Also in the following, Q will denote an arbitrary compact separable T_1-space and " $'$ " will denote 2-sided differentiation along any continuously differentiable (C^1) vector field on Q. Further, for $\vec{v} = (v_1,\ldots,v_n) \in V^n$ we write $|\vec{v}|^2 = \Sigma v_i^2 = \vec{v}\cdot\vec{v}$. Finally " c " will denote "constant" and "pcw." will abbreviate "piecewise."

ISBN 0-12-481820-X

In [3] and [4] (see also [6]) are developed the variational equations for minimal projections:

THEOREM (Variational Equations (VE)). *Let* $L^p(Q) \supset X \supset V = [v_1,\ldots,v_n] = [\vec{v}]$, $1 \le p \le \infty$. *Then* $L^q(Q) \supset X^* \supset P_{min} = [g_1,\ldots,g_n] = [\vec{g}]$ *where* $\exists$ *nonzero* $\vec{v}_{(1)}, \vec{v}_{(2)} \in V^n$, $\vec{v}_{(k)} \in V^n_{\pm}$ *(see §3),* $3 \le k \le n$, *such that*

$$\text{(VE)} \quad \begin{cases} (*) \quad \frac{1}{p}\vec{g}\,'\cdot\vec{v}_{(1)} = \frac{1}{q}\vec{g}\cdot\vec{v}\,'_{(1)} & V \subset C^1_{pcw.}(Q) \\ (o) \quad \vec{g}\cdot\vec{v}_{(k)} = c_{pcw.}, & 2 \le k \le n \ (independent). \end{cases}$$

Note 1. The theorem applies to any separable Banach space X, identified as usual with a subspace of $L^\infty(Q)$, where Q is the set of extreme points of the ball of the dual.

The (VE) in general determine P_{min} up to a "small" number of constants which can then be determined in those situations where necessary and sufficient conditions (NASC) for minimal projections are known. That is, knowledge of the (VE) reduces the problem to a finite dimensional minimum problem which can then be solved successfully via known NASC.

In the following will be presented simple formulas solving the (VE) and yielding P_{min}.

II. SUPER-ORTHOGONAL PROJECTIONS

Consider $X \subset L^p$, $1 \le p \le \infty$. Any orthogonal projection (see e.g. [7]) $P_\perp$ can be written $[\vec{g}] = [w\vec{v}]$ where $\vec{v}$ is an arbitrary n-tuple of independent functions. (See also [10].)

DEFINITION 1 ([3]). $P_\perp$ is called *super-orthogonal* if $w = \frac{E}{\vec{v}_0\cdot\vec{v}}$ for some $\vec{v}_0 \in V^n$, where $E = e^{p\int^t \vec{v}_0\,'\cdot\vec{v}\,/\,\vec{v}_0\cdot\vec{v}}$.

THEOREM 1 [3]. *If* $P_\perp$ *is minimal among all projections, then* $P_\perp$ *is super-orthogonal,* $1 \le p < \infty$.[†]

This theorem is proved by establishing the (*)-equation for P_{min} and showing that if P_{min} is orthogonal, then it must be super-orthogonal. (To check that a super-orthogonal projection satisfies the (*)-equation, see §3.) Note that the n-1 (o)-equations are automatically satisfied by $P_\perp$:

$$g_{k-1}v_k - g_k v_{k-1} = 0 \qquad (2 \le k \le n).$$

Note 2. If $P_\perp$ is super-orthogonal with $\vec{v}_0 = \vec{v}$, then $w = |\vec{v}|^{p-2}$ and if further $\vec{v}$ is chosen (see [1], [14]) so that $\langle v_i, wv_j \rangle = \delta_{ij}$, we obtain the "natural" projection P_p where $\|P_p\| \le n^{\left|\frac{1}{2} - \frac{1}{p}\right|}$, $1 \le p \le \infty$.

EXAMPLE. The Fourier projections, $1 \le p \le \infty$, are examples of orthogonal (in fact natural) projections which are minimal.

Note 3. The super-orthogonal projections form an n(n-1) parameter family $(\vec{v}_0 = A_{n \times n}\vec{v})$. In fact if V is symmetric we can restrict our attention to *symmetric* super-orthogonal projections which form an n(n-1)/2 parameter family.

Because of Theorem 1 and Notes 2-3, it is clear that the super-orthogonal projections form a relatively good (in the sense of having projections with "small" norm) and simple (in the sense of form) subfamily of projections from which a minimum may be sought.

PROPOSITION 1 ([3]). *In the case* $X = L^\infty$, $V = [1, t, \ldots, t^{n-1}]$, *the super-orthogonal projections coincide with the interpolating projections.*

COROLLARY 1 ([8],[11]-[13]). *In the case of Proposition 1 and* Q = [a,b] , *the minimal super-orthogonal projection is that which has an equi-oscillating Lebesgue function.*

[†] If $P_\perp(p)$ is minimal (and therefore super-orthogonal), $1 \le p < \infty$, then $P_\perp(\infty) = \text{(subsequential)} \lim_{p\to\infty} P_\perp(p)$ is minimal (and super-orthogonal (Definition 1 (p = ∞))).

III. FORMULAS FOR MINIMAL PROJECTIONS

There have been very few known examples of P_{min} which are structurally nontrivial. The three examples which provide insight sufficient to lead to the establishment of the general variational equations (VE) are listed below.

Example 1 (Hölder). $n = 1$, $X \subset L^p$, $V = [v]$.

Example 2b (Cheney-Franchetti [9]). $n = 2$, $X = L^1$, $V = [1,t]$.

Example 3a (Chalmers-Metcalf [5]). $n = 3$, $X = L^\infty$, $V = [1,t,t^2]$.

Recently, a fourth example has been worked out by use of the (VE) and the "Lebesgue function = const." part of the NASC criteria for minimal L^1-projections established in [9]:

Example 3b (Chalmers [2]). $n = 3$, $X = L^1$, $V = [1,t,t^2]$.

Consequently a relatively simple general formula becomes apparent which at once solves the (VE) and provides the form for P_{min} in all known cases, $1 \le p \le \infty$, where $1 \in V$ if $n \ge 2$.

We introduce the following notation which will be made clearer by the examples.

DEFINITION 2. $V_\pm = V$ with $n-1$ possible "sign changes."

THEOREM 2. *(1) below solves the* (VE), $1 \le p \le \infty$, *and, in all known cases (assuming* $1 \in V$ *if* $n \ge 2$ *),* $\exists\, P_{min} = [\vec{g}]$ *where*

$$\vec{g} = \{(1+\delta)\frac{E}{\vec{v}_\pm \cdot \vec{v}}\}\vec{v}_\pm + \vec{c}_\pm . \tag{1}$$

Here $\vec{v}_\pm \in V_\pm^n$, $\vec{v} \in V^n$, $\vec{c}_\pm = \vec{c}_{pcw.}$,

$$E = e^{p\int^t \vec{v}_\pm \cdot \vec{v}\,'/\vec{v}_\pm \cdot \vec{v}},$$

$$\delta = (p-1)\int^t \vec{c}_\pm \cdot \vec{v}\,'/E ;$$

and $\exists$ *(independent)* $\vec{v}_{(2)} \in V^n$ *and* $\vec{v}_{(k)} \in V_{\pm}^n$ $(3 \le k \le n)$ *such that* $\vec{g} \cdot \vec{v}_{(k)} = c_{pcw.}$ $(2 \le k \le n)$.

Proof. The last phrase above states that the $n-1$ (o)-equations are satisfied. The (*)-equation also holds:

$$\vec{g}' = E\left\{\left[\frac{\delta'}{\vec{v}_{\pm}\cdot\vec{v}} + \frac{(1+\delta)p\vec{v}_{\pm}\cdot\vec{v}'}{(\vec{v}_{\pm}\cdot\vec{v})^2} - (1+\delta)\frac{(\vec{v}_{\pm}'\cdot\vec{v}+\vec{v}_{\pm}\cdot\vec{v}')}{(\vec{v}_{\pm}\cdot\vec{v})^2}\right]\vec{v}_{\pm} + \frac{(1+\delta)\vec{v}_{\pm}'}{\vec{v}_{\pm}\cdot\vec{v}}\right\} .$$

Hence $\vec{g}'\cdot\vec{v} = (p-1)\vec{c}_{\pm}\cdot\vec{v}' + (p-1)\dfrac{(1+\delta)E\vec{v}_{\pm}\cdot\vec{v}'}{\vec{v}_{\pm}\cdot\vec{v}} = (p-1)\vec{g}\cdot\vec{v}'$ and the (*)-equation follows by noting that $p-1 = \frac{p}{q}$. To see that (1) holds in all known cases see the examples below. □

Example 1. $n = 1$, $L^p \supset X \supset V = [v]$, $1 \le p \le \infty$. Then $v_{\pm} = v$, $c_{\pm} = 0$, whence $E = (v \cdot v)^{p/2}$ and

$$g = |v|^{p-2}v .$$

Example 2. $n = 2$, $1 \le p \le \infty$, $L^p[-1,1] = X \supset V = [1,t]$. Then $\vec{v}_{\pm} = \vec{v} = (1,\alpha t)$, $\vec{c}_{\pm} = (0,\pm c)$, " $\pm$ " denotes 1 sign change (at $t_1 = 0$); hence $E = |\vec{v}|^p$ and

$$\vec{g} = (1+\delta)|\vec{v}|^{p-2}\vec{v} + (0,\pm c)$$

with $\delta = \pm(p-1)c\alpha\int_0^t |\vec{v}|^{-p}$. Check that a (o)-equation holds with $\vec{v}_{(2)} = (-\alpha t,1)$. More particularly, we have in the following subexamples:

Example 2a (well known). $p = \infty$. $c = 0$, $\alpha > 0$, $\vec{g} = \lim_{p\to\infty}(1+\alpha^2t^2)^{\frac{p-2}{2}}(1,\alpha t)$ concentrates all its mass at $\{-1,1\}$ yielding the interpolating projection at $\{-1,1\}$.

Example 2b ([9]). $1 = p$. $\vec{g} = \dfrac{(1,\alpha t)}{\sqrt{1+\alpha^2t^2}} + (0,\pm c)$.

Example 3. $n = 3$, $1 \le p \le \infty$, $L^p[-1,1] = X \supset V = [1,t,t^2]$. Then, on

$[0,1]$ (extend to $[-1,0]$ by symmetry),

$$\vec{v}_\pm = (a_1 \pm b_1 t + c_1 t^2, \pm a_2 + b_2 t \pm c_2 t^2, a_3 \pm b_3 t + c_3 t^2) \in V_\pm^3 ,$$
$$\vec{v} = (\alpha_1 + \gamma_1 t^2, \beta_2 t, \alpha_3 + \gamma_3 t^2) \in V^3 ,$$
$$\vec{c}_\pm = (c_\pm^{(1)}, c_\pm^{(2)}, c_\pm^{(3)}) ,$$

" $\pm$ " denotes 2 sign changes (at $t_1 = -t_0$ and $t_2 = t_0$ for some $t_0 \in (0,1)$). Thus

$$\vec{g} = w\vec{v}_\pm + \vec{c}_\pm ,$$

where $w = (1+\delta)E/\vec{v}_\pm \cdot \vec{v}$ and $\vec{v}_\pm$ and $\vec{c}_\pm$ are connected by the existence of $\vec{v}_{(2)}, \vec{v}_{(3)}$ given in Theorem 2.

More particularly, we have the following subexamples:

Example 3a ([5]). $p = \infty$. On $[0,1]$, $\vec{v}$ is such that $\vec{v}_\pm \cdot \vec{v} = (b \pm t)^4$, and $\vec{c}_\pm = \vec{0}$. Now $w = w_1 + w_2$, where $w_1 = \dfrac{E}{\vec{v}_\pm \cdot \vec{v}}$ and $w_2 = \dfrac{\delta E}{\vec{v}_\pm \cdot \vec{v}}$ are respectively the discrete part and the continuous (L^1) part of the weight w . In fact δ and E cancel each other (i.e., $\delta E \to c_{pcw.}$), as $p \to \infty$, while E concentrates all its mass at $\{-1,0,1\}$ (as $p \to \infty$). Thus $w_2 = \dfrac{c_{pcw.}}{(b \pm t)^4}$ and w_1 is a measure supported on $\{-1,0,1\}$. Clearly $\vec{v}_\pm \cdot \vec{v} = (b \pm t)^4$ yields the first (o)-equation $(\vec{v}_{(2)} = \vec{v})$ and, writing $\vec{v}_\pm = (v_1, v_2, v_3)$, we have that a second (o)-equation is given by $\vec{v}_{(3)} = (0, -v_3, v_2)$.

Note 4. Since the (*)-equation in Example 3a is $g_2 = t(-\dfrac{2\gamma_1}{\beta_2} g_1 - \dfrac{2\gamma_3}{\beta_2} g_3)$, $\vec{v}_\pm$ can be rewritten as $\vec{v}_\pm = (d \pm t, t(d \pm t), t(\pm c_1 + t))$. Note that this form has two more parameters than the form given for this example in [6] where it was implied erroneously that $c_1 = d = b$.

Example 3b ([2]). $p = 1$. On $[0,1]$, $\vec{v}$ is such that $\vec{v}_\pm \cdot \vec{v} = (1 \pm \alpha t + \beta t^2)^2$ and $\vec{v}_\pm \cdot \vec{v}\,' = \vec{v}_\pm' \cdot \vec{v}$ whence $w = \dfrac{1}{1 \pm \alpha t + \beta t^2}$, and $\vec{c}_\pm = (c_1, \pm c_2, c_3)$. (Note that P_{min} is independent of $\vec{c}_\pm$ and hence without loss we may take $\vec{c}_\pm = \vec{0}$.) The first (o)-equation arises by choosing $\vec{v}_2 = (a+bt^2, \varepsilon t, c+dt^2)$ so that $\vec{v}_\pm \cdot \vec{v}_2 = 1 \pm \alpha t + \beta t^2$ and the second (o)-equation is analogous to that in Example 3a.

OTHER EXAMPLES. One can check that Theorem 2 holds in all other known cases. For example, if P_{min} is orthogonal, $1 \le p < \infty$ (see §2), then $\vec{v}_\pm$ has no sign changes (i.e., $\vec{v}_\pm \in V^n$) and $\vec{c}_\pm = \vec{0}$, i.e., the form for super-orthogonal projections obtains. In particular, if $p = 2$, then also $\vec{v}_\pm \cdot \vec{v}' = \vec{v}_\pm' \cdot \vec{v}$ (or just take $\vec{v}_\pm = \vec{v}$) and $\vec{g} = \vec{v}_\pm$.

IV. CONCLUSIONS

If $1 \notin V$ then formula (1) must be modified:

Example 4. $n = 2$, $L^1[-1,1] = X \supset V = [v_1, v_2]$, where v_1 is symmetric and nonvanishing and v_2 is antisymmetric. Then the (VE)-equations are

$$(*) \quad g_1' v_1 + \alpha g_2' v_2 = 0$$

$$(o) \quad g_1 v_2 - g_2 v_1 = c$$

yielding $\vec{g} = F|\vec{v}|^{-1}\vec{v} + (0, \pm\frac{c}{v_1})$, $F = 1 + \alpha\int_0^t (\frac{v_2}{v_1})(\frac{1}{v_1})'$.

Formula (1) determines P_{min} up to a finite number of parameters which may then be determined by applying part of known NASC if such are known. Such conditions are available for $p = 1, \infty$ since the extremals for P can be obtained (by use of the Lebesgue function). For example, the result of Cheney-Franchetti [9] that for $p = 1$ the Lebesgue function must be constant says that all δ-functions are extremal, and analogous statements are true (locally) for $p = \infty$. Thus the r constants remaining in formula (1) may be found, in the cases $p = 1, \infty$, by setting the Lebesgue function equal at $r+1$ (arbitrary) points and solving the resulting r (nonlinear) equations. For the specific values of the constants in Examples 2a, 3a-b, see the references.

One notes that a very general solution to the (VE) holds by observing that the $n-1$ (o)-equations are linear in $g_1,\ldots,g_n$ with coefficients in $V_\pm$. Thus one can write $\vec{g}_2 = U_{n-1}^{-1}(\vec{c}_{pcw.} - g_1\vec{v}_2)$ for some $(n-1)\times(n-1)$ matrix U_{n-1} with entries in $V_\pm$, where $\vec{g}_2 = (g_2,\ldots,g_n)$ and $\vec{v}_2 \in V_\pm^{n-1}$, and substitute in the (*)-equation to get a first order differential equation for g_1 with coefficients being rational functions of members of $V_\pm$. Formula (1) says in effect that the matrix U_{n-1} has (i.e., the (o)-equations have) a simple form.

SUMMARY. Formula (1) effectively brings the determination of a minimal projection to the point where known NASC can be successfully applied to find the values of the remaining (finitely many) parameters. Indeed, this has been the case for example in the situation of Examples 3a and 3b where the remaining constants have been readily determined by numerical procedures.

REFERENCES

1. Chalmers, B. L., "A natural simple projection with norm $\leq \sqrt{n}$," submitted for publication.

2. ______, "A minimal projection from $L^1[a,b]$ onto the quadratics," in preparation.

3. ______, "The (*)-equation for minimal projections and super-orthogonal projections," in preparation.

4. Chalmers, B. L. and F. T. Metcalf, "The variational equations for minimal projections," submitted for publication.

5. ______, "A minimal projection from $C[a,b]$ onto the quadratics," in preparation.

6. ______, "Multiplicative variations lead to the variational equations for minimal projections," to appear in: *Approximation Theory III*, E. W. Cheney, ed. (Proc. of Conf., Austin, TX, January, 1980).

7. Cheney, E. W. and K. H. Price, "Minimal projections," in: *Approximation Theory*, A. Talbot, ed., Academic Press, London, 1970, 261-289.

8. deBoor, C. and A. Pinkus, "Proof of the conjectures of Bernstein and Erdös concerning the optimal nodes for polynomial interpolation," *J. Approx. Theory 24* (1978), 289-303.

9. Franchetti, C. and E. W. Cheney, "The problem of minimal projections in L_1-spaces," in: *Approximation Theory II*, G. G. Lorentz, et al, ed., Academic Press, New York, 1976, 365-368. (Also CNA Report 106, Univ. of Texas, Austin, 1975.)

10. ______, *Orthogonal Projections in Spaces of Continuous Functions*, CNA Report 113, Univ. of Texas, Austin, 1976.

11. Kilgore, T. A., "Optimization of the norm of the Lagrange interpolation operator," *Bull. Amer. Soc. 83*(1977), 1069-1071.

12. ______, "A characterization of Lagrange interpolating projections with minimal Tchebycheff norm," *J. Approx. Theory 24*(1978), 273-288.

13. Kilgore, T. A. and E. W. Cheney, "A theorem on interpolation in Haar subspaces," *Aeguationes Math. 14*(1976), 391-400.

14. Lewis, D. R., "Finite dimensional subspaces of L^p," *Studia Math. 63*(1978), 207-212.

TWO-PART SPLITTING AND ADI-CONVERGENCE

John de Pillis

University of California, Riverside

I. INTRODUCTION

The solution vector x, to the linear system $Ax = b$, where x and b are n-vectors and A is a sparse $n \times n$ matrix, is often a discrete approximation to a solution of a differential equation. For example, consider the Laplacian operator, $\Delta u = 0$, where $u: \mathbb{R} \times \mathbb{R} \to \mathbb{R}$ is constrained in advance by its values on the boundary of some region D. We may impress a grid on D and then approximate u_{xx} and u_{yy} by difference equations, relative to the grid points. The approximate values of u at the n grid points, say, are then realized as the coordinates of an n-vector x, which is the solution to the linear equation $Ax = b$. By strategic ordering on the grid points, $n \times n$ matrix A can be made to have certain useful properties (e.g., positive semi-definiteness, narrow bandwidth).

Since sparity of A is often an intrinsic property, direct methods of solution, like Gaussian elimination, which tend to induce fill-in, are not always appropriate for machine solution. A more practical choice often is an iterative method which preserves sparseness. These include the so-called stationary first-degree Jacobi, Gauss-Seidel, Successive Over Relaxation (SOR) and the Alternating Direction Implicit (ADI) methods. For an exposition of these techniques, see Varga [5].

In this paper, we concentrate on these iterative sparse-preserving techniques. There follows a unified overview of first-degree methods

ISBN 0-12-481820-X

(Section 2), recent developments in second-degree methods (Section 3), second-degree methods as an acceleration of first-degree methods (Section 4), and second degree methods in tandem with ADI, particularly as applied to a generalized nonsymmetric Laplacian differential equation (Section 5).

II. FIRST-DEGREE STATIONARY METHODS: OVERVIEW

Consider the linear system $Ax = b$ where A is an $n \times n$ matrix with entries in some field and x and b are appropriate n-vectors. Since matrix A may not be easy to invert, we pull off an easy-to-invert term, A_0, and write

$$A = A_0 - A_1 , \tag{2.1a}$$

an additive splitting. This A_0 also induces a multiplicative splitting,

$$A = A_0(I-B) \quad \text{where}$$
$$B = I - A_0^{-1}A = A_0^{-1}A_1 . \tag{2.1b}$$

Matrix B is called the *iteration matrix*.

Each of these splittings has its use. The additive splitting (2.1a) defines a sequence $\{x_k\}$ for arbitrary fixed x_0 via

$$A_0 x_{k+1} - A_1 x_k = b , \tag{2.2a}$$

or

$$x_{k+1} = Bx_k + A_0^{-1}b, \qquad b=0,1,2,\ldots . \tag{2.2b}$$

Since $A = A_0 - A_1$, it is easy to see (from 2.2a, say) that if x_k converges at all, it must converge to the solution vector x, where $Ax = b$. Notice the additive splitting (2.1a) defines the sequence (2.2a), while the multiplicative splitting (2.1b)(which leads to an equivalent definition) also carries information on how fast the sequence $\{x_k\}$ converges. In fact, if $\rho(B)$ is the spectral radius of iteration matrix B, then the *asymptotic rate of convergence*, R_x, is defined by

$$R_x = -\log_{10}(\rho(B)) . \tag{2.3}$$

It turns out that $1/R_x$ represents, asymptotically, the number of iterations in (2.2a) or (2.2b) which suffice to produce one additional decimal place of accuracy in the x_k's. The splitting(s) of (2.2a) and (2.2b) are called *stationary* since there is no altering of parameters from iteration to iteration. Since each x_{k+1} depends only on one previous vector, x_k, the method is said to be *first-degree*, or a *one-part* splitting.

With the additive/multiplicative splitting idea above, we are poised to present an overview for stationary *second*-degree methods, or two-part splittings.

III. SECOND-DEGREE STATIONARY METHODS: OVERVIEW

Throughout, we consider $A = A_0 - A_1$ as before, where A_0 is the *same* easy-to-invert matrix component of $n \times n$ matrix A, and $Ax = b$. We now extend the one-part splittings (2.1a) and (2.1b). The new additive splitting is

$$A = A_0 - C_1 - C_2 , \tag{3.1a}$$

and the new multiplicative splitting is

$$A = A_0(I - B_1)(I - B_2) . \tag{3.1b}$$

The connection between (3.1a) and (3.1b) is given by

$$\begin{aligned} C_1 &= A_0(B_1 + B_2) , \\ C_2 &= -A_0(B_1 B_2) . \end{aligned} \tag{3.1c}$$

These ideas were first introduced in [2].

Analogously to the development in Section 2, we use the additive splitting (3.1a) for arbitrary initial y_0, y_1 to define the sequence $\{y_k\}$ by

$$A_0 y_{k+2} - C_1 y_{k+1} - C_2 y_k = b , \tag{3.2a}$$

or

$$y_{k+2} = (B_1 + B_2) y_{k+1} - (B_1 B_2) y_k + A_0^{-1} b \quad \text{for all } k = 0,1,2,\ldots . \tag{3.2b}$$

Again, since $A = A_0 - C_1 - C_2$, we see from (3.2a) that if $\{y_k\}$ converges at all, it necessarily converges to solution vector x, where $Ax = b$.

As has been recently shown [2],[3], the multiplicative splitting (3.1b) contains the information for computing asymptotic convergence rate, R_y, for $\{y_k\}$. In fact,

$$R_y = \min\{-\log(\rho(B_1)), -\log(\rho(B_2))\}. \tag{3.3}$$

R_y carries the interpretation: $1/R_y$ is asymptotically the number of iterations of (3.2b) which suffice to produce one more decimal place of accuracy in the y_k's.

Notice that "fast" convergence of one-part scheme (2.2a) and two-part scheme (3.2a) results whenever the spectral radius of the iteration matrices B, B_1, B_2 are "small." This fact carries an interesting interpretation; namely, that in devising our splittings, (2.1b) and (3.1b), we must create matrices B, B_1, B_2 whose spectra are embraced or captured by the smallest possible *circle*. In [4] is developed the idea of writing $A = A_0(I-B)$ and capturing B with the smallest possible *ellipse* symmetric with respect to the origin. From this, we construct a two-part scheme which flows from the one-part splitting. This connection is developed in the following section.

IV. SECOND-DEGREE METHODS AS ACCELERATION OF FIRST-DEGREE METHODS

In (3.1a), (3.1b) and (3.1c), we have already given the connection between one-part and two-part methods. In fact, the same invertible A_0 appears in scheme (2.1a) and (2.1b). The interlinking is further analyzed in [4] in the following sense: Choose a one-part splitting $A = A_0 - A_1 = A_0(I-B)$. This gives rise to sequence

$$x_{k+1} = Bx_k + A_0^{-1}b \tag{4.1}$$

with convergence rate $R_x = -\log(\rho(B))$. (See (2.2), (2.3).) Then, with iteration matrix B (and A_0) in hand, choose invertible U with the property that for some scalars θ and λ,

$$B = \theta(U+\lambda U^{-1}) .$$

(Such a matrix U can always be found [4].) Now in (3.1b), set

$$B_1 = \mu U, \qquad B_2 = \mu\lambda U^{-1} .$$

Substituting the above into (3.1b) yields the two-part stationary sequence.

$$y_{k+2} = (1+\lambda\mu^2)By_{k+1} - \lambda\mu^2 y_k + (1+\lambda\mu^2)A_0^{-1}b \qquad k=0,1,2,\ldots . \tag{4.2}$$

As it turns out the convergence rate $R_y = -\log(\mu)$ [2]. There are several features to note in the creation of the two-part sequence:

- No new operators are needed in (4.2) *viz-a-viz* (4.1).
- Method (4.2) does require one extra vector in storage. That is, each vector y_{k+2} depends on the previous two vectors.

Features of (4.2) which are not obvious are,

- The optimal scalars μ and λ which are to be chosen, easily derive from the configuration of $\sigma(B)$, the spectrum of B.
- As long as $\sigma(B)$ lies inside the infinite vertical strip $Z = \{z: -1 < \mathrm{Re}(z) < 1\}$, convergence of (4.2) is assured. In contrast, $\sigma(B)$ must lie within the (more restrictive) unit circle in order for (4.1) to converge. Moreover, as long as $\sigma(B)$ is captured by an ellipse, symmetric about the origin, then convergence of (4.2) is always faster than that of (4.1) (i.e., $R_y > R_x$) whenever the embracing ellipse is not a circle.

We now present the straight-forward algorithm which reveals optimal scalars μ,λ, once the operator $A = A_0(I-B)$ is known and once (estimates of) the spectrum of B is known.

THEOREM ([4]).

$$\left(\begin{array}{c}\textit{ALGORITHM for}\\ \textit{DERIVING (4.2) from (4.1)}\end{array}\right)$$

A1. Capture $\sigma(B)$ *in an ellipse which is symmetric about the origin. Let* M_r *and* M_i *be the respective lengths of the real and imaginary semi-axes.*

A2. Define $\lambda = (M_r - M_i)/(M_r + M_i)$.

A3. Define μ *to be the unique root in the unit interval of the quadratic* $(M_r + M_i)(1+\lambda\mu^2) = 2\mu$.

Then $\{y_k\}$ *of (4.2) converges with asymptotic convergence rate* $R_y = -\log(\mu)$ *, whereas* $\{x_k\}$ *of (4.1) converges with rate* $R_x = -\log(\rho(B))$. *As we have noted,* $R_y > 0$ *whenever* $\sigma(B) \in Z$ *and* $R_y > R_x$ *whenever* $M_i \neq M_r$.

Since we are now able to capture the spectrum of B with an ellipse (so that (4.2) improves upon (4.1)), are there any realistic situations where this technique may be applied? The next section illustrates this in conjunction with the ADI method for a generalized Laplacian.

V. SECOND-DEGREE METHODS AND ADI

First, a brief word about the ADI technique (cf. [5],[6]). The standard setting usually considers a linear system $Ax = b$ where $n\times n$ matrix A decomposes in some natural way as $A = (H+V+E)$; positive definiteness and commutivity is often hypothesized. But a generic way of describing the setting for ADI might be:

Given $Ax = b$ for $n\times n$ matrix A . Design (by some means) two $n\times n$ invertible matrices Δ_1 and Δ_2 . Write

$$\begin{aligned} \Delta_1 z_{k+\frac{1}{2}} &= (-A+\Delta_1)z_k + b \\ \Delta_2 z_{k+1} &= (-A+\Delta_2)z_{k+\frac{1}{2}} + b \ . \end{aligned} \tag{5.1a}$$

These equations are clearly equivalent to the "usual" first-order system (substitute for $z_{k+\frac{1}{2}}$)

$$z_{k+1} = \Delta_2^{-1}(A+\Delta_2)\Delta_1^{-1}(A+\Delta_1)z_k + \tilde{b} \ . \tag{5.1b}$$

The iteration matrix

$$B = \Delta_2^{-1}(A+\Delta_2)\Delta_1^{-1}(A+\Delta_1) \tag{5.1c}$$

of (5.1b) is called the Peaceman-Rachford iteration matrix.

To particularize, consider the generalized Laplacian over the unit square. Then

$$-\Delta u + au_x + bu = 0$$

where $u(1,y) = 0$ for all $y \in [0,1]$. By setting a uniform grid over the unit square, and by ordering the grid points lexicographically ($(x_1,y_1) < (x_2,y_2)$ if $x_1 < x_2$ or $x_1 = x_2$ and $y_1 < y_2$) finite difference approximations for the derivatives produce the linear system $Ax = b$ where $A = H+V$, where H is "almost" symmetric and V is a skew-symmetric plus a positive diagonal. In any case, we may follow the ADI format (5.1a), to produce

$$\begin{aligned}(H+\rho I)u_{k+\frac{1}{2}} &= (\rho I-V)u_k + c \\ (\rho' I+V)u_{k+1} &= (\rho' I-H)u_{k+\frac{1}{2}} + C\ ,\end{aligned} \qquad (5.2a)$$

for scalars ρ and ρ', which leads to an equivalent first-order system (5.1b) with Peaceman-Rachford iteration matrix

$$T_{\rho,\rho'} = (\rho' I+V)^{-1}(\rho' I-H)(H+\rho I)^{-1}(\rho I-V)\ . \qquad (5.2\ )$$

Success (or rapid convergence) of the ADI method (5.2a) depends on a small spectral radius for $T_{\rho,\rho'}$ of (5.2). As Ray Chin [1] has shown, scalars ρ and ρ' may be chosen to produce spectra of $T_{\rho,\rho'}$ which live in the infinite vertical strip $Z = \{z: -1 < Re(z) < 1\}$ and which are captured by symmetric ellipses with high eccentricity. In this case, we can accelerate (5.2a) by using this fortuitous Peaceman-Rachford iteration matrix $T_{\rho,\rho'}$ of (5.2) in conjunction with (4.2). (In (4.2), use our $T_{\rho,\rho'}$ for B.) The result is a second-degree system (4.2) whose convergence rate exceeds that of the ADI system (5.2a). Details of specific quantities will appear in a future paper.

REFERENCES

1. Chin, R., *Spectra of Certain ADI Systems and A Generalized Laplacian*, Lawrence Livermore Lab technical report, 1980.

2. de Pillis, J. and M. Neumann, "Iterative methods with k-part split-ings ings," *IMA J. Numer. Anal.* (to appear).

3. ______, "A noncommutative spectral theorem for operator entried companion matrices," *Linear and Multilinear Alg.* (to appear).

4. de Pillis, J., "How to embrace your spectrum for faster iterative results," *Linear and Multilinear Alg.* (to appear).

5. Varga, R., *Matrix Iterative Analysis*, Prentice-Hall, Englewood Cliffs, New Jersey, 1962.

6. Wachspress, E., *Iterative Solution of Elliptic Systems and Applications to the Neutron Diffusion Equations of Reactor Physics*, Prentice-Hall, Englewood Cliffs, New Jersey, 1966.

UNIFORM σ-ADDITIVITY AND UNIFORM CONVERGENCE OF CONDITIONAL EXPECTATIONS IN THE SPACE OF BOCHNER OR PETTIS INTEGRABLE FUNCTIONS

Nicolae Dinculeanu

University of Florida, Gainesville

I. INTRODUCTION

Let (X,Σ,μ) be a measure space, E a Banach space, L^1_E the space of Bochner μ-integrable functions $f:X\to E$ with norm $\|f\|_1 = \int|f|d\mu$, and $\mathcal{L}^1_E$ the space of strongly measurable, Pettis μ-integrable functions $f:X\to E$ with norm

$$(f)_1 = \sup\{\int|\langle f,x^*\rangle|d\mu;x^* \in E^*_1\} .$$

Here we denoted by E^* the conjugate space of E and by E^*_1 the unit ball of E^*. (Similarly, E_1 denotes the unit ball of E.)

In §2 we characterize uniform σ-additivity of the indefinite integrals $\{\int_{(\cdot)}|f|d\mu;f\in K\}$ for sets $K\subset L^1_E$, by means of convergence of conditional expectations $E_\pi f\to f$, for the topology $\sigma(L^1_E,L^\infty_{E^*})$, *uniformly* for $f\in K$, where π are finite partitions of X. In §3 we give a similar characterization of uniform σ-additivity of the indefinite integrals $\{\int_{(\cdot)} fd\mu;f\in K\}$ for sets $K\subset\mathcal{L}^1_E$.

II. UNIFORM σ-ADDITIVITY IN THE SPACE L^1_E

For each function f from L^1_E (or from $\mathcal{L}^1_E$) we denote by $f\mu$ the measure with density f, defined on Σ by $(f\mu)(A) = \int_A fd\mu$, for $A\in\Sigma$.

ISBN 0-12-481820-X

If $K \subset L_E^1$ (or $K \subset \mathcal{L}_E^1$) we denote

$$K\mu = \{f\mu ; f \in K\} = \{\int_{(\cdot)} f d\mu ; f \in K\}$$

and

$$|K|\mu = \{|f|\mu ; f \in K\} = \{\int_{(\cdot)} |f| d\mu ; f \in K\} .$$

We remark that $K\mu$ (or $|K|\mu$) is uniformly σ-additive, if and only if, for every countable subset $K_0 \subset K$, the set $K_0\mu$ (or $|K_0|\mu$) is uniformly σ-additive. Further, for any function $g \in L^\infty_{E^*}$ we write $\langle K,g \rangle = \{\langle f,g\rangle ; f \in K\}$, where $(\langle f,g\rangle)(x) = \langle f(x),g(x)\rangle$ for $x \in X$. With these notations we can state the following lemma.

LEMMA 1. *(a) Let* $K \subset L_E^1$. *Then the set* $|K|\mu = \{\int_{(\cdot)} |f| d\mu ; f \in K\}$ *is uniformly σ-additive, if and only if, for every function* $g \in L^\infty_{E^*}$, *the set* $\langle K,g\rangle\mu = \{\int_{(\cdot)} \langle f,g\rangle d\mu ; f \in K\}$ *is uniformly σ-additive.*

(b) Let S *be a set and* $f_n, f : S \to L_E^1$, $n=1,2,\ldots$ *functions such that* $f_n(s) \to f(s)$ *for the topology* $\sigma(X_E^1, L^\infty_{E^*})$, *uniformly for* $s \in S$. *If for each* n , *the set* $f_n(S)$ *is bounded in* L_E^1 *and the set* $|f_n(S)|\mu$ *is uniformly σ-additive, then* $f(S)$ *is bounded in* L_E^1 *and* $|f(S)|\mu$ *is uniformly σ-additive.*

Proof. (a) The first implication follows from the inequalities $\int_A |\langle f,g\rangle| d\mu \le \|g\|_\infty \int_A |f| d\mu$ for $A \in \Sigma$, $f \in K$ and $g \in L^\infty_{E^*}$.

Conversely, deny that $|K|\mu$ is uniformly σ-additive: there is $\varepsilon > 0$, a sequence (f_n) from K , and a sequence (A_n) of disjoint sets from Σ , such that

$$\int_{A_n} |f_n| d\mu > \varepsilon \text{ , for each } n .$$

For each n , there is a Σ-step function $g_n : X \to E^*$ with $|g_n| \le 1$, such that ([4], Thm. 5, p. 288):

$$\int_{A_n} \langle f_n, g_n \rangle d\mu > \varepsilon \ .$$

If we set $g = \sum_{1 \le n < \infty} \varphi_{A_n} g_n$, then $|g| \le 1$, hence $g \in L^{\infty}_{E^*}$, and

$$\int_{A_n} \langle f_n, g \rangle d\mu = \int_{A_n} \langle f_n, g_n \rangle d\mu > \varepsilon \ , \text{ for each } n \ ,$$

hence $\langle K, g \rangle \mu$ is not uniformly σ-additive.

(b) To prove (b), let $g \in L^{\infty}_{E^*}$. Then, for each n , the set $\langle f_n(S), g \rangle$ is bounded in L^1 and the set $\langle f_n(S), g \rangle \mu$ is uniformly σ-additive; hence the set $\langle f_n(S), g \rangle$ is relatively weakly compact in L^1 (see [1]). Also, $\langle f_n(s), g \rangle \to \langle f(s), g \rangle$ in the weak topology of L^1 , uniformly for $s \in S$. By Lemma 6 in [1], the limit set $\langle f(S), g \rangle$ is relatively weakly compact in L^1 ; hence the set $\langle f(S), g \rangle$ is bounded in L^1 and the set $\langle f(S), g \rangle \mu$ is uniformly σ-additive. Then the set $f(S)$ is bounded in L^1_E and, by the first part of the lemma, the set $|f(S)|\mu$ is uniformly σ-additive.

We shall apply the above lemma in case $S = K \subset L^1_E$, f is the identity mapping of K , and f_n are conditional expectations E_{π_n} determined by finite partitions π_n . If X is a locally compact abelian group with the Haar measure, other particularizations of f_n can be used; for example, convolution with an approximate unit, and translations. In these particular cases of the functions f_n , we are able to give also a partial converse of part (b) of Lemma 1.

THEOREM 2. *Let* $K \subset L^1_E$ *be a bounded set.*

(i) If for each countable subset $K_0 \subset K$ *there is a sequence* (π_n) *of finite partitions such that* $E_{\pi_n} f \to f$ *in* L^1_E *for the topology* $\sigma(L^1_E, L^{\infty}_{E^*})$ *, uniformly for* $f \in K_0$ *, then the set*

$$|K|\mu = \{ \int_{(\cdot)} |f| d\mu ; f \in K \}$$

is uniformly σ-additive.

Conversely:

(ii) If $|K|\mu$ *is uniformly σ-additive, then* $E_\pi f \to f$ *in* L^1_E *for the topology* $\sigma(L^1_E, L^\infty_{E^*})$ *, uniformly for* $f \in K$ *, the limit being taken along the net* (π) *of all finite partitions.*

(iii) If $|K|\mu$ *is uniformly σ-additive and if every countable set* $K_0 \subset K$ *contains a countable subset* $K' \subset K_0$ *such that*

$$\sup\{|f(t)|; f \in K'\} < \infty, \quad \mu\text{-a.e.}\ ,$$

then for each countable subset $K_0 \subset K$ *there is a sequence* (π_n) *of finite partitions such that* $E_\pi f \to f$ *in* L^1_E *for the topology* $\sigma(L^1_E, L^\infty_{E^*})$ *, uniformly for* $f \in K_0$.

Proof. To prove (i), let $g \in L^\infty_{E^*}$ and $\pi = (E_1, \ldots, E_n)$ a finite partition (consisting of disjoint sets $E_i \in \Sigma$ with $0 < \mu(E_i) < \infty$; but the union of the sets E_i may be different from X). For every E_i , we have $g\varphi_{E_i} \in L^1_{E^*}$ and

$$\mu(E_i)^{-1} \int_{(\cdot)} |g|\varphi_{E_i} d\mu$$

is σ-additive. Let $M = \sup\{\|f\|_1; f \in K\} < \infty$ and $A \in \Sigma$. Then, for every $f \in K$ we have

$$\begin{aligned}
|\int_A \langle E_\pi f, g\rangle d\mu| &= |\int_A \sum_{1\le i\le n} \mu(E_i)^{-1} \langle \int_{E_i} f d\mu, g\varphi_{E_i}\rangle d\mu| \\
&= |\sum_{1\le i\le n} \mu(E_i)^{-1} \langle \int_{E_i} f d\mu, \int_A g\varphi_{E_i} d\mu\rangle| \\
&\le \sum_{1\le i\le n} \mu(E_i)^{-1} \int_{E_i} |f| d\mu \int_A |g|\varphi_{E_i} d\mu \\
&\le M \sum_{1\le i\le n} \mu(E_i)^{-1} \int_A |g|\varphi_{E_i} d\mu \ ,
\end{aligned}$$

therefore, the set

$$\langle E_\pi K, g\rangle\mu = \{\int_{(\cdot)} \langle E_\pi f, g\rangle\mu ; f\in K\}$$

is uniformly σ-additive. From Lemma 1 it follows that the set

$$|E_\pi K|\mu = \{\int_{(\cdot)} |E_\pi f|\, d\mu ; f\in K\}$$

is also uniformly σ-additive. Since $\|E_\pi f\|_1 \le \|f\|_1$, the set $E_\pi K$ is bounded in L^1_E .

Now let $K_0 \subset K$ be countable and let (π_n) be a sequence of finite partitions such that $E_{\pi_n} f \to f$ in L^1_E for $\sigma(L^1_E, L^\infty_{E^*})$, uniformly for $f \in K_0$. Since for each n , the set $E_{\pi_n} K$ is bounded in L^1_E and the set $|E_{\pi_n} K|\mu$ is uniformly σ-additive, from Lemma 1(b) we deduce that $|K_0|\mu$ is uniformly σ-additive. Then the set $|K|\mu$ is also uniformly σ-additive.

Part (ii) has been proved in [1], step "K" of the proof of Theorem 1. For part (iii) see the proof of Theorem 1 in [2].

Remark. If in (i) we impose, in addition, that for every set $A \in \Sigma$, the set $K(A) = \{\int_A f d\mu ; f\in K\}$ is conditionally weakly compact in E , then we deduce that K is conditionally compact in L^1_E for the topology $\sigma(L^1_E, L^\infty_{E^*})$.

III. UNIFORM σ-ADDITIVITY IN THE SPACE L^1_E

A set $F \subset E^*_1$ is said to be *norming* for a set of functions $K \subset L^1_E$ if for every $f \in K$ we have

$$|f(t)| = \sup\{|\langle f(t), x^*\rangle| ; x^* \in F\}, \qquad \mu\text{-a.e.}$$

We remark first that if $K \subset L^1_E$ and if $F \subset E^*_1$ is norming for K , then the set $K\mu = \{\int_{(\cdot)} f d\mu ; f\in K\}$ is uniformly σ-additive, if and only if the set

$$\langle K, F\rangle\mu = \{\int_{(\cdot)} \langle f, x^*\rangle d\mu ; f\in K, x^*\in F\}$$

is uniformly σ-additive. Using this remark, we can prove the following analog of Lemma 1(b) for sets from L^1_E.

LEMMA 3. *Let* S *be a set,* $f_n, f: S \to L^1_E$, $n=1,2,\ldots$ *functions and* $F \subset E^*_1$ *a norming set for the sequence* $\{f, f_1, f_2, \ldots\}$. *Assume that*

$$\langle f_n(s), x^*\rangle \to \langle f(s), x^*\rangle \quad \textit{weakly in } L^1,$$

uniformly for $s \in S$ *and* $x^* \in F$.

If for each n, *the set* $f_n(S)$ *is bounded in* L^1_E *and the set* $f_n(S)\mu$ *is uniformly* σ*-additive, then the set* $f(S)$ *is bounded in* L^1_E *and the set* $f(S)\mu$ *is uniformly* σ*-additive.*

For the proof, we apply Lemma 1(b) to the functions φ_n and φ defined on $S\times F$ by $\varphi_n(s,x^*) = \langle f_n(s), x^*\rangle$ and $\varphi(s,x^*) = \langle f(s), s^*\rangle$.

Remark. The assumption in the statement of Lemma 3 is satisfied, for example, if $f_n(s) \to f(s)$ *weakly* in L^1_E, uniformly for $s \in S$.

We state now an analog of Theorem 2, for sets from L^1_E.

THEOREM 4. *Let* $K \subset L^1_E$ *be a bounded set.*

(i) If for each countable set $K_0 \subset K$, *there is a countable set* $F \subset E^*_1$, *norming for* K_0, *and a sequence* (π_n) *of finite partitions such that* $\langle E_{\pi_n} f, x^*\rangle \to \langle f, x^*\rangle$, *weakly in* L^1, *uniformly for* $f \in K_0$ *and* $x^* \in F$, *then the set* $K\mu = \{\int_{(\cdot)} f d\mu; f \in K\}$ *is uniformly* σ*-additive.*

Conversely:

(ii) If $K\mu$ *is uniformly* σ*-additive, then* $\langle E_\pi f, x^*\rangle \to \langle f, x^*\rangle$ *weakly in* L^1, *uniformly for* $f \in K$ *and* $x^* \in E^*_1$, *where* (π) *is the net of all finite partitions.*

(iii) If $K\mu$ *is uniformly* σ*-additive and if every countable set* $K_0 \subset K$ *contains a countable subset* $K' \subset K_0$ *such that*

$$\sup\{|f(t)|; f \in K'\} < \infty, \quad \mu\text{-a.e.},$$

then for every countable set $K_0 \subset K$ *, there is a sequence* (π_n) *of finite partitions and a countable set* $F \subset E_1^*$ *, norming for* K_0 *, such that* $\langle E_{\pi_n} f, x^* \rangle \to \langle f, x^* \rangle$ *weakly in* L^1 *, uniformly for* $f \in K_0$ *and* $x^* \in F$ *.*

Proof. (i) We apply Theorem 2 and deduce that $\langle K,F\rangle\mu$ is uniformly σ-additive; then $K\mu$ is also uniformly σ-additive.

(ii) If $K\mu$ is uniformly σ-additive, then $\langle K,E_1^*\rangle\mu$ is uniformly σ-additive and we apply Theorem 2(ii) to the set $\langle K,E_1^*\rangle \subset L^1$.

(iii) Let $K_0 \subset K$ be countable. We can assume that the functions of K_0 have their values in a separable subspace $E_0 \subset E$. Let $F \subset E_1^*$ be a countable set, norming for K_0 . The set $\langle K_0,F\rangle = \{\langle f,x^*\rangle; f \in K_0, x^* \in F\}$ is countable and bounded in L^1 , and the set

$$\langle K_0,F\rangle\mu = \{ \int_{(\cdot)} \langle f,x^*\rangle \mu ; f \in K_0, x^* \in F\}$$

is uniformly σ-additive. We then apply Theorem 2(iii).

REFERENCES

1. Brooks, J. K. and N. Dinculeanu, "Weak compactness in spaces of Bochner-integrable functions and applications," *Advances in Math.* *24*(1977), 172-188.

2. ______, "On weak compactness in spaces of Bochner-integrable functions," *Advances in Math.* (to appear).

3. ______, "Weak and strong compactness in the space of Pettis-integrable functions," in: *Contemporary Mathematics, Vol. 2,* Amer. Math. Soc., 1980 (to appear).

4. Dinculeanu, N., *Vector Measures*, Pergamon Press, Oxford, 1967.

5. Lewis, D. R., "Conditional weak compactness in certain inductive tensor products," *Math. Ann.* *201*(1973), 201-209.

JONES'S LEMMA AND INACCESSIBLE CARDINALS

Eric K. van Douwen[1)]

Institute for Medicine and Mathematics,
Ohio University, Athens, Ohio

The following lemma is essentially due to Jones, [J], and now bears his name.

JONES'S LEMMA. *If* X *is normal, then* $2^{|D|} \leqq 2^{d(X)}$ *for each closed discrete* $D \subseteq X$.

We here consider the question of whether Jones's Lemma can be strengthened in a natural way. We need some definitions.

Let κ, λ, μ denote cardinals. Define cardinal functions d , density, e , extent, and $\hat{e}$ (in Juhász' notation, [Ju]) by

$$d(X) = \min\{|A| : A \subseteq X \text{ is dense}\}$$

$$e(X) = \sup\{|D| : D \subseteq X \text{ is closed discrete}\}$$

$$\hat{e}(X) = \min\{\kappa : |D| < \kappa \text{ for every closed discrete } D \subseteq X\} .$$

Also, define the weak power $2^{<\kappa}$ of κ by

$$2^{<\kappa} = \sup_{\lambda<\kappa} 2^{\lambda} .$$

Clearly Jones's Lemma is equivalent to the statement that

if X is normal then $2^{<\hat{e}(X)} \leqq 2^{d(X)}$.

This reduces Jones's Lemma to one single inequality (rather than $|\{\kappa : \kappa < \hat{e}(X)\}|$ many). Admittedly, this reduction is nothing but an unsatisfactory linguistic trick. However, it suggests the following

1) Research supported by NSF Grant MCS 78-09484

ISBN 0-12-481820-X

question:

If X is normal, then is $2^{e(X)} \leqq 2^{d(X)}$?

While the answer to this question is rather easy, we find it interesting since it involves inaccessible cardinals: Recall that a cardinal is called *inaccessible* if it is an uncountable regular limit cardinal, and that if I is the statement

I: there is an inaccessible cardinal

then $\mathrm{CON(ZFC)} \Longleftrightarrow \mathrm{CON(ZFC} + \neg I)$, and $\mathrm{CON(ZFC} + I)$ is properly stronger than $\mathrm{CON(ZFC)}$ (hence $\mathrm{CON(T)}$ means that T is consistent); however, most set theorists believe that ZFC does not imply $\neg I$.

We now answer our question as follows.

THEOREM. *Consider the following statements*

A. *There is a normal* X *with* $2^{e(X)} > 2^{d(X)}$.

B. *There is an infinite cardinal* δ *and a limit cardinal* ε *such that* $2^{\varepsilon} > 2^{\delta}$ *but* $2^{\kappa} = 2^{\delta}$ *if* $\delta \leqq \kappa < \varepsilon$.

Then

1° $A \Longleftrightarrow B$;

2° *Any* ε *like in* B *is inaccessible; and*

3° $\mathrm{CON(ZFC} + A) \Longleftrightarrow \mathrm{CON(ZFC} + I)$

Proof. 1° $A \Rightarrow B$. Let $\varepsilon = e(X)$ and $\delta = d(X)$. If $\delta < \omega$ then $\delta = \varepsilon = |X|$, hence $\varepsilon > \delta \geqq \omega$. By Jones's Lemma no closed discrete set of X has cardinality ε , hence ε is a limit. If $\delta \leqq \kappa < \varepsilon$, then there is closed discrete $D \subseteq X$ with $\kappa \leqq |D| < \varepsilon$, hence with $|D| = \kappa$. Then clearly $2^{\kappa} \geqq 2^{\delta}$, while $2^{\kappa} \leqq 2^{\delta}$ by Jones's Lemma. □

$B \Rightarrow A$. For any space Y , and $A \subseteq Y$, we say that the space Z is obtained from Y by making A discrete if Z has the same underlying set as Y and has

$$\{U \cup B: U \text{ open in } Y, B \subseteq A\}$$

as topology.

Let $[0,\varepsilon]$, the set of ordinals $\leq \varepsilon$, carry the usual order topology. Let E be the space obtained from $[0,\varepsilon]$ by making $[0,\varepsilon)$ discrete. Clearly, $e(E) = \varepsilon$, since ε is a limit. Let Π denote the product of 2^δ copies of $\{0,1\}$ in the discrete topology. Recall that $d(\Pi) \leq \delta$, [E, 2.3.15].

We claim that βE can be embedded into Π. To this end we have to prove that βE is zero-dimensional and has weight at most 2^δ, [E, 6.2.16]. βE is zero-dimensional since ε has a neighborhood basis of clopen neighborhoods in E, hence in βE, whose complements are discrete. Since βE is infinite, it follows that $w(\beta E)$ is the number of clopen sets of E. This number clearly equals $2 \cdot \Sigma_{\lambda<\varepsilon} 2^\lambda$. As $\Sigma_{\lambda<\kappa} 2^\lambda = 2^{<\kappa}$ for all infinite κ it follows that $w(\beta E) = 2^{<\varepsilon} = 2^\delta$, as required.

Assume $\beta E \subseteq \Pi$. Clearly E is nowhere dense since it has a dense set of isolated points. Since $d(\Pi) \leq \delta$, we can choose a dense $A \subseteq \Pi$ with $|A| = \delta$ that misses βE. Let X be the space obtained from the subspace $E \cup A$ of Π by making A discrete. Clearly $d(X) = \delta$, and $e(X) = e(E) = \varepsilon$ (since $\delta < \varepsilon$). To see that X is normal consider disjoint closed $F, G \subseteq X$. Since the points of A are isolated we may assume without loss of generality that $F \cup G \subseteq E$. Since E is normal, F and G have disjoint closures in βE, hence in Π. Therefore F and G have disjoint neighborhoods in Π, hence in the subspace $E \cup A$ of Π, hence in X. □

2° It suffices to recall that Bukovský, [Bu], and Hechler, [He], have proved that

(*) if λ is singular, and if there is $\mu < \lambda$ such that $2^\kappa = 2^\mu$ for $\mu \leq \kappa < \lambda$, then $2^\lambda = 2^\mu$.

For completeness sake we include the simple proof: There is a decomposition $\mathcal{D}$ of $[0,\lambda)$ with $\mu \leq |\mathcal{D}| < \lambda$ such that $\mu \leq |D| < \lambda$ for $D \in \mathcal{D}$. Then

$$2^\lambda = |P([0,\lambda)]| = \Pi_{D \in \mathcal{D}}|P(D)| = (2^\mu)^\mu = 2^\mu .$$

3° Since $A \Longleftrightarrow B$ by 1° and $B \Rightarrow I$ by 2°, it suffices to prove CON(ZFC + I) $\Rightarrow$ CON(ZFC + B). The usual proof of CON(ZFC + I) $\Rightarrow$ CON(ZFC + 2^ω is inaccessible) establishes just this: One first notes that CON(ZFC + I) $\Rightarrow$ CON(ZFC + I + GCH), so we can start with a model M of ZFC + GCH which has an inaccessible cardinal ε. Then $\varepsilon^\lambda = \varepsilon$ for $\lambda < \varepsilon$ because of GCH, hence if N is the model obtained from M by adding ε Cohen reals one has $2^\lambda = \varepsilon$ for $\omega \leq \lambda < \varepsilon$, and ε is still weakly inaccessible, [K, VII, 5.17 and G.1], but of course $2^\varepsilon > \varepsilon$.

REMARKS. 1° Suppose we modify the proof of $B \Rightarrow A$ by letting E be discrete, with $|E| = \varepsilon$. Then βE embeds in Π if $2^\varepsilon = 2^\delta$. If $\varepsilon = \delta = 2^\omega$ then the space one gets from Π by making Π_E discrete is Bing's example G, [B], in a disguise brought to my attention by Ryszard Engelking in personal communication, while if $\varepsilon = \omega_1$ and $\delta = \omega$ then the space one gets from $A \cup E$ by making A discrete is Heath's example of [H].

2° In our proof of 3° we have $\varepsilon = 2^\delta$. One can also have $2^\varepsilon > 2^\delta > \varepsilon$. [Of course $2^\delta < \varepsilon$ is impossible.]

REFERENCES

[B] Bing, R. H., "Metrization of topological spaces," *Canad.J. Math.* 3 (1951) 175-186.

[Bu] Bukovský, L., "The continuum problem and the powers of alephs," *Comm. Math. Univ. Car.* 6 (1965) 181-197.

[E] Engelking, R.H., *General Topology*, Polish Scientific Publishers, 1977.

[H] Heath, R. W., "Separability and $\aleph_1$-compactness," *Coll. Math* 12 (1964) 11-14.

[He] Hechler, S. H., "Powers of singular cardinals and strong form of the negation of the generalized continuum hypothesis," *Zeitschr. f. Math. Logik und Grundlagen der Math. 19*(1973), 83-84.

[J] Jones, F. B., "Concerning normal and completely normal spaces," *Bull. AMS 43*(1937), 671-677.

[Ju] Juhász, I., *Cardinal functions, 20 years later*, Mathematisch Centrum, Amsterdam.

[K] Kunen, K., *Set theory*, North Holland.

BOREL-ADDITIVE FAMILIES AND BOREL MAPS IN METRIC SPACES

R. W. Hansell

University of Connecticut, Storrs

0. PRELIMINARIES

This talk is intended to survey several problems in the descriptive theory of sets in nonseparable metric spaces (some dating back to 1935), and to announce some results I have recently obtained regarding them. For background, we refer the reader to the textbook of Kuratowski [K_1], and the papers [K_2], [S_1], and [H_1]. However, we recall here the standard notation and terminology used in describing the various classes of Borel sets and Borel mappings. The G_0 sets of a space X are simply the open sets, and for $0 < \alpha < \omega_1$, a G_α set is a countable union (if α is even) or countable intersection (if α is odd) of sets from $\bigcup_{\beta<\alpha} G_\beta$. The F_α sets are defined similarly except here F_0 stands for the closed sets, and we take countable intersections for even α and countable unions for odd α (limit ordinals are even). Another useful classification is to let $\Sigma_\alpha =$ G_α (for α even) or F_α (for α odd) and say these sets are "of additive class α"; interchanging G_α and F_α, we obtain the family Π_α of sets "of multiplicative class α." We will also use the symbol Σ_α (or $\Sigma_\alpha(X)$) to denote the class of all Σ_α sets in X, and similarly for the classes Π_α, G_α, and F_α. Finally, we recall that a mapping $f:X \to Y$ of topological spaces is a Borel map if $f^{-1}(U)$ is a Borel set in X for every open $U \subset Y$; it is of (Baire) class α provided

ISBN 0-12-481820-X

$f^{-1}(U) \in \Sigma_\alpha(X)$ for every open $U \subset Y$.

For metrizable X, one has $G_\alpha \subset F_{\alpha+1}$ and $F_\alpha \subset G_{\alpha+1}$, and that $\bigcup_{\alpha<\omega_1} G_\alpha = \bigcup_{\alpha<\omega_1} F_\alpha$ = the family of all Borel subsets of X. Since most of the results presented below are known only for the case when X is metrizable, we simplify the presentation by stipulating that *all spaces considered henceforth are to be metrizable.*

I. BOREL-ADDITIVE FAMILIES AND HOW THEY ARISE

To motivate the main concept with which the present talk is concerned, let us consider any Borel map f from a space X to a space Y. If $\mathcal{U}$ is any family of open sets of Y, then the family $\mathcal{B} = f^{-1}(\mathcal{U})$ (the inverse images of members of $\mathcal{U}$) has the property that the union of any subfamily is a Borel set in X; in short, we say that $\mathcal{B}$ is a *Borel-additive family in* X. If, for some fixed ordinal $\alpha < \omega_1$, the union of every subfamily of $\mathcal{B}$ is of class Σ_α (or F_α, etc.) in X, then we will say that $\mathcal{B}$ is Σ_α-*additive* (respectively F_α-*additive*, etc.) in X. We observe that if there is a countable open family $\{U_n | n \in \omega\}$ such that each member of $\mathcal{U}$ is a union of members from $\{U_n\}$, then $f^{-1}(\mathcal{U})$ will be Σ_α-additive, where $\alpha = \sup\{\alpha_n\}$ and $f^{-1}(U_n) \in \Sigma_{\alpha_n}$ ($n \in \omega$). In particular, if Y is a *separable* (metric) space, then every Borel map $f: X \to Y$ is of bounded class. A fundamental question in descriptive set theory is whether this property of Borel maps continues to hold for general (metric) spaces.

II. SOME BASIC PROBLEMS CONCERNING BOREL-ADDITIVE FAMILIES AND SOME HISTORY

The above question (of whether a Borel map between metric spaces must have a class) was first raised by A. H. Stone [S_1] for the case of a Borel isomorphism (a bijection such that both it and its inverse are Borel maps); the general question was later raised by Stone in [S_2]. Since every metric space has a σ-discrete base, Stone's question leads naturally to the following fundamental question regarding Borel-additive families:

Question 1. Is every disjoint Borel-additive family in a (metric) space Σ_α-additive for some $\alpha < \omega_1$?

Using Montgomery's theorem [Mo], that a subset of a metric space X which is locally of class Σ_α (or Π_α, $\alpha > 0$) is globally of that class in X , it is easily seen that a σ-discrete (or σ-locally countable) family of Σ_α sets in X is Σ_α-additive in X . More generally, call a family $\{E_a | a \in A\}$ *σ-discretely decomposable* (abbreviated σdd) if there is a family $\{E_{an} | a \in A, n \in \omega\}$, discrete for fixed n , such that each $E_a = \bigcup_{n\in\omega} E_{an}$.[1] In [H_1, Lemma 7] we observed that every σdd family $\subset \Sigma_\alpha(X)$ is Σ_α-additive in X . Moreover, we proved the following.

2.1 THEOREM [Hansell, 1969]. *If* X *is absolutely analytic, then every disjoint analytic-additive (hence Borel-additive) family in* X *is* σdd .[2]

2.2 COROLLARY. *If* X *is absolutely analytic, then every disjoint Borel-additive family* $\subset \Sigma_\alpha(X)$ *is* Σ_α*-additive in* X .

Although Theorem 2.1 enabled me to extend to general (metric) spaces a substantial part of the "separable theory" of Borel maps, for the case when the domain is absolutely analytic (see [H_1,H_2]), I was unable to answer Question 1 in this case. What was needed was the following:

2.3 THEOREM [Preiss, 1972]. *If* X *is a disjoint Borel-additive family*

---

[1]Alternatively, one can show that $\{E_t | t \in T\}$ is σdd in (the metric space) X iff there is a point-countable open family $\{U_t | t \in T\}$ in X such that $E_t \subset U_t$ for all $t \in T$.

[2]Recall that a (metric) space X is absolutely analytic if it is an analytic set in every (equivalently, some) complete metric space which contains it topologically. An analytic set is one which can be obtained from the closed sets using Souslin's operation -A . Of course, every Borel set in X is analytic in X .

in a (metric) space X *, then there is an* $\alpha < \omega_1$ *such that* $X \subset \Sigma_\alpha(X)$ *(see [P]).*

Combining 2.2 and 2.3 we immediately obtain the following positive partial answer to Question 1 (cf. also Lemma 2.8 of the paper by W. Fleissner $[F_1]$).

2.4 THEOREM. *If* X *is absolutely analytic, then every disjoint Borel-additive family in* X *is* Σ_α*-additive for some* $\alpha < \omega_1$ *. Consequently, every Borel map from* X *to a (metric) space* Y *is of bounded class.*

About the same time that Theorem 2.1 was proved, it was also discovered that it is consistent with the usual axioms of set theory (ZFC) for there to exist an uncountable set of reals every subset of which is a relative F_σ set; these are called "Q-sets." Note that the family of singletons of a Q-set is F_σ-additive, but is not σdd since "σ-discrete" in separable metric spaces is equivalent to "countable." Thus we cannot omit the assumption that X is absolutely analytic in Theorem 2.1. Ironically, it turns out (see §III) that the existence of a Q-set implies that this same assumption *can* be omitted from Theorem 2.4.

In their joint paper [K-P], J. Kaniewski and R. Pol observed that *point-finite* Borel-additive families arose naturally as the inverse image of a discrete (or locally-finite) open family under a lower Borel measurable compact-valued (multi-) map.[3] Furthermore, they showed that Theorem 2.1 continues to hold with "disjoint" replaced by "point-finite" (and that it fails in general for "point-countable"). This was used to prove the following extension of the Kuratowski and Ryll-Nardzewski selection theorem: If X is absolutely analytic and Y is any (equivalently, complete) metric space, then every compact-valued map $F: Y \to 2^Y$ of lower class

[3] A set-valued map $F: X \to 2^Y$ is *lower Borel measurable* if $F^{-1}(U) = \{x \in X \mid F(x) \cap U \neq \emptyset\}$ is a Borel set in X for every open $U \subset Y$; it is of *lower class* α if $F^{-1}(U) \in \Sigma_\alpha(X)$ for every open $U \subset Y$.

$\alpha < \omega_1$ admits a selector of class α. It was also observed that in this theorem "class α" could be generalized to "Borel measurable" provided the following question (cf. Theorem 2.3) had an affirmative answer:

Question 2. Given a point-finite Borel-additive family $\mathcal{X}$ in some (absolutely analytic or general metric) space X, does there exist an $\alpha < \omega_1$ such that $\mathcal{X} \subset \Sigma_\alpha(X)$?

We observed above that it is consistent with ZFC for Theorem 2.1 to be false if we omit the assumption that X is absolutely analytic. In the other direction, W. Fleissner [F_1] has shown that if there is a model of set theory with a supercompact cardinal, then there is a model in which every point-finite analytic-additive family in an *arbitrary* metric space is σdd. In particular, in such models (if they exist) Question 1 is answered in the affirmative. Fleissner also asked in [F_1] whether the following natural extension of Question 1 could be answered (with or without assuming the existence of large cardinals).

Question 3. Is every point-finite Borel-additive family in a metric space Σ_α-additive for some $\alpha < \omega_1$?

We point out that the special case of Question 3 when the members of the family are singletons has been answered in the affirmative by A. Miller in the separable case, and extended to arbitrary metric spaces by K. Kunen (see [Mi, Thm. 48]).

To motivate our final question concerning Borel-additive families, let us consider the following problem posed by K. Kuratowski in 1935 and which has remained open until now: Given (metric) spaces X, Y, and Z, and maps $f: X \to Y$, $g: X \to Z$ of classes α and β respectively, does it follow that the map $(f,g): X \to Y \times Z$, where $(f,g)(x) = (f(x),g(x))$, is Borel measurable (of bounded class)? (See §8, Prob. 2 of [K_2], where the problem is stated in a somewhat more restrictive form.) Using Theorem 2.1 it can be shown that (f,g) is of class $\gamma = \max(\alpha,\beta)$ when X is absolutely analytic (see [H_1, Thm. 4] and also [S_2, Thm. 3]). To see what the problem is in general, let us outline a proof of this: Suppose U is open in $Y\times Z$, and let $\{V_{na} | a\in A_n, n\in\omega\}$ be a σ-discrete open base for Y. For each $a \in A_n$ we define $W_{na} = \cup\{W | W$ is open in Z, and $V_{na}\times W \subset U\}$. Then, for each fixed $n \in \omega$, $\{f^{-1}(V_{na}) | a\in A_n\}$ is a disjoint Σ_α-additive family in X and $\{g^{-1}(W_{na}) | a\in A_n\} \subset \Sigma_\beta(X)$ (note that the latter family will not even be point-countable in general; it is Σ_β-additive, but this is of no consequence here). Finally, we observe that

$$(f,g)^{-1}(U) = \bigcup_{n\in\omega} \bigcup_{a\in A_n} f^{-1}(V_{na}) \cap g^{-1}(W_{na}) . \tag{1}$$

Now, if X is absolutely analytic, then by Theorem 2.1 the family $\{f^{-1}(V_{na}) | a\in A_n\}$ is σdd. Hence the family $\{f^{-1}(V_{na}) \cap g^{-1}(W_{na}) | a\in A_n\}$ is also σdd and $\subset \Sigma_\gamma(X)$ where $\gamma = \max(\alpha,\beta)$; consequently, by Corollary 2.2, it is Σ_γ-additive. It follows that $(f,g)^{-1}(U) \in \Sigma_\gamma(X)$ proving that (f,g) is of class γ.

An analysis of the above proof leads one naturally to the following definition and question:

DEFINITION. An indexed family $\{X_a | a\in A\}$ of subsets of a space X is said to be *hereditarily-Borel-additive* if, whenever $\{B_a | a\in A\} \subset \Sigma_\alpha(X)$ for some $\alpha < \omega_1$, then $\{X_a \cap B_a | a\in A\}$ is Σ_β-additive in X for some $\beta < \omega_1$.

Question 4. Is every point-finite Σ_α-additive family in a (metric) space X hereditarily-Borel-additive?

III. NEW RESULTS ON BOREL-ADDITIVE FAMILIES AND ANSWERS TO (MOST OF) THE ABOVE QUESTIONS

Theorem 2.3 was proven by Preiss in the general setting of abstract σ-fields. Although our answer to Question 2 (and, in particular, the following lemma) can be proven with the same generality, for the sake of brevity we will state it here only for the σ-field of Borel sets of a metric space.

3.1 LEMMA. *If* S *is any separable metric space and* $\{X_s \mid s \in S\}$ *is a point-finite Borel-additive family in the (metric) space* X, *then the family* $\{X_s \times \{s\} \mid s \in S\}$ *is disjoint and Borel-additive in* $X \times S$.

We can now give the following affirmative answer to Question 2.[1]

3.2 THEOREM. *If* $\mathcal{X}$ *is a point-finite Borel-additive family in a (metric) space* X, *then there is an* $\alpha < \omega_1$ *such that* $\mathcal{X} \subset \Sigma_\alpha(X)$.

Sketch of Proof. For each $\alpha < \omega_1$ suppose $\mathcal{X} \not\subset \Sigma_\alpha(X)$ and choose $X_{s(\alpha)} \in \mathcal{X} - \Sigma_\alpha(X)$. Treating $S = \{s(\alpha) \mid \alpha < \omega_1\}$ as a subset of the reals (say), we may apply the lemma to conclude that

$$H = \cup\{X_{s(\alpha)} \times \{s(\alpha)\} \mid \alpha < \omega_1\} \in \Sigma_{\beta+1}(X \times S)$$

for some $\beta < \omega_1$. Since $X \times \{s(\alpha)\}$ is closed in $X \times S$, it follows that $H \cap (X \times \{s(\alpha)\}) = X_{s(\alpha)} \times \{s(\alpha)\} \in \Sigma_{\beta+1}(X \times S)$ for each $\alpha < \omega_1$. But then $X_{s(\alpha)} \in \Sigma_{\beta+1}(X)$ for each $\alpha < \omega_1$, giving the desired contradiction.

The task of completely answering Question 3 has been more formidable. We have several affirmative partial answers which seem to indicate that the complete answer should be an unqualified "yes." The proof of the following lemma (and that of Lemma 3.1 above) is a straightforward induction argument using the monotone class lemma.

3.3 LEMMA. *Let* $\{X_s \mid s \in S\}$ *be a point-finite Borel-additive family in the (metric) space* X, *and let* T *be an arbitrary nonempty set. Then*

$$\cup\{X_s \times \{t\} \mid (s,t) \in M\} \in \sigma(\mathcal{B} \times T)$$

for all $M \in \sigma(S \times T)$.

Here $\sigma(S \times T)$ stands for the σ-field on $S \times T$ generated by the rectangles $\mathcal{R}_0 = \{A \times B \mid A \subset S, B \subset T\}$ (and similarly for $\sigma(\mathcal{B} \times T)$ except here the

---

[1]Professor Roman Pol has recently informed me that his student S. Spahn has also proven Theorem 3.2 using a different technique.

first factor ranges over the Borel σ-field of X). Recall that $\sigma(S\times T) = \bigcup_{\alpha<\omega_1} R_\alpha$ where R_α is defined (inductively) to be the family of all countable unions (intersections) of sets from $\bigcup_{\beta<\alpha} R_\beta$ if α is even (respectively odd).

We now have the following affirmative partial answer to Question 3.

3.4 THEOREM. *Let* $\{X_s \mid s\in S\}$ *be point-finite and Borel-additive in the (metric) space* X . *If there is an uncountable set* T *such that* $2^{S\times T} = \sigma(S\times T)$, *then* $\{X_s \mid s\in S\}$ *is* Σ_α*-additive for some* $\alpha < \omega_1$.

The general idea of the proof is similar to the one sketched above for Theorem 3.2, except we use 3.3 in place of 3.1.

Conditions under which the equality

$$(*) \quad 2^{S\times T} = \sigma(S\times T)$$

holds have been studied by various authors, and we cite Kunen [K], Rao [R_1,R_2], Bing, Bledsoe, and Mauldin [B-B-M], Mauldin [M], and most recently, Miller [Mi]. In particular, it is known that (*) is consistent with ZFC for $|S|,|T| \le c$ (see [K]). Specifically, (*) will hold in the case $|S|,|T| \le \omega_1$ ([K],[R_1,R_2]), and also when $|S|,|T| \le c$ if Martin's Axiom is assumed ([K]); and (*) does not hold when $c < |S| \le 2^{|T|}$ ([K],[R_2]). (See also §4 of [Mi].) Also, by modifying the proof of Theorem 3 of [B-B-M], we can show that (*) holds if $|T| < c$ and the following axiom is assumed.

(B) There exists an uncountable set W such that 2^W is countably generated.

Note that CH (continuum hypothesis) implies the negation of (B), and that MA + ¬CH implies (the existence of a Q-set, and hence) (B). Finally, if X is separable and $x_s \in X_s$ (in the hypothesis of 3.4), then it can be shown that $W = \{x_s \mid s\in S\}$ will satisfy (B) whenever $|S| > \omega_0$. We summarize this as follows:

3.5 THEOREM. *If* $\{X_s | s \in S\}$ *is a point-finite Borel-additive family in the (metric) space* X *, then* $\{X_s \; s \in S\}$ *is* Σ_α*-additive in* X *for some* $\alpha < \omega_1$ *provided any one of the following hold (under the given axiom): (i)* X *is separable; (ii)* $|S| \le \omega_1$ *; (iii) (MA)* $|S| \le c$ *; (iv) (B)* S *arbitrary.*

The fact that Question 4 can be answered in the affirmative without equivocation is remarkable in two respects. Firstly, that it *can* be answered without making further assumptions, and secondly, that the (relatively straightforward) proof has gone unnoticed until now.

3.6 THEOREM. *Every point-finite* Σ_α*-additive family in a (metric) space* X *is hereditarily-Borel-additive. More precisely, if* $\{X_a | a \in A\}$ *is* Σ_α*-additive and* $\{B_a | a \in A\} \subset G_\beta(X)$ *, then* $\{X_a \cap B_a | a \in A\}$ *is* $\Sigma_{\alpha+\beta}$*-additive (if* β *is even) or* $\Pi_{\alpha+\beta}$*-additive (if* β *is odd).*

Sketch of Proof. For the case when $\beta = 0$, let $\{U_{ns} | s \in S_n, n \in \omega\}$ be a σ-discrete open base for X , and define $X_{ns} = \cup\{X_a | U_{ns} \subset B_a\}$. Then one easily sees that

$$(1) \quad \bigcup_{a \in A} X_a \cap B_a = \bigcup_{n \in \omega} \bigcup_{s \in S_n} X_{ns} \cap U_{ns} .$$

Since $\{X_{ns} \cap U_{ns} | s \in S_n\}$ is discrete in X and $\subset \Sigma_\alpha(X)$, it follows that the set on the right of (1) is Σ_α in X . But the same argument applies to any $A' \subset A$, and thus we conclude that $\{X_a \cap B_a | a \in A\}$ is Σ_α-additive.

The proof is now completed by induction on β using the following general properties:

$$(2) \quad \text{if each } B_a = \bigcup_{n \in \omega} B_{an} \text{, then } \bigcup_{a \in A} X_a \cap B_a = \bigcup_{n \in \omega} \bigcup_{a \in A} X_a \cap B_{an} ;$$

$$(3) \quad \text{if each } B_a = \bigcap_{n \in \omega} B_{an} \text{ where } B_{an+1} \subset B_{an} \text{ for all } n \text{, then}$$

$$\bigcup_{a\in A} X_a \cap B_a = \bigcap_{n\in\omega} \bigcup_{a\in A} X_a \cap B_{an} .$$

We now cite as corollaries two consequences of Theorem 3.6 (cf. the discussion in §II).

3.7 COROLLARY. *If* $f:X \to Y$ *and* $g:X \to Z$ *are of classes* α *and* β *respectively, then the diagonal map* $f,g:X \to Y\times Z$ *is of class* $\min(\alpha+\beta,\beta+\alpha)$.

3.8 COROLLARY. *If* $F:X \to 2^Y$ *is compact-valued and lower Borel measurable, then* F *has a Borel selector of bounded class.*

REMARK. The bound on the class of f,g in 3.7 cannot in general be sharpened to $\max(\alpha,\beta)$ as in the case when X is absolutely analytic (see §II). W. Fleissner [F_2] has shown that it is consistent for there to exist a Q-set X such that X^2 is not a Q-set. Thus if Y is the set X with the discrete topology and $f,g:X^2 \to Y$ are defined by $f(x,y) = x$ and $g(x,y) = y$, then it is easily verified that f and g are of class 1 but that f,g is not of class 1 (it must be, of course, of class 2).

We conclude with two examples which show why the assumption of point-finiteness is necessary in Theorems 3.2, 3.5, and 3.6.

3.9 *EXAMPLE.* In the space $\mathbb{R}$ of real numbers, there exists a point-countable F_σ-additive family which is not hereditarily-Borel-additive: By a theorem of Hausdorff [K_1, p. 484] there exists a strictly decreasing transfinite sequence $\{E_\alpha\}_{\alpha<\omega_1}$ of F_σ sets in $\mathbb{R}$ having an empty intersection; and so we may choose an $x_\alpha \in E_\alpha - E_{\alpha+1}$ for each $\alpha<\omega_1$. It follows that $\{E_\alpha\}_{\alpha<\omega_1}$ is point-countable and F_σ-additive, but the disjoint family $\{E_\alpha \cap \{x_\alpha\}\}_{\alpha<\omega_1}$ cannot even be analytic-additive in $\mathbb{R}$ by Theorem 2.1.

3.10 *EXAMPLE.* In the space $\mathbb{R}$ of real numbers, there exists a point-countable Borel-additive family $\not\subset \Sigma_\alpha(\mathbb{R})$ for any $\alpha < \omega_1$: There exists a disjoint family $\{B_\alpha\}_{\alpha<\omega_1}$ of Borel sets of $\mathbb{R}$ of unbounded classes with $\mathbb{R} = \bigcup_{\alpha<\omega_1} B_\alpha$ [K_1, p. 484]. Letting $C_\beta = \bigcup_{\alpha\leq\beta} B_\alpha$, it is easy to see that $\{C_\beta\}_{\beta<\omega_1}$ is of unbounded class, and hence that $\{\mathbb{R}-C_\beta\}_{\beta<\omega_1}$ is point-countable, Borel-additive and $\not\subset \Sigma_\alpha(\mathbb{R})$ for any $\alpha < \omega_1$.

REFERENCES

[B-B-M] Bing, R. H., W. W. Bledsoe and R. D. Mauldin, "Sets generated by rectangles," *Pacific J. Math.* *51*(1974), 27-36.

[F_1] Fleissner, W. G., "An axiom for nonseparable Borel theory," *Trans. Amer. Math. Soc.* *251*(1979), 309-328.

[F_2] ______, "Squares of Q sets," *Notices Amer. Math. Soc.* *195*(1979), A-475.

[H_1] Hansell, R. W., "Borel measurable mappings for nonseparable metric spaces," *Trans. Amer. Math. Soc.* *161*(1971), 145-169.

[H_2] ______, "On Borel mappings and Baire functions," *Trans. Amer. Math Soc.* *194*(1974), 195-211.

[K-P] Kaniewski, J. and R. Pol, "Borel-measurable selectors for compact-valued mappings in the nonseparable case," *Bull. Acad. Polon. Sci. Ser. Sci. Math. Astronomy, Phys.* *23*(1975), 1043-1050.

[Kn] Kunen, K., *Inaccessibility Properties of Cardinals*, Ph.D. Thesis, Stanford Univ., 1968.

[K_1] Kuratowski, K., *Topology, Vol. 1*, Academic Press, New York; PWN, Warsaw, 1966.

[K_2] ______, "Quelques problèmes concernant espaces métriques nonséparables," *Fund. Math.* *25*(1935), 532-545.

[M] Mauldin, R. B., "On rectangles and countably generated families," *Fund. Math.* *95*(1977), 129-139.

[Mi] Miller, A. W., "On the length of Borel hierarchies," *Ann. Math. Logic* *16*(1979), 233-267.

[Mo] Montgomery, D., "Nonseparable metric spaces," *Fund. Math.* *25*(1935), 527-534.

[P] Preiss, D., "Completely additive disjoint systems of Baire sets is of bounded class," *Comment. Math. Univ. Carolinae* *15*(1972), 341-344.

[R_1] Rao, B. V., "On discrete Borel spaces and projective sets," *Bull. Amer. Math. Soc.* *75*(1969), 614-617.

[R_2] Rao, B. V., "On discrete Borel spaces," *Acta Math. Acad. Sci. Hungaricae* 22(1971), 197.

[S_1] Stone, A. H., "Nonseparable Borel sets," *Rozprawy Mat.* 28(1962).

[S_2] ______, *Some Problems of Measurability*, Lecture Notes in Math. No. 375, Springer-Verlag, Berlin, 1974.

THE USE OF SET-THEORETIC HYPOTHESES IN THE STUDY OF MEASURE AND TOPOLOGY

Donald A. Martin*

University of California, Los Angeles

I. INTRODUCTION

We shall consider set-theoretic hypotheses not provable from the usual axioms of set theory: in particular, Martin's Axiom and large cardinal axioms. We give two examples showing how proofs using these hypotheses can lead to outright proofs.

The use of such hypotheses in solving mathematical problems has become familiar in recent year. Of course, this is really not a new phenomenon: It has for many years been a standard practice to answer questions by assuming the continuum hypothesis. In recent years, however, a few - often mutually incompatible - set-theoretic hypotheses have proven extremely fruitful not just in answering a few isolated questions but rather in producing coherent general pictures in certain parts of mathematics.

Most prominent among these hypotheses are the Axiom of Constructibility, Martin's Axiom, and large cardinal axioms. The word "axiom" is somewhat misleading in all three cases. Their status is - at least in a sociological sense - quite unlike that of the Zermelo-Fraenkel axioms. The latter presumably codify certain more or less evident principles about the iterative notion of sets. In any event, a theorem proved using the Z-F axioms is simply a theorem. One does not have to mention which axioms

*Partially supported by NSF Grant #MCS 78-02989

ISBN 0-12-481820-X

one has used (except, occasionally, the axiom of choice). The three sorts of axioms mentioned at the beginning of this paragraph are not treated in this way, and in fact are treated somewhat differently one from the other.

The Axiom of Constructibility does not seem to be an evident fact about the concept of set. Rather it might be though of as an attempt to *complete* the concept of set. It says that there are no sets except (a) ordinal numbers and (b) those other sets required by the Z-F axioms. The only thing about sets it does not precisely delimit is the nature of the ordinal numbers themselves. Until recently it has appear that the Axiom of Constructibility is strong enough to answer every natural mathematical question not explicitly about ordinals (i.e., all questions except those such as "Do inaccessible cardinals exist?"). Recent results of Harvey Friedman suggest that this is not quite true, however.

The main practical use of the Axiom of Constructibility is nevertheless to get consistency results. Since Gödel showed that the Axiom is relatively consistent with the Z-F axioms, we know that anything proved from the Axiom cannot be refuted using the Z-F axioms. In recent years, using the Axiom to prove such consistency results has become possible for non-logicians, since Ronald Jensen has proved several powerful purely combinatorial consequences of the Axiom (e.g., the "diamond" principle). The reader interested in this should consult [7]. [3] in the same volume is a general introduction to the Axiom of Constructibility.

Martin's Axiom's status is almost entirely that of a device for generating consistency proofs. It is by no means obviously true, and it is not at all a completion of the concept of set. But it is provably consistent (see [2]), and any assertion which can be proved consistent by a certain general method can be proved consistent by simply deducing it from Martin's Axiom (henceforth denoted by "MA"). Thus, for example, general topologists with no knowledge of logic can generate consistency proofs in their own discipline using MA, and this has become a fairly common occurrence. See [11] for an introduction to the Axiom and some of its basic consequences.

Large cardinal axioms are quite different from the two "axioms" already discussed. They have not been shown relatively consistent with the Z-F axioms, and it is even a theorem that they cannot be proved relatively consistent by any methods now understood. On the other hand, a number of set theorists think not only that they are consistent but also that it can be argued, with various degrees of plausibility, that they are *true* of the concept of set. The most surprising property of large cardinal axioms, first discovered by Solovay, is that while they assert the existence of monstrously large sets, they nevertheless have consequences in the world of small sets, e.g. in the realm of the real numbers.

All three kinds of hypotheses occasionally have another kind of use - one which has nothing to do with independence results or with the "axiomatic" status of the hypotheses. It is sometimes possible to prove an outright theorem by (1) first proving the theorem assuming one of our hypotheses and (2) then eliminating the hypothesis. In the rest of this paper, two very different examples of such proofs will be given, one using MA and the other using large cardinals.

II. MARTIN'S AXIOM AND A PROBLEM ABOUT RECTANGLES

The first theorem proved by the method of using and then eliminating MA was probably Baumgartner and Hajnal [1], a result in infinitary combinatorics. The theorem to be discussed below was first proved by a direct construction [4], but people unaware of [4] occasionally raise the question and reanswer it. This is how a proof via MA arose. (The problem was posed to the author by N. Linial; the author solved it using

MA; then Linial and E. Straus found a non-MA proof. After the lecture on which this paper is based, J. Mycielski told the author about [4].) In any event it is a good example of how one might use MA to prove an outright theorem.

The problem is as follows: Does every planar set of positive measure contain a rectangle $X \times Y$ with X uncountable and Y of positive measure? (There are counterexamples to both X and Y being of positive measure.)

Our method for solving the problem (solved already in [4], as we have remarked) is:

(1) We replace the assertion we wish to prove by a stronger and more combinatorial assertion.

(2) We prove the stronger assertion from MA.

(3) By [2] we know there is a Boolean-valued model $V^{\mathcal{B}}$ of set theory in which MA holds. Thus the stronger assertion - and so the theorem to be proved - holds in $V^{\mathcal{B}}$.

(4) We show that the assertion to be proved is *absolute:* it is true in $V^{\mathcal{B}}$ if and only if it is true in the real world.

Let us begin by recalling the statement MA. Suppose P is partially ordered by $\leq$. $p,q \in P$ are *compatible* if there is an $r \in P$, $p \geq r$, $q \geq r$. An *antichain* in P is a set of pairwise incompatible elements of P. P has the *c.a.c.* if every antichain in P is countable. $D \subseteq P$ is *dense* if for each $p \in P$ there is a $q \in D$ with $q \leq p$. A non-empty $G \subseteq P$ is a *filter* if (1) for every $p,q \in G$ there is an $r \in G$ with $r \leq p$ and $r \leq q$ and (2) if $p \in G$ and $q \geq p$ then $q \in G$. If κ is a cardinal number MA_κ says that for every P with the c.a.c. and every family F of size $\leq \kappa$ of dense subsets of P, there is a filter G such that G meets each member of F. (MA is the assertion $MA_{<2^{\aleph_0}}$.) In [13] (see also [2]) it is shown that there is a complete Boolean algebra $\mathcal{B}$ such that $MA_{\aleph_1}$ is true in the Boolean model $V^{\mathcal{B}}$.

LEMMA 1. *Assume* $MA_{\aleph_1}$. *Let* H *be an uncountable family of linear sets of positive measure. There is an uncountable subfamily* H' *of* H *such that the intersection of all the members of* H' *has positive measure.*

Proof. (The lemma may be due to Solovay. It is an easy extension of a result of [10]). Let P be the set of all infinite sequences

$$F_1, F_2, \ldots$$

where each F_i is a closed linear set, all but finitely many F_i are the whole real line, and F_i has measure $> \frac{1}{i}$. Partially order P by

$$F_1, F_2, \ldots \leq F_1', F_2', \ldots$$

if and only if, for every i, $F_i \subseteq F_i'$.

We first show that P has the c.a.c. Let $\{p^i : i \in I\}$ be an uncountable family of pairwise incompatible elements of P. Let $p^i = F_1^i, F_2^i, \ldots$. Passing to a subfamily if necessary, we may assume that there is a fixed n such that F_m^i is the whole line for all $m > n$ and all $i \in I$.

Each F_j^i has measure $> \frac{1}{j}$. Passing to a subfamily if necessary, we may assume that there is a fixed $\varepsilon > 0$ such that each F_j^i has measure $> \frac{1}{j} + \varepsilon$. For each i and j let E_j^i be a finite intersection of complements of open rational intervals such that $E_j^i \supseteq F_j^i$ and $E_j^i - F_j^i$ has measure $< \frac{\varepsilon}{2}$. Since there are only countably many finite intersections of complements of open rational intervals, we may - by passing to a subfamily - assume that, for each $j \leq n$, all the E_j^i are the same set E_j.

For each $i, i' \in I$, there is some $j \leq n$ such that $F_j^i \cap F_j^{i'}$ has measure $\leq \frac{1}{j}$. Otherwise

$$F_1^i \cap F_1^{i'}, F_2^i \cap F_2^{i'}, \ldots$$

would be an element of P witnessing that the two sequences are compatible. But $F_j^i \cap F_j^{i'} \supseteq (E_j^i \cap E_j^{i'}) - [(E_j^i - F_j^i) \cup (E_j^{i'} - F_j^{i'})] = E_j - [(E_j^i - F_j^i) \cup (E_j^{i'} - F_j^{i'})]$. Thus the measure of $F_j^i \cap F_j^{i'}$ is at least the measure of E_j minus $\left(\frac{\varepsilon}{2} + \frac{\varepsilon}{2}\right)$. Since F_j^i has measure $> \frac{1}{j} + \varepsilon$, E_j has measure $> \frac{1}{j} + \varepsilon$. This shows that $F_j^i \cap F_j^{i'}$ has measure $> \frac{1}{j}$, a contradiction.

We now return to the proof of the lemma. Passing to a subfamily if necessary, let $\{Y_i, i \in I\}$ be a family of positive measure sets, with I of size $\aleph_1$. For each $i \in I$, let

$$D_i = \{F_1, F_2, \ldots : \text{ some } F_j \subseteq Y_i\} .$$

To see that each D_i is dense, note that Y_i has measure $> \frac{1}{j}$ for some j . Given $F_1, F_2, \ldots$ choose j big enough so that F_j is the whole line and Y_i has measure $> \frac{1}{j}$. Let $F_j^* \subseteq Y_i$ be closed with measure $> \frac{1}{j}$. $F_1, F_2, \ldots, F_{j-1}, F_j^*, F_{j+1}, \ldots$ is the desired element of $P \cap D_i$.

By $MA_{\aleph_1}$ let G be a filter meeting each D_i . Let

$$F_1, F_2, \ldots = \bigcap_G F_1', \bigcap_G F_2', \ldots .$$

Each F_j is the intersection of a family of closed sets every finite subfamily of which has intersection with measure $> \frac{1}{j}$. Thus F_j has measure $\geq \frac{1}{j}$. Each Y_i is a superset of some F_j . Thus some $F_j \subseteq Y_i$ for uncountably many i .

LEMMA 2. *If* $MA_{\aleph_1}$ *, then every planar set of positive measure contains a rectangle* $X \times Y$ *with* X *uncountable and* Y *of positive measure.*

Proof. By Fubini's Theorem, the set of x whose section Y_x has positive measure is uncountable. By Lemma 1, there is an uncountable X such that

$$\bigcap_{x \in X} Y_x \text{ has positive measure.}$$

We may then choose $Y = \bigcap_{x \in X} Y_x$.

We now turn to the problem of eliminating $MA_{\aleph_1}$. Note that the assertion we wish to prove is unaffected if we require our planar set to be closed, since every set of positive measure contains a closed set of positive measure. Note also that, if we can find X and Y as desired then we may find *closed* X and Y , since the closure of a rectangle is a rectangle, and since we are assuming our planar set closed. Thus $MA_{\aleph_1}$ implies the following assertion (equivalent to the assertion we are trying to prove):

For every closed set A in the plane, of positive measure, there are closed sets X and Y such that X is uncountable, Y has positive measure, and $X \times Y \subseteq A$.

This is what logicians call a Π^1_3 statement. It begins with a universal quantifier over the reals (since closed sets may be coded as reals). This is followed by existential quantifiers over the reals (X exists, Y exists, and quantifiers to say that X is uncountable and Y has positive measure). Next occur universal quantifiers over the reals (to say $(\forall x)(\forall y)(x \in X \ \& \ y \in Y \rightarrow (x,y) \in A)$). All conditions on the reals mentioned in the quantifiers are Borel. By a lemma of J. R. Shoenfield [11], we have the following absoluteness principle:

Every Π^1_3 statement true in a Boolean valued model $V^{\mathcal{B}}$ is true in the real world.

This is proved by noting that, if the Π^1_3 statement is false in the real world, then some real z satisfies the corresponding Π^1_2 condition. Since $V^{\mathcal{B}}$ is an *extension* of the set-theoretic universe, this real z belongs to $V^{\mathcal{B}}$. As Shoenfield [12] shows, a Π^1_2 condition is equivalent to a certain linear ordering's being a well-ordering. This same linear ordering is present in $V^{\mathcal{B}}$ and is a well-ordering there. Thus z is a counterexample to the Π^1_3 statement in $V^{\mathcal{B}}$.

THEOREM (Eggleston [4]). *Every planar set of positive measure contains a rectangle* $X \times Y$ *with* X *uncountable and* Y *of positive measure.*

III. THE WADGE ORDERING

We wish to classify subsets of topological spaces in terms of their complexity. Though the theory we give below applies, with minor changes, to many spaces (e.g., the reals), the definitions take their simplest form in Baire space, and so we restrict ourselves to it. *Baire space* $\mathcal{N}$ is the product of countably many copies of the discrete space consisting of the positive integers. In other words it is the set of all infinite sequences of positive integers, endowed with the topology with basic open sets those sets determined by finite sequences of positive integers. $\mathcal{N}$ is homeomorphic to the irrationals.

If $X, Y \subseteq \mathcal{N}$ and X is a continuous preimage of Y, then X is no more complicated than Y, in any reasonable sense. Thus we say that $X \leq_W Y$ just in case there is a continuous $f : \mathcal{N} \to \mathcal{N}$ with $X = f^{-1}(Y)$. The relation ($X \leq_W Y$ and $Y \leq_W X$) is an equivalence relation. The *Wadge degree* $d(X)$ of X is the union of the equivalence class of X and the equivalence class of $\mathcal{N} - X$. (We treat a set and its complement as being of the same complexity.) The Wadge degrees are partially ordered by

$$d(X) \leq d(Y) \Longleftrightarrow (X \leq_W Y \text{ or } X \leq_W \mathcal{N} - Y).$$

If we wish to deal with, for example, the reals, we have to vary the definition by slightly weakening the continuity condition on f. A surprising fact is that for many spaces (Polish spaces, countable products of discrete spaces) the Wadge degrees of Borel sets are linearly ordered, indeed are well-ordered.

To indicate how this is proved, we introduce a finer notion of degrees. If X and Y are subsets of N, we define a two person game $G(X,Y)$ as follows. Players I and II take turns listing positive integers. I lists m_1, then II lists n_1, then I lists m_2, etc. II *wins* the resulting play of $G(X,Y)$ just in case

$$m_1, m_2, \ldots \in X \iff n_1, n_2, \ldots \in Y .$$

The notion of a *winning strategy* for I or II for $G(X,Y)$ is defined in the obvious way. We say that

$$X \leq_{\ell} Y \iff \text{II has a winning strategy for } G(X,Y) .$$

We define the *ℓ-degrees* from $\leq_{\ell}$ just as we defined the Wadge degrees from $\leq_W$. Since a strategy for II induces a continuous $f : N \to N$ and a winning strategy for II induces such an f with $f^{-1}(Y) = X$,

$$X \leq_{\ell} Y \Rightarrow X \leq_W Y .$$

We say that $G(X,Y)$ is *determined* if I or II has a winning strategy.

LEMMA 1 (W. Wadge). *If* $G(X,Y)$ *is determined, then* $X \leq_{\ell} Y$ *or* $Y \leq_{\ell} N - X$.

Proof. If II has a winning strategy for $G(X,Y)$, then $X \leq_{\ell} Y$. If I has a winning strategy for $G(X,Y)$, then II has a winning strategy for $G(Y, N - X)$. II simply plays the given strategy for I with a one-move delay.

The lemma shows that, in so far as games are determined, the ℓ-degrees - and so the Wadge degrees - are linearly ordered.

In fact, these degrees are *well-ordered*.

LEMMA 2 (Martin [8] based on an idea of L. Monk). *Suppose* $X_1, X_2, \ldots$ *are subsets of* N *and that, for every-continuous* g : *the Cantor set*

$\to N$, $g^{-1}(X_1)$ *is measurable. (The Cantor set is the product of countably many copies of* $\{0,1\}$. *The measure in question is the product of the obvious measure on* $\{0,1\}$.) *Then there is an* i *such that* I *does not have a winning strategy for both* $G(X_i, X_{i+1})$ *and* $G(X_i, N - X_{i+1})$.

Remarks. The hypothesis is of course satisfied if X_1 is Borel. If all the games are determined, the lemma just says that $d(X_1), d(X_2), \ldots$ is not an infinite descending chain of ℓ-degrees.

Proof. Suppose that S_i^0 is a winning strategy for I for $G(X_i, X_{i+1})$ and that S_i^1 is a winning strategy for I for $G(X_i, N - X_{i+1})$. For each $\varepsilon_0, \varepsilon_1, \ldots$ in the Cantor set, consider players 1, 2, 3,... and let player i play $S_i^{\varepsilon_i}$ against player $i+1$. Since all of the strategies are strategies for the first player, this produces infinite sequences $x_1, x_2, \ldots$, where x_i is the play of player i. The function

$$\varepsilon_1, \varepsilon_2, \ldots \to x_1$$

is continuous. Thus p_1 = the probability that $x_1 \in X_1$ is defined. It can also be seen that $p_i(\varepsilon_1, \ldots, \varepsilon_n)$ = the probability that $x_i \in X_i$, given $(\varepsilon_1, \ldots, \varepsilon_n)$, is defined. Note that

$$p_{n+1}(\varepsilon_1, \ldots, \varepsilon_n) = \tfrac{1}{2} p_{n+2}(\varepsilon_1, \ldots, \varepsilon_n, 1) + \tfrac{1}{2}(1 - p_{n+2}(\varepsilon_1, \ldots \varepsilon_n, 0)),$$

since

$$x_{n+1} \in X_{n+1} \iff (x_{n+2} \in X_{n+2} \ \&\ \varepsilon_{n+1} = 1 \text{ or } x_{n+2} \notin X_{n+2} \ \&\ \varepsilon_{n+1} = 0).$$

Note also that $p_{n+2}(\varepsilon_1, \ldots, \varepsilon_{n+1})$ is independent of $(\varepsilon_1, \ldots, \varepsilon_{n+1})$, since $x_{n+2}, x_{n+3}, \ldots$ are independent of $(\varepsilon_1, \ldots, \varepsilon_{n+1})$. Thus, if $p = p_{n+2}(\varepsilon_1, \ldots, \varepsilon_{n+1})$,

$$p_{n+1}(\varepsilon_1, \ldots, \varepsilon_n) = \frac{1}{2} p + \frac{1}{2}(1 - p) = \frac{1}{2}.$$

Now $p_1(\varepsilon_1, \ldots, \varepsilon_n)$ is either $p_{n+1}(\varepsilon_1, \ldots, \varepsilon_n)$ or $1 - p_{n+1}(\varepsilon_1, \ldots, \varepsilon_n)$, so $p_1(\varepsilon_1, \ldots, \varepsilon_n) = \frac{1}{2}$. This contradicts the 0-1 law, and the lemma is proved.

We have seen that, in so far as the relevant games are determined, we can classify the subsets of N in a transfinite sequence of levels. But are the games determined? Using the axiom of choice, one can easily construct sets X and Y such that $G(X,Y)$ is not determined. If the Axiom of Constructibility holds, there is a non-determined game $G(X,Y)$ with X and Y *analytic*, i.e., a continuous image of N.

LEMMA (H. Friedman [5], improving a result of Martin). *If a measurable cardinal exists, then* $G(X,Y)$ *is determined for every analytic* X *and* Y.

Since we shall not give the proof, we omit the definition of a measurable cardinal. In outline, the proof consists of replacing $G(X,Y)$ by another game G^* played in a space of the size of the measurable cardinal, but a space whose winning condition is *open*. G^* is determined, by a result of Gale and Stewart [6]. A winning strategy for $G(X,Y)$ is then produced by integrating the strategy for G^* with respect to certain measures whose existence is guaranteed by the cardinal's measurability.

THEOREM (W. Wadge). *If a measurable cardinal exists, then the Wadge degrees of Borel sets are well-ordered.*

Can we eliminate the use of the large cardinal axiom from this thoerem? The answer is yes. Martin [9] proves (with no hypotheses) the determinacy of all two-person games with Borel winning conditions. (The winning condition may depend in an arbitrary Borel fashion on the pair $\langle m_1, m_2, \ldots ; n_1, n_2, \ldots \rangle$; also N may be replaced by any countable product of discrete spaces.) Thus we get

THEOREM. *The Wadge degrees of Borel sets are well-ordered.*

Note that the elimination of the large cardinal axiom here was quite different from the elimination of $MA_{\aleph_1}$ from the proof in the last section. In that case we directly applied Shoenfield's absoluteness theorem and the existence of Boolean models of $MA_{\aleph_1}$. Here we proceeded differently. It was necessary to replace the proof, from a measurable cardinal, of the determinacy of Borel games by a quite different outright proof.

What then was the value of the large cardinal axiom? The answer is that has *predicted* a number of results, results which were then proved outright. The proofs of the determinacy theorems from the large cardinal axiom and the proofs about Wadge degrees (a) were fairly short and (b) were done before the outright theorem that Borel games are determined. Thus work with large cardinal axioms, or - more directly - work with "axioms" asserting that many games are determined, led with relative ease to a substantial body of theory (the theory of Wadge degrees, which goes beyond the results mentioned here). Part of this theory (the Wadge degrees of Borel sets) was put on solid ground by the outright determinancy proof, but it might not have existed at all without the use of the "axioms."

Let me finish with a clarification. Though I have been stressing the value of large cardinal axioms in generating outright proofs, I don't wish to suggest that this is their main importance. I feel that the most interesting consequences of large cardinal axioms are those for which the axioms are essential.

REFERENCES

1. Baumgartner, J., and A. Hajnal, "A proof (involving Martin's axiom) of a partition relation," *Fund. Math.* *78*(1973), 193-203.

2. Burgess, J.,"Forcing," *Handbook of Mathematical Logic*, J. Barwise, ed., North-Holland, Amsterdam - New York - Oxford, (1977), 453-489.

3. Devlin, K.,"Constructibility," *Handbook of Mathematical Logic*, J. Barwise, ed., North-Holland, Amsterdam-New York-Oxford,(1977),453-489.

4. Eggleston, H., "Two measure properties of Cartesian product sets," *Quart. J. Math. Oxford*(2), 5(1954), 108-115.

5. Friedman, H., "Determinateness in the low projective hierarch," *Fund. Math. 72*(1971), 79-95.

6. Gale, D., and F. Stewart, "Infinite games with perfect information," *Contributions to the Theory of Games, Annals of Math. Studies, Vol.28*, Princeton University Press, Princeton, N.J., 245-266.

7. Juhaśz, "Consistency results in topology," *Handbook of Mathematical Logic*, J. Barwise, ed., North-Holland, Amsterdam-New-Oxford, (1977), 503-522.

8. Kechris, A., and D. Martin, "Infinite games and effective descriptive set theory," to appear in the Proceedings of the London Conference on Analytic Sets, C. A. Rogers and J. Jayne, eds.

9. Martin, D., "Borel determinacy," *Annals of Math. 102*(1975), 363-371.

10. Martin, D., and R. Solovay, "Internal Cohen extensions," *Annals of Math. Logic 2*(1970), 143-178.

11. Rudin, M., "Martin's axiom," *Handbook of Mathematical Logic*, J. Barwise, ed., North-Holland, Amsterdam-New York-Oxford, (1977), 491-501.

12. Shoenfield, J. R., "The problem of predicativity," *Essays on the Foundations of Mathematics*, Magnes Press, Jerusalem, (1961), 132-139.

13. Solovay, R., and S. Tennenbaum, "Iterated Cohen extensions and Souslin's problem," *Annals of Math. 94*(1970), 201-245.

PROBLEMS ON FINITELY ADDITIVE INVARIANT MEASURES

Jan Mycielski

University of Colorado
Boulder, Colorado 80309

0. I will discuss here some old but still unsolved problems of the theory of finitely additive invariant measures and mention recent results (especially of J.M. Rosenblatt) pertaining to some of them. This subject is as old as measure theory but our story really begins with Hausdorff's proof [8] of the nonexistence of a finitely additive universal invariant measure in $\mathbb{R}^3$ or in the 2-sphere S^2. Then there comes the work of Banach and Tarski [1, 2, 17] and many others (see Tarski [18], Cohen and Fickett [4], and my surveys [9, 10, 11, 14]).

1.1. *Problem* (E. Marczewski alias Szpilrajn, circa 1930). Does there exist a finite sequence $A_1,\dots,A_n$ of disjoint open sets in the southern hemisphere and rotations $\rho_1,\dots,\rho_n$ of S^2 such that $\rho_1(A_1)\cup\dots\cup\rho_n(A_n)$ is dense in S^2?

Of course, the conjecture is that the answer is no, but open sets A_i with boundaries of positive measure are difficult to study. I will state a few equivalent forms of 1.1 and indicate the reasons for their equivalence.

1.2 Do there exist for every open non empty set $V \subseteq S^2$ a finite sequence of disjoint open subsets $A_1,\dots,A_n$ of V and rotations $\rho_1,\dots,\rho_n$ such that $\rho_1(A_1)\cup\dots\cup\rho_n(A_n)$ is dense in S^2?

1.3 Does there exist a paradoxical decomposition of S^2 with parts

ISBN 0-12-481820-X

having the property of Baire? (The definition of paradoxical decompositions is given below; a set has the property of Baire iff it is of the form $(A \cup K_1) \setminus K_2$, where A is open and K_1 and K_2 are meager i.e. of the first category.)

1.4 Does there exist a nonnegative finitely additive rotation invariant measure μ over the algebra of subsets of S^2 having the property of Baire which satisfies $\mu(S^2) = 1$?

In fact 1.3 was the form of the problem which Marczewski told to me. He found that the method of Banach [1] yields a negative answer for the circle S^1 and for the spaces $\mathbb{R}^1$ and $\mathbb{R}^2$ (see [10] for simplified proofs). The equivalence of 1.4 with the negation of 1.3 follows from a remark and a theorem of Tarski which I will state now.

1.5 *Remark* (Tarski). If μ is a measure as required in 1.4 then μ vanishes over all meager sets.

This follows from the existence of a paradoxical decomposition of S^2 . In fact we can pack S^2 by finite decompositions into any nonempty open subset of S^2 (see [2,18]). In particular we can pack every meager set into sets of arbitrarily small diameters. Since all subsets of meager sets must have the property of Baire (being meager) it follows that meager sets have measure zero.

To state the theorem of Tarski we need the following concepts. Let $\mathbb{B}$ be a Boolean algebra and G a group of automorphisms of $\mathbb{B}$. We say that an element $a \in \mathbb{B}$ has a paradoxical decomposition in $(\mathbb{B}, G)$ iff there exist elements

$$a_1, \ldots, a_m, b_1, \ldots, b_n$$

in $\mathbb{B}$ satisfying

$$a_i \leq a, b_k \leq a, \ a_i \wedge a_j = b_k \wedge b_\ell = a_i \wedge b_k = \underline{0}$$

for $i,j = 1,\ldots,m,\ i \neq j$ and $k,\ell = 1,\ldots,n,\ k \neq \ell,$ and elements

$$g_1, \ldots, g_m, h_1, \ldots, h_n$$

of G such that

$$g_1(a_1) \vee \ldots \vee g_m(a_m) = h_1(b_1) \vee \ldots \vee h_n(b_n) = a .$$

1.6. THEOREM (Tarski). *The following conditions are equivalent:*

(i) *There exists a nonnegative finitely additive measure* μ *over* $\mathbb{B}$ *which is invariant under* G *and satisfies* $\mu(a) = 1$.

(ii) a *has no paradoxical decompositions in* $(\mathbb{B}, G)$

For a short proof of this theorem see [11].

Now, the equivalence of 1.4 and the negation of 1.3 is an immediate consequence of 1.6. The equivalence of 1.1, 1.2 and 1.3 also follows from 1.6 and the following remarks. We look at the quotient algebra $\mathfrak{B}$ of the Boolean algebra of subsets of S^2 having the property of Baire modulo the ideal of meager sets. We observe that

1.7. Every coset in $\mathfrak{B}$ has a (unique) regular open representative. (A set $A \subseteq S^2$ is regular open iff A equals the interior of its closure.)

1.8 Let μ be a nonnegative finitely additive measure over $\mathfrak{B}$ which is invariant under the natural action of the rotations of S^2 over $\mathfrak{B}$. Then for every $a \in \mathfrak{B}$, $a \neq \underline{0}$ the following conditions are equivalent:

(i) $0 < \mu(\underline{1}) < \infty$,

(ii) $0 < \mu(a) < \infty$.

It is clear that 1.5, 1.6, 1.7 and 1.8 yield the equivalence of 1.1, 1.2 and 1.3.

Problems similar to 1.4 are also open for the spaces $\mathbb{R}^n (n \geq 3)$, if we require $\mu(\text{unit cube}) = 1$ or $\mu(\mathbb{R}^n) = 1$, and for all S^n $(n \geq 2)$. The difficulty lies in the nonamenability of the groups of isometries of these spaces. And if we require only the invariance of μ relative to an amenable subgroup (e.g., in the case of $\mathbb{R}^n$,the group of translations) then the answer is yes (see [10]). But some more intricate results are possible (see [10], § 5, 6 and 7).

2.1 *Problem* (Banach and Ulam [3]). Let C be a compact metric space. Does there exist a nonnegative finitely additive measure over the algebra of all Borel subsets of C such that $\mu(C) = 1$ and $\mu(A) = \mu(B)$ whenever A is isometric to B ?

Problem 2.1 remains open already for the Hilbert cube $[0,1]^{\omega}$ with a metrization $(\Sigma a_i^2(x_i - y_i)^2)^{1/2}$, where $a_i > 0$ and $\Sigma a_i^2 < \infty$. But, if a_i tends to 0 sufficiently fast, then the answer is yes (see [6]). Also with some other metrizations of $[0,1]^{\omega}$, e.g. $\max|a_i(x_i - y_i)|$ with $0 < a_i < c\alpha^i$ for some constants c and $\alpha < 1$ the answer is yes (see [13]). For these positive cases the measure μ is countably additive and hence equal to the standard product measure in $[0,1]^{\omega}$. Countable additivity is not possible in general, e.g. if C is countably infinite. But in this case we have

2.2 THEOREM (Roy O. Davies and A. J. Ostaszewski [5]). *If* C *is countable then 2.1 has a positive answer.*

By a modification of the construction of Haar's measure we have proved the following

2.3 THEOREM (J. Mycielski [12, 13]). *If one restricts the last clause of 2.1 to open* A's *and* B's *then the answer is yes and* μ *can be countably additive.*

In a weak sense there exists a negative answer to the problem 2.1:

2.4 THEOREM (Rosenblatt [15]). *There exists a rotation invariant ideal in the Boolean algebra of all Borel subsets of* S^2 *such that* S^2 *has paradoxical decompositions modulo this ideal.*

3.1 *Problem* (S. Ruziewicz [1]). Let μ be a nonnegative finitely additive invariant under isometries measure over the algebra of Lebesgue measurable subsets of $\mathbb{R}^3$ with $\mu(\text{unit cube}) = 1$. Must μ be equal to the Lebesgue measure?

Similar questions for $\mathbb{R}^n$ $(n \geq 3)$ and S^n $(n \geq 2)$ are also open. For $\mathbb{R}^n$, $n \geq 5$, and S^n , $n \geq 4$, the answer to Problem 3.1 is yes. [D. Sullivan, to appear].

For S^1, $\mathbb{R}^1$ and $\mathbb{R}^2$ the answer is no (Banach, see [16]).

By an argument similar to the proof of 1.5 we can show that if μ satisfies the suppositions of 3.1 then it vanishes over all null sets. Therefore the following theorem comes close to a solution of 3.1.

3.2 THEOREM (J.M. Rosenblatt [16]). *If* $n \geq 2$ *and* μ *is a non-negative finitely additive measure over the class of bounded Lebesgue measurable subsets of* $\mathbb{R}^n$, $\mu([0,1]^n) = 1$, μ *vanishes over all null sets, and* μ *is invariant under the three transformations*

$$(x_1,\ldots,x_n) \to (x_1 + 1, x_2,\ldots,x_n),$$

$$(x_1,\ldots,x_n) \to (x_1, x_1 + x_2, x_3,\ldots,x_n)$$

$$(x_1,\ldots,x_n) \to (x_2,x_3,\ldots,x_n,(-1)^{n+1}x_1)$$

then μ *agrees with the Lebesgue measure.*

It is not yet clear whether a better understanding of the method of Robsenblatt can lead to a solution of 3.1.

4.1. *Problem.* Let ρ be a metrization of the real line $\mathbb{R}$ which generates the usual topology and satisfies $\rho(x + z, y + z) = \rho(x,y)$ for all $x,y,z \in \mathbb{R}$. Let $A,B \subseteq \mathbb{R}$ be Borel sets which are isometric relative to ρ. Must $\lambda(A) = \lambda(B)$, where λ is the Lebesgue measure?

For many natural metrizations, e.g. if $\rho(0,x)$ is increasing for $x > 0$, the answer is yes. It is yes in general if A and B are open. Those facts are proved in [7], but the methods of [7] are insufficient for a complete solution.

REFERENCES

1. Banach, S.,"Sur le problème de la mesure," *Fund. Math.* 4(1923), 7-33. (Reprinted in S. Banach, Oeuvres vol. I, PWN Warszawa 1967, pp. 66-89.

2. Banach, S., and A.Tarski,'Sur la decomposition des ensembles de points en parties respectivement congruentes," *Fund. Math.* 6(1924), 244-277. (Reprinted in S. Banach, ibidem, pp. 118-148.)

3. Banach, S. and S.Ulam, *Problem 2, 17.VII. 1935,* Scottish Book (to appear), and Problem P 34, *Coll. Math.* 1 (1948), pp. 152-153.

4. Cohen, R., and J. Fickett, "The max norm in $\mathbb{R}^n$ - isometries and measure," *Coll. Math.* (to appear).

5. Davies, Roy O., and A. J. Ostaszewski, Denumberable compact spaces admit isometry-invariant finitely additive measures, *Mathematika* (to appear).

6. Fickett, J., Thesis, University of Colorado 1978, (to appear in Studia Math.)

7. Fickett, J., and Jan Mycielski, "A problem of invariance for Lebesgue measure," *Coll. Math.* *42*(1979), 123-125.

8. Hausdorff, F., *Crundzüge der Mengenlehre*, 1914, reprinted by Chelsea 1965, 469-472.

9. Mycielski, Jan, *Commentaries to some papers of S. Banach*, in S. Banach, Oeuvres, Vol. 1, PWN Warszawa 1967, 318-327.

10. ____________, "Finitely additive invariant measures (I)," *Coll. Math.* *42* (1979), pp. 309-318.

11. ____________, "Finitely additive invariant measures (III)," *Coll. Math.* (to appear).

12. ____________, "Remarks on invariant measures in metric spaces," *Coll. Math.* *32*(1974), 105-112.

13. ____________, "A conjecture of Ulam on the invariance of measure in Hilbert's cube," *Studia Math.* *60*(1977), 1-10.

14. ____________, "Two problems on geometric bodies," *Amer. Math. Monthly* *84*(1977), 116-118.

15. Rosenblatt, J., "Finitely additive invariant measures," *Coll. Math.* *42* (1979), pp. 361-363.

16. ____________, "Uniqueness of invariant means for measure-preserving transformations," *Transactions of the AMS* (to appear).

17. Tarski, A., "Algebraische Fassung des Massproblems," *Fund. Math.* *31* (1938), 47-66.

18. ____________, "Cardinal Algebras," *Oxford Univ. Press*, New York 1949.

INTERSECTION NUMBERS AND SPACES OF MEASURES

I. Namioka

University of Washington
Seattle, Washington

This is a short report on a paper written jointly with G. Mägerl. The paper will appear in Mathematische Annalen.

Let $(S,\leqq)$ be a partially ordered set. Given a finite sequence $(s_1,\ldots,s_n)$ in S (where the terms are not necessarily distinct), let $i_S(s_1,\ldots,s_n) = {}^m/n$ where m is the largest integer such that there exists a subsequence $(s_{i_1},\ldots,s_{i_m})$ of length m that admits a lower bound in S, i.e. for some t in S, $t \leqq s_{i_k}$ for $k = 1,\ldots,m$. For a non-void subset A of S, let

$$I_S(A) = \inf\{i_S(s_1,\ldots,s_n):(s_1,\ldots,s_n) \text{ is a finite sequence in } A\}.$$

Following Kelley [2], we call $I_S(A)$ the *intersection number* of A in S. Clearly $0 \leqq I_S(A) \leqq 1$, and $I_S(A) \leqq I_S(B)$ whenever $\emptyset \neq B \subset A \subset S$.

We say that a partially ordered set S satisfies *condition* (C) (resp. condition $(C\alpha)$, where $0 < \alpha \leqq 1$) if there is a sequence $\{A_n : n \in \mathbb{N}\}$ of subsets of S such that $S = \cup\{A_n : n \in \mathbb{N}\}$ and $I_S(A_n) > 0$ (resp. $I_S(A_n) \geqq \alpha$) for each n in $\mathbb{N}$. There is quite a lot of redundancy in the above classification of partially ordered sets as the following theorem shows:

THEOREM 1. *If a partially ordered set* S *satisfies* $(C\,\alpha)$ *for some* α *in* $(0,1)$, *then* S *satisfies* $(C\,\beta)$ *for all* β *in* $(0,1)$.

Let X be a compact Hausdorff space, and let $M(X) = C(X)^*$, $M^+(X) =$

ISBN 0-12-481820-X

$\{\mu \in M(X): \mu \geq 0\}$, $M_1(X) = \{\mu \in M(X): \|\mu\| \leq 1\}$, and $M_1^+(X) = M^+(X) \cap M_1(X)$. Of course $M_1^+(X)$ is the space of regular Borel probability measures on X. A *π-base* P for the space X is a collection of non-void open subsets of X with the property that each non-void open subset of X contains a member of P. Clearly, each base for the topology of X is a π-base for X provided the base does not include the empty set. A π-base is partially ordered by the inclusion.

In the next three theorems, P is a fixed π-base for a compact Hausdorff space X. These theorems relate properties of P qua a partially ordered set to topological and measure theoretic properties of X.

THEOREM 2. *(Kelly [2], Herbert-Lacey [1]). The partially ordered set* P *has property* (C) *if and only if there exists a member of* $M_1^+(X)$ *that is supported on* X.

THEOREM 3. *The following conditions are equivalent:*

(a) *The partially ordered set* P *satisfies* ($C\alpha$) *for some* α *in* $(0,1)$.

(b) *The partially ordered set* P *satisfies* ($C\beta$) *for all* β *in* $(0,1)$.

(c) $M^+(X)$ *is weak** - *separable.*

(d) $M_1(X)$ *is weak** - *separable.*

(e) $M_1^+(X)$ *is weak** - *separable.*

THEOREM 4. *The partially ordered set* P *satisfies* (C1) *if and only if* X *is separable.*

The equivalence of (a) and (b) in Theorem 3 is, of course, a special case of Theorem 1. We remark on the weak* - separability of $M(X)$ at the end of this note.

In the proofs of Theorems 2 and 3, the information on intersection numbers must be converted to that of measures. This is accomplished by the next two lemmas. The ideas behind these lemmas are mainly due to

Kelley [2]. If X is a set, we denote by $\mathcal{X}$ the family of all non-void subsets of X. The family $\mathcal{X}$ is partially ordered by inclusion. A *mean* m on the set X is a positive linear functional on $\ell^\infty(X)$ such that $m(1) = 1$.

LEMMA 1. *Let* X *be a set, let* $\mathcal{A}$ *be a non-empty subfamily of* $\mathcal{X}$, *and let* $0 < \alpha \leqq 1$. *Then* $I_X(\mathcal{A}) \geqq \alpha$ *if and only if there is a mean* m *on* X *such that* $m(\chi_A) \geqq \alpha$ *for each* A *in* $\mathcal{A}$.

Proof (a sketch). First assume that there is a mean m on X such that $m(\chi_A) \geqq \alpha$ for each A in $\mathcal{A}$. Let $(A_1,\ldots,A_n)$ be a sequence in $\mathcal{A}$. Then $\|\Sigma\{\chi_{A_k} : k = 1,\ldots,n\}\|_\infty = n \cdot i_X(A_1,\ldots,A_n)$. Hence
$n\,\alpha \leqq m(\Sigma\{\chi_{A_k} : k = 1,\ldots, n\}) \leqq n \cdot i_X(A_1,\ldots,A_n)$, or $\alpha \leqq i_X(A_1,\ldots,A_n)$.
It follows that $I_X(\mathcal{A}) \geqq \alpha$.

Conversely suppose that $I_X(\mathcal{A}) \geqq \alpha$. Let C be the convex hull of $\{\chi_A : A \in \mathcal{A}\}$ in $\ell^\infty(X)$. Then the assumption implies that $\|f\|_\infty \geqq \alpha$ for each f in C. Let $D = \{f \in \ell^\infty(X) : f \leqq \alpha\}$. Then $(\mathrm{Int}\, D) \cap C = \emptyset$. Hence by the separation theorem, there is an m in $\ell^\infty(X)^*$ such that $m \neq 0$ and $\sup\{m(f) : f \in D\} \leqq \inf\{m(g) : g \in C\}$. The last inequality implies that $m \geqq 0$; hence we may assume that m is a mean on X. It is then easy to show that $M(\chi_A) \geqq \alpha$ for each A in $\mathcal{A}$.

LEMMA 2. *Let* X *be a compact Hausdorff space, let* $\mathcal{A}$ *be a family of non-void closed subsets of* X, *and let* $0 < \alpha \leqq 1$. *Then* $I_X(\mathcal{A}) \geqq \alpha$ *if and only if there is a member* μ *of* $M_1^+(X)$ *such that* $\mu(A) \geqq \alpha$ *for each* A *in* $\mathcal{A}$.

Proof. Assume that $I_X(\mathcal{A}) \geqq \alpha$. By Lemma 1, there is a mean m on X such that $m(X_A) \geqq \alpha$ for each A in $\mathcal{A}$. Let μ be the restriction of m to $C(X)$. Then $\mu \in M_1^+(X)$. If $A \in \mathcal{A}$, then $\mu(f) \geqq \alpha$ whenever $f \in C(X)$ and $\chi_A \leqq f$. By the regularity of μ, this implies that $\mu(A) \geqq \alpha$. The converse is proved as in Lemma 1.

We conclude with a few remarks concerning the weak*- separability of

(M(X) where X is a compact Hausdorff space. Clearly if $M^+(X)$ is weak*-separable, M(X) is weak* - separable. However, Talagrand [3] recently constructed with the continuum hypothesis a compact Hausdorff space X such that M(X) is weak* - separable and $M^+(X)$ is not. Can one construct such a space without the continuum hypothesis? It would be also interesting to give a topological characterization of those compact Hausdorff spaces X for which M(X) is weak* - separable. If X and Y are compact Hausdorff spaces such that C(X) and C(Y) are isomorphic as Banach spaces, then clearly M(X) is weak* - separable if and only if M(Y) is. Surprisingly, under the same condition, it is also true that $M^+(X)$ is weak* - separable if and only if $M^+(Y)$ is. This follows from Theorem 3.

REFERENCES

1. Herbert, D. J., Lacey, H.E., "On supports of regular Borel measures," *Pacific J. Math.*, *27*(1968), 101-118.

2. Kelley, J. L., "Measures in Borlean algebras," *Pacific J. Math.*, *9* (1959), 1165-1177.

3. Talagrand, M., "Séparabilite vague dans l'espaces de measures," (to appear).

AXIOMS, THEOREMS, AND PROBLEMS RELATED TO THE JONES LEMMA

Peter J. Nyikos

University of South Carolina, Columbia

The subject of this paper is the Jones Lemma and its numerous repercussions in metrization theory and axiomatic set theory. Probably the most famous is the "normal Moore space problem," which now has more set-theoretic axioms bearing on it than any other single problem in topology. I will describe some of those axioms, and also mention two other problems where the Jones Lemma has shed a great deal of light.

The Jones Lemma, which is really the first four theorems of [6], is the contrapositive of the case $\kappa = \aleph_0$, $\lambda = \aleph_1$ or c of the following general fact:

LEMMA. *Let* X *be a space with a discrete subspace* S *of cardinality* λ *and a dense subspace* D *of cardinality* κ .

a. If X *is hereditarily normal, then* $2^{\kappa} \geq 2^{\lambda}$.

b. If X *is normal and* S *is closed, then* $2^{\kappa} \geq 2^{\lambda}$.

Of course, if $\lambda = c$, and $\kappa = \aleph_0$, the conclusions are false and thus we have a handy way of proving certain spaces are not normal or not hereditarily normal. Among them are the Sorgenfrey plane and the tangent disk space. But the most famous consequence was proved by Jones in the same paper:

THEOREM 1. *If* $2^{\aleph_0} < 2^{\aleph_1}$ *, then every separable normal Moore space is metrizable.*

Also in [6], Jones introduced the problem of whether every normal Moore

ISBN 0-12-481820-X

space is metrizable. I will return to this theme after mentioning some less well known consequences of the Jones Lemma.

In 1948, M. Katetov published a proof of the following:

THEOREM 2 ([8]). *Let* X *be a compact space. The following are equivalent:*

1. X^3 *is hereditarily normal;*
2. X^2 *is perfectly normal;*
3. X *is metrizable.*

The obvious question that arises from this is whether the first two conditions can be replaced by " X^2 is hereditarily normal." I showed in 1977 that it is consistent that they cannot: if one splits $\aleph_1$ points of the closed unit interval, one obtains a compact space which is not metrizable (it does not have a countable base) but whose square is hereditarily normal under the axiom $MA + \neg CH$ [9]. On the other hand, the following mathematical joke suggests that it is consistent to include the condition " X^2 is hereditarily normal."

Begin with a compact space such that X^2 is hereditarily normal. By another result of Katetov in the same paper, X is perfectly normal, hence hereditarily Lindelöf. Under $MA + \neg CH$, X is hereditarily separable [12, p. 16]. Thus X^2 is separable. By Jones's lemma and $2^{\aleph_0} < 2^{\aleph_1}$, every discrete subspace of X^2 is countable. Now a result of Szentmiklóssy assures us, if we assume $MA + \neg CH$, that X^2 is hereditarily Lindelöf, hence perfectly normal. Hence by Theorem 2, X is metrizable.

Of course, we have used the mutually incompatible axioms $MA + \neg CH$ and $2^{\aleph_0} < 2^{\aleph_1}$ in arriving at our conclusion, but the joke does have its serious side. It shows that if we could arrive at the $MA + \neg CH$ results by assuming an axiom compatible with the Jones axiom $2^{\aleph_0} < 2^{\aleph_1}$, then we would show that Theorem 2 could consistently be strengthened as stated above. If that cannot be done, it still shows that if there is a "real" example of a compact nonmetrizable space whose square is hereditarily normal, it must have very different properties under different axioms. If we assume MA + ¬CH , its square must contain an uncountable discrete subspace, and be separable. If we assume $2^{\aleph_0} < 2^{\aleph_1}$, it must either not be separable, or every discrete subspace of its square must be countable. In the former case, it would have to be an L-space: a hereditarily Lindelöf regular space which is not hereditarily separable; in the latter case, its square must at least contain an S-space (a hereditarily separable regular space which is not hereditarily Lindelöf). Moreover, if X^2 is not itself a (compact) S-space, then it must also contain an L-space. This is because of the following fact, which is so well known it is hardly ever mentioned, and so easy to prove that no proof seems to have appeared in print:

LEMMA. *Let* X *be a regular space in which every discrete subspace is countable.*

a. If X *is not hereditarily separable, it contains an L-space.*

b. If X *is not hereditarily Lindelöf, it contains an S-space.*

To give you some idea of how far we are from a "real" counterexample to the improvement of Theorem 2: we still do not know whether $2^{\aleph_0} < 2^{\aleph_1}$ is enough to imply the existence of an S or L space, although the stronger axiom CH does imply it.

The mathematical joke is also good enough to establish:

THEOREM 3 [MA + ¬CH]. *If* X *is a compact space such that* X^2 *is hereditarily collectionwise normal, then* X *is metrizable.*

Indeed, the only place where we assumed $2^{\aleph_0} < 2^{\aleph_1}$ in the joke is when we showed X^2 separable → every discrete subspace of X^2 is countable. Hereditary collectionwise normality gives this in ZFC :

LEMMA. *Let* X *be a separable space.*

a. If X *is hereditarily collectionwise Hausdorff, then every discrete subspace of* X *is countable.*

b. If X *is collectionwise Hausdorff, then every closed discrete subspace of* X *is countable.*

I do not know whether "normal" can be weakened to "Hausdorff" in Theorem 3.

A similar mathematical joke can be told about a question on which several Russian topologists have worked: Does every compact, hereditarily normal space contain a nontrivial convergent sequence? Here, too, we have a "consistent counterexample," though at the opposite extreme from the one above: under the axiom $\diamond$, which implies CH, Fedorchuk has constructed a compact, hereditarily normal, hereditarily separable space which has no infinite zero-dimensional subspaces at all! [4]

On the other hand, look at an arbitrary countably infinite subset Q of a compact, hereditarily normal space. By Jones's lemma, every discrete subspace of $\bar{Q}$ must be countable if we assume $2^{\aleph_0} < 2^{\aleph_1}$. By Szentmiklóssy's result, $MA + \neg CH$ implies that $\bar{Q}$ is hereditarily Lindelöf, hence first countable! So in fact there is a sequence from Q converging to any point in $\bar{Q}$ and we can see that the whole space is sequentially compact. If we assume further that our compact space has countable tightness (that is, if A is a subset and $x \in \bar{A}$, then there is a countable $Q \subset A$ such that $x \in \bar{Q}$), then the space is Fréchet-Urysohn.

Here again, the joke gives us a connection with the " S and L problem." From the lemma there follows:

THEOREM 4 [$2^{\aleph_0} < 2^{\aleph_1}$]. *Let* X *be compact and hereditarily normal. At least one of the following is true.*

(1) X *is sequentially compact.*

(2) X *contains an L-space and an S-space.*

(3) X *contains a compact S-space of cardinality* $\geq 2^{\aleph_1} > 2^{\aleph_0}$.

Indeed, if the $\bar{Q}$ we obtain above is *not* hereditarily Lindelöf (which would make X sequentially compact, as above), then it must contain an S-space. If it does not also contain an L-space, it must be itself an S-space. (So the hereditary separability of Fedorchuk's example is not quite the frosting on the cake that it may seem to be.) If $\bar{Q}$ is not sequentially compact, it must be of cardinality $\geq 2^{\aleph_1}$ [5].

But the most famous application of the Jones lemma was found by Jones himself:

THEOREM 1 ([6]). *If* $2^{\aleph_0} < 2^{\aleph_1}$, *then every separable normal Moore space is metrizable.*

This was probably the first application of the axiom " $2^{\aleph_0} < 2^{\aleph_1}$ " outside of pure set theory, and elsewhere I have suggested that it be called "Jones's Hypothesis" for this reason [11]. In the same paper, Jones posed:

The Normal Moore Space Problem: Is every normal Moore space metrizable?

With the possible exception of the S and L problem, this is the most important unsolved problem of set-theoretic topology (at least if one combines it with some very closely related problems, like the one of whether every first countable normal space is collectionwise normal). It is also probably the oldest. I have already written a paper on F. B. Jones's main mathematical contributions to the problem [11]; there and in F. Tall's survey paper [14] and thesis [13] and in M. E. Rudin's notes [12] one can get a good feel for the history of the problem. In this paper I will focus on some of the unexpected twists and turns of the problem.

First, the main "consistency counterexamples" seem to fly in the face of what F. Burton Jones did in [6]. Even before the existence of a Q-set of reals (an uncountable subspace of the real line, every subset of which is a G_δ in the relative topology) was shown to be consistent, R. H. Bing showed that if one joins the upper-half plane with the Euclidean topology to a Q-set of reals with the tangent disk topology, one obtains a separable, nonmetrizable normal Moore space [1]. In 1967, it was shown that $MA + \neg CH$ implies there is a Q-set of reals. Ten years later, Devlin and Shelah showed it was consistent that CH (and hence "Jones's hypothesis" $2^{\aleph_0} < 2^{\aleph_1}$) be true and there be a nonmetrizable normal Moore space [3]. Of course, their example was not separable. In the summer of 1980 their result was superseded by W. Fleissner's remarkable construction showing CH *alone is enough to imply the existence of a nonmetrizable normal Moore space!* It would be ironic indeed if it could be shown that "Jones's hypothesis," which has been so effective in demolishing examples of nonmetrizable normal Moore spaces, were itself enough to imply that one exists.

This demolition has gone further than Jones himself imagined it could. At the same time Jones obtained the results of [6], he was looking at what he called "tin can spaces" (what we have come to call Aronazajn trees) and the closely related "road spaces" [7],[11] as possible examples of "real" nonmetrizable normal Moore spaces. He never dreamed that his axiom $2^{\aleph_0} < 2^{\aleph_1}$ would imply that the Moore ones are not normal! This was first shown by Devlin and Shelah [3] and then more neatly by A. Taylor [15]. Taylor's proof involved showing that the following two axioms are equivalent to "Jones's hypothesis":

Φ: Suppose the nodes of the full binary tree of height ω_1 are colored by two colors. Then at each level of the tree it is possible to pick a color in such a way that every branch of the tree has nodes of the picked color on stationary many levels.

Θ: Suppose the branches of the full binary tree of height ω_1 are colored by the $2^{\aleph_0}$ colors of the rainbow, with all nodes at limit levels colored in the same way on all the branches. Then there is a branch B which has a "typical coloring": on stationary many levels there is another branch which forks off from B at that level and is colored in the same way as B up to that level.

In stating Θ we are assuming that the way a node is colored on one branch has no influence on the way it is colored on other branches.

Of course, it is immediate that Θ implies Jones's hypothesis: if $2^{\aleph_0} = 2^{\aleph_1}$ we could simply color each of the $2^{\aleph_1}$ branches of the binary tree with a different color! It was only because Devlin and Shelah gave Φ and Θ much more abstract statements that this implication escaped their notice [2]. [The first person to notice the implication was Y. Avraham.] In particular, their formulation of Φ resembles closely the usual statement of $\diamond$. Conversely, if one substitutes "pick a node" for "pick a color" and "picked nodes" for "nodes of the picked color" in the statement of Φ above, one obtains a restatement of $\diamond$.

[Aside: it is somewhat more difficult to prove that the restatement of Θ with "two colors" replacing "$2^{\aleph_0}$ colors" is also equivalent to "Jones's hypothesis." The key is to set up, if $2^{\aleph_0} = 2^{\aleph_1}$, a 1-1 correspondence between the branches of the full binary tree of height ω_1, and that of height ω. Then we color the first ω nodes of a branch of the former according to whether there is a 0 or 1 at that node on the branch which is associated with it. Then given any two branches of the former tree, there will be a node on a finite-integer level at which their colorings disagree.]

Then there is the unexpected turn taken in the direction of an axiom implying all normal Moore spaces are metrizable. At first the results seemed to fit the pattern established by Theorem 5: W. Fleissner showed in 1974 that if $V = L$ (an axiom implying CH and hence Jones's hypothesis), then every normal space of character $\leq \aleph_1$ is collectionwise Hausdorff, and F. Tall had a number of similar results in certain models of CH [14]. But my breakthrough in 1977 came at the opposite end of the spectrum, in a model of set theory in which $2^{\aleph_0}$ is weakly inaccessible (and then some!):

THEOREM 6 ([10]). *If* PMEA , *then every normal space of character* $< c$ *is collectionwise normal and hence every normal Moore space is metrizable.*

The reason this does not settle the normal Moore space problem is that PMEA implies the consistency of there being a measurable cardinal (in fact, a proper class of measurable cardinals) and this is something whose consistency cannot be shown by simply assuming the consistency of ZFC . In fact there seems to be no plausible way of ever ruling out the possibility that ZFC implies there are no measurable cardinals.

Even after Theorem 6 was obtained, it was conjectured in [14] that if there is a model in which all normal Moore spaces are metrizable, there is also such a model in which the Generalized Continuum Hypothesis holds. Fleissner's 1980 result pretty well kills this idea and all other ideas of [14] in the direction of obtaining a "good" model where the conclusion of Theorem 6 holds. We seem to be further than ever from a final solution of the normal Moore space problem.

REFERENCES

1. Bing, R. H., "Metrization of topological spaces," *Canadian J. Math. 3* (1951), 175-186.

2. Devlin, K. and S. Shelah, "A weak version of ◇ which follows from $2^{\aleph_0} < 2^{\aleph_1}$," *Israel J. Math. 29*(1978), 239-247.

3. ______, "A note on the normal Moore space conjecture," to appear.

4. Fedorchuk, V. V., "Fully closed mappings and the consistency of some theorems of general topology with the axioms of set theory," *Math. USSR Sbornik 28*(1976), 1-26.

5. Franklin, S. P., "Spaces in which sequences suffice I and II," *Fund. Math.* *57*(1965), 107-115 and *Fund. Math.* *61*(1967), 51-56.

6. Jones, F. B., "Concerning normal and completely normal spaces," *Bull. Amer. Math. Soc.* *43*(1937), 671-677.

7. ______, "Remarks on the normal Moore space metrization problem," *Ann. Math. Studies* *60*(1966), 115-120.

8. Katetov, M., "Complete normality of Cartesian products," *Fund. Math.* *36*(1948), 271-274.

9. Nyikos, P. J., "A compact nonmetrizable space P such that P^2 is completely normal," *Topology Proc.* *2*(1977), 359-363.

10. ______, "A provisional solution to the normal Moore space problem," *Proc. Amer. Math. Soc.* *78*(1980), 429-435.

11. ______, "F. Burton Jones's contributions to the normal Moore space problem," to appear.

12. Rudin, M. E., *Lectures on Set Theoretic Topology*, AMS Regional Conf. Series in Math. No. 23, Providence, 1975.

13. Tall, F. D., "Set theoretic consistency results and topological theorems concerning the normal Moore space conjecture and related problems," *Rozprawy Mat.* *148*(1977).

14. ______, "The normal Moore space problem," in: *Topological Structures II, Part 2*, Math. Centre Amsterdam, 1980, 243-261.

15. Taylor, A., "Diamond principles, ideals and the normal Moore space problem," *Canadian J. Math.* (to appear).

THE SPACES P(S) OF REGULAR PROBABILITY MEASURES WHOSE TOPOLOGY IS DETERMINED BY COUNTABLE SUBSETS

Roman Pol (*)

University of Washington and The Polish Academy of Sciences

I. THE QUESTION

We say that a topological space X is determined by countable sets at a point $x \in X$, or, in short, that X has countable tightness at x, if for each set $A \subset X$ such that $x \in \bar{A}$ there is a countable set $C \subset A$ with $x \in \bar{C}$; if X has countable tightness at each point, we say that X has countable tightness, see [E].

Given a compact space S we denote by $P(S)$ the space of all regular probability measures on S endowed with the weak - star topology. In other words, we consider $P(S)$ as the subspace of the dual space $C(S)^*$ of the Banach space $C(S)$ of continuous real functions on S, and $\mu_\sigma \to \mu$ means that $\mu_\sigma(f) \to \mu(f)$ for each $f \in C(S)$, see [S].

The question we are going to discuss is: when does the space $P(S)$ have countable tightness?

This question can also be formulated in the following way: for what compact spaces S is it true if $f : S \to R^\alpha$ is a continuous map into the product of real lines, then the closed convex hull conv $f(S)$ of $f(S)$ has countable tightness (where R^α carries the pointwise linear structure)?[1]

(*) This paper was written while the author was a Visiting Assistant Professor at the University of Washington.

[1] The equivalence follows from the fact that the map f can always be extended continuously to a surjection $f^* : P(S) \to$ conv $f(S)$, cf. [S].

ISBN 0-12-481820-X

The details will be published in [P].

II. HAYDON'S EXAMPLE

The following example due to R. Haydon shows that under the Continuum Hypothesis (in short CH) the space $P(S)$ may fail to have countable tightness for a compact space S with countable tightness.

Example 2.1 ([H]). Assuming CH, there exists a compact space X such that each point in X has a countable base of neighbourhoods, X is non - separable, and X is the support of a regular probability measure μ.[2]

The space $P(S)$ fails to have countable tightness at the point μ. Indeed, μ is in the closure of the set $A = \{\lambda \in P(X)$: the support of λ is finite $\}$, but μ does not belong to the closure of any countable subset of A, since the support of μ is non - separable.

Question 2.2. Is it possible in the realm of the usual set theory to construct a compact space S with countable tightness such that $P(S)$ fails to have countable tightness?

Also, the Haydon example does not answer the following question.

Question 2.3. If $\mu \in P(S)$ is a purely atomic measure (see [S]) in the compact space S with countable tightness, is it true that the space $P(S)$ has countable tightness at the point μ?

III. THE CONVEX ANALOGUE TO LINDELÖF PROPERTY

The question we discuss is related to the following property (C) of Banach spaces defined by H.H. Corson [C] (we restrict ourselves here to function spaces).

A function space $C(S)$ has property (C), provided that each family of closed and convex sets in $C(S)$ with empty intersection contains a

[2] W. Fleissner and E. van Douwen [F -vD] have shown that such spaces X exist in various models of set theory, and, on the other hand, it is known that under Martin's axiom and $\neg$CH the space X with these properties does not exist.

countable subfamily with empty intersection.

The property (C) can be viewed as a convex analogue to the Lindelöf property of the weak topology of the function space $C(S)$. The dual property (C^*) defined below turns out to be a convex analogue to countable tightness of the space $P(S)$:

THEOREM 3. *The function space* $C(S)$ *has property* (C) *if and only if the space* $P(S)$ *has the following property* (C^*) : *If* $\mu \in \bar{A}$, *with* $A \subset P(S)$, *then* $\mu \in \text{conv } C$ *for some countable set* $C \subset A$.

Question 3.2 Is property (C) of $C(S)$ equivalent to countable tightness of $P(S)$? In other words, is the convex analogue to countable tightness (C^*) equivalent to countable tightness in $P(S)$?

The following two theorems contain some positive results in this direction.

THEOREM 3.3 *For compact scattered spaces* S *(more generally - for countable products of such spaces) all three properties: countable tightness of* S , *countable tightness of* $P(S)$, *and property* (C) *of* $C(S)$, *are equivalent.*

THEOREM 3.4 *If* S *is a compact space, then* $C(S^N)$ *has property* (C) *if and only if* $P(S^N)$ *has countable tightness.*[3]

In connection with Theorem 3.4 the following question arises (cf. Theorem 4.1 and Question 4.2).

Question 3.5 If $C(S)$ has property (C), does $C(S \times S)$ have this property? If $P(S)$ has countable tightness, does $P(S \times S)$ have countable tightness?

IV. MEASURES WITH METRIZABLE - LIKE SUPPORT

We shall say that a regular probability measure μ on a compact space S has metrizable - like support, provided that there exists a sequence $\underset{\sim}{A}_1, \underset{\sim}{A}_2, \ldots$ of finite disjoint families of compact sets in S such that for each open cover $\underset{\sim}{U}$ of S and $\varepsilon > 0$ there exists $\underset{\sim}{A}_i$ which refines $\underset{\sim}{U}$ and $\mu(S \setminus \bigcup \underset{\sim}{A}_i) < \varepsilon$.

Let us denote by $\underset{\sim}{S}$ the class of all compact spaces S such that each $\mu \in P(S)$ has metrizable-like support.

The class $\underset{\sim}{S}$ is stable under the operation of countable product, under continuous images, and also under the operation $P(S)$ (i.e. if $S \in \underset{\sim}{S}$, then $P(S) \in \underset{\sim}{S}$).

The class $\underset{\sim}{S}$ also contains many "classical" compact spaces:compact scattered spaces, weakly compact subsets of Banach spaces, the space of real functions on the interval whose variation is less or equal 1 with pointwise topology (cf. also Question 4.2).

THEOREM 4.1 *If* $S \in \underset{\sim}{S}$ *then property* (C) *of* $C(S)$ *is equivalent to countable tightness of* $P(S^N)$.

Question 4.2 Is it true that if $P(S)$ has countable tightness, then $S \in \underset{\sim}{S}$?

It would be interesting to find the answer in the following special case. Let $B_1(N^N)$ be the space of functions of the first Baire class on the space of irrationals N^N endowed with the pointwise topology. It is a deep theorem of H. Rosenthal [R] and J. Bourgain, D.H. Fremlin and M. Talagrand [B - F - T] that each compact set $K \subset B_1(N^N)$ is a Fréchet space (see [E]), i.e. if $f \in \bar{A}$ with $A \subset K$, then $f = \lim f_n$, where $f_n \in A$. G. Godefroy and M. Talagrand [G] proved that if a compact space K embeds in $B_1(N^N)$ then $P(K)$ also embeds in $B_1(N^N)$ and, in particular, $P(K)$ has countable tightness.

Question 4.3. Is it true that compact subspaces of the space $B_1(N^N)$ belong to the class $\underset{\sim}{S}$?

A very special result in this direction is that, as we noticed before, the compact sets of functions of bounded variation on the interval are in.

REFERENCES

[B-F-T] Bourgain, J., D. H. Fremlin, M. Talagrand, "Pointwise compact sets of Baire-measurable functions," *Amer. J. Math. 100*(1978), 845-886.

[C] Corson, H. H., "The weak topology of a Banach space," *Trans. Amer. Math. Soc. 101*(1961), 1-15.

[E] Engelking, R., *General Topology*, Warszawa 1977.

[F-vD] Fleissner, W. G., E. K. van Douwen, *The definable forcing axiom: an alternate to Martin's axiom* (preprint).

[G] Godefroy, G., "Compacts de Rosenthal," *Pac. J. Math.* (to appear).

[H] Haydon, R., "On dual L^1 - spaces and injective bidual Banach spaces," *Isr. J. Math. 31*(1978), 142-152.

[P] Pol, R., *Note on the spaces P(S) of regular probability measures whose topology is determined by countable subsets* (preprint).

[R] Rosenthal, H., "Pointwise compact subsets of the first Baire class," *Amer. J. Math. 99*(1977), 362-378.

[S] Semadeni, Z., *Banach spaces of continuous functions*, Warszawa 1970.

STRUCTURE AND CONVEXITY OF ORLICZ SPACES OF VECTOR FIELDS

M. M. Rao

University of California, Riverside

I. INTRODUCTION

A study of vector fields is motivated by the theory of group representations. Their significance and a systematic investigation of the spaces of vector fields, of L^p-type, have been recognized and undertaken by Godement [5] when the vectors are confined to Hilbert spaces. If the vectors are Banach space valued, then a detailed analysis of the structure of the spaces was presented in ([4], Ch. I) extending the work of [5]. The theory has been developed from an abstract point of view by Dinculeanu, and a final account of his results was given in ([2], Ch. VII) for the vector fields of L^Φ-type, to be defined below. The following results will generally complement the latter work, and they are needed for some recent study of problems in [8], where some of them were stated without details. Only their abstract structure theory and convexity properties will be treated here.

If (Ω,Σ,μ) is a general measure space, for each $\omega\in\Omega$, let X_ω be a real or complex Banach space with norm $|\cdot|$. If $F = \{\underset{\sim}{x}:\underset{\sim}{x} = [x_\omega \in X_\omega, \omega\in\Omega]\} = \underset{\omega\in\Omega}{\times} X_\omega$, then the elements of F are called vector fields (= champs de vecteurs, in [5]). Evidently F is a vector space under componentwise operations. [For definiteness, the scalar field of each X_ω is taken as the complex numbers, $\mathbb{C}$.] A subvector space $F_0 \subset F$ is called

ISBN 0-12-481820-X

fundamental (cf. [2],[5]) if the following conditions are met.

(i) $\{\underset{\sim}{x}(\omega):\underset{\sim}{x}\in F_0\} \subset X_\omega$ is dense for each $\omega\in\Omega$,

(ii) the function $\omega \mapsto |\underset{\sim}{x}(\omega)|\chi_A$ is μ-measurable for each $A\in\Sigma$, and $\underset{\sim}{x}\in F_0$,

(iii) given $\varepsilon > 0$, $\underset{\sim}{x}\in F$, $\{A,B\} \subset \Sigma$, $A \subset B \Rightarrow \exists\ \underset{\sim}{y}\in F_0$ such that

$$\underset{\sim}{x}\chi_A = \underset{\sim}{y}\chi_A\ ,\ \|\underset{\sim}{y}\|_B = \sup_{\omega\in B}|\underset{\sim}{y}(\omega)| \le \|\underset{\sim}{x}\|_A + \varepsilon\ .$$

[Hereafter $\|\underset{\sim}{y}\| = \|\underset{\sim}{y}\|_\Omega$, and as usual $(\underset{\sim}{x}\chi_A)(\omega) = \underset{\sim}{x}(\omega)\chi_A(\omega)$.]

In what follows $\bar{F}_0$ denotes the completion of F_0 under the norm: $\underset{\sim}{x} \mapsto \|\underset{\sim}{x}\|$. For topological spaces Ω, condition (iii) was not used in [4] and [5]. Instead, (i) was strengthened to continuity (Σ being the Borel algebra) which in the abstract case is replaced by the (weaker) (i) and (ii) here. In either case, F_0 is a $B(\Omega)$-module where $B(\Omega)$ is the complex function algebra. The spaces $L^\Phi_{F_0}(\mu)$ will be introduced, after recalling the relevant integration concepts for vector fields.

Consider $S_{F_0}(\Sigma) = \{\underset{\sim}{x} = \sum_{i=1}^n \underset{\sim}{x}_i\chi_{A_i}\ ,\ \underset{\sim}{x}_i\in F_0\ ,\ A_i\in\Sigma\ ,\ \text{disjoint}\}$, and

$$M_{F_0}(\Sigma) = \{\underset{\sim}{x}\in F\colon \exists\ \underset{\sim}{x}_n\in S_{F_0}(\Sigma)\ ,\ |x-\underset{\sim}{x}_n| \to 0 \text{ in measure}\}\ .$$

These are called spaces of *step* and *totally measurable* vector fields. The integrals are now introduced first for $\underset{\sim}{x}\in S_{F_0}(\Sigma)$ as

$$\int_A \underset{\sim}{x}d\mu = \sum_{i=1}^n \underset{\sim}{x}_i\mu(A\cap A_i)\ ,\ \underset{\sim}{x} = \sum_{i=1}^n \underset{\sim}{x}_i\chi_{A_i}\ ,\ A_i, A\in\Sigma, \mu(A) < \infty\ , \tag{1}$$

and then $\underset{\sim}{x}\in M_{F_0}(\Sigma)$ is termed integrable if there exist $\underset{\sim}{x}_n\in S_{F_0}(\Sigma)$ integrable in the sense that the integral in (1) exists (as element of F_0), $\underset{\sim}{x}_n \to \underset{\sim}{x}$ a.e.(μ) , and $\int_A |\underset{\sim}{x}_n - \underset{\sim}{x}_m| d\mu \to 0$ as $n,m \to \infty$ for all $A\in\Sigma$. Then, by definition,

$$\int_A \underset{\sim}{x}d\mu = \lim_n \int_A \underset{\sim}{x}_n d\mu \quad (\ \in\bar{F}_0\ ),\ A\in\Sigma\ . \tag{2}$$

One shows by a standard computation that the integrals in (1) and (2) are well defined and do not depend on the representation or the sequence used.

Further $\underset{\sim}{x} \mapsto \int_A \underset{\sim}{x} d\mu$, $A \in \Sigma$ defines a linear mapping of the space of integrable vector fields $L_{F_0}(\mu)$ into $\bar{F}_0$.

Let X_ω^*, $\omega \in \Omega$, be the adjoint space of X_ω , and denote by F^* $(= \times_{\omega\in\Omega} X_\omega^*)$ and F_0^* $(\subset F^*)$, the latter being a fundamental field, defined analogously. For the following analysis F_0 and F_0^* are fixed, but arbitrary. If $X_\omega = X$, $\omega \in \Omega$, then one can (and does) identify F_0 with X by taking constant mappings from Ω to X .

If $\langle \cdot , \cdot \rangle : X_\omega \times X_\omega^* \to \mathbb{C}$ is the duality pairing, then for each $\underset{\sim}{x} \in S_{F_0}(\Sigma)$, $\underset{\sim}{x}^* \in S_{F_0^*}(\Sigma)$, the functions $\omega \mapsto \langle \underset{\sim}{x}(\omega), \underset{\sim}{x}^*(\omega)\rangle$, $\omega \mapsto |\underset{\sim}{x}(\omega)|$, and $\omega \mapsto |\underset{\sim}{x}^*(\omega)|$ are μ-measurable. With this set up in the next section, Orlicz spaces of vector fields $L^\Phi_{F_0}(\mu)$ are introduced and their structure investigated, including the integral representations. The last section is devoted to the study of duality and uniform convexity (and smoothness) of these spaces. It is this aspect that is of interest in nonabelian harmonic analysis, which generalizes the work of [5]. There Ω will be a locally compact group and μ a left Haar measure. Since the latter is not necessarily σ-finite, the study should be general enough for this application. Thus μ *will only be assumed hereafter to have the finite subset property*, i.e., $A \in \Sigma, \mu(A) > 0 \Rightarrow \exists\ B \in \Sigma$, $B \subset A$ and $0 < \mu(B) < \infty$, or the (slightly stronger) *localizability property* which, however, is more general than σ-finiteness.

II. ORLICZ SPACES OF VECTOR FIELDS: STRUCTURE

Let $\Phi : \mathbb{R} \to \bar{\mathbb{R}}^+$ be a symmetric convex function, $\Phi(0) = 0$, $\Phi(\mathbb{R}) \neq \{0,\infty\}$, and define $\Psi : \mathbb{R} \to \bar{\mathbb{R}}^+$ by: $\Psi(u) = \sup\{|u|v - \Phi(v) : v \geq 0\}$. Then Ψ is also a convex function and one can redefine the pair $\{\Phi, \Psi\}$ so that

$$\Phi(1) + \Psi(1) = 1 , \quad \text{and of course } xy \leq \Phi(x) + \Psi(y) . \tag{3}$$

(See [11], p. 173 on this point.) The pair $\{\Phi,\Psi\}$ is termed a complementary "modified" Young functions. Also Φ is called a continuous Young function, if it is continuous and $\Phi(u) = 0$ iff $u = 0$. With this, the Orlicz class $\tilde{L}^{\Phi}_{F_0}(\mu)$ and space $L^{\Phi}_{F_0}(\mu)$ can be introduced as follows:

$$\tilde{L}^{\Phi}_{F_0}(\mu) = \{\underset{\sim}{x} \in M_{F_0}(\Sigma) : \int_\Omega \Phi(|\underset{\sim}{x}(\omega)|)d\mu < \infty\},$$

and

$$L^{\Phi}_{F_0}(\mu) = \{\underset{\sim}{x} \in M_{F_0}(\Sigma) : \alpha\underset{\sim}{x} \in \tilde{L}^{\Phi}_{F_0}(\mu) \text{ for some } \alpha > 0\}.$$

Then $L^{\Phi}_{F_0}(\mu)$ is a vector space while $\tilde{L}^{\Phi}_{F_0}(\mu)$ is a convex (generally nonlinear) subset of the former. For analysis on these spaces, consider the following functionals on $M_{F_0}(\Sigma)$ defined by:

$$\rho_\Phi(\underset{\sim}{x}) = \int_\Omega \Phi(|\underset{\sim}{x}(\omega)|)d\mu(\omega), N_\Phi(\underset{\sim}{x}) = \inf\{\alpha > 0 : \rho_\Phi(\tfrac{1}{\alpha}\underset{\sim}{x}) \le \Phi(1)\},$$

$$\|\underset{\sim}{x}\|_\Phi = \sup\{\int_\Omega |\langle \underset{\sim}{x}(\omega), \underset{\sim}{x}^*(\omega)\rangle| d\mu(\omega) : \rho_\Psi(\underset{\sim}{x}^*) \le 1, \underset{\sim}{x}^* \in M_{F_0^*}(\Sigma)\},$$

$$\|\underset{\sim}{x}\|'_\Phi = \sup\{\int_\Omega |\underset{\sim}{x}(\omega)||\underset{\sim}{x}^*(\omega)| d\mu(\omega) : \rho_\Psi(|\underset{\sim}{x}^*|) \le 1, \underset{\sim}{x}^* \in M_{F_0^*}(\Sigma)\}.$$

It is readily seen that $N_\Phi(\cdot)$, $\|\cdot\|_\Phi$, $\|\cdot\|'_\Phi$ are norms when the vector fields differing only on μ-null sets are identified. Further $\|\underset{\sim}{x}\|_\Phi \le \|\underset{\sim}{x}\|'_\Phi$ and hence $L^{\Phi}_{F_0}(\mu)$, the space of equivalence classes of vector fields in $L^{\Phi}_{F_0}(\mu)$ is a normed linear space under either of the norms. Let $\tilde{L}^{\Phi}_{F_0}(\mu)$ be similarly defined. The relations between these norms, as well as the basic structure of these spaces can be described precisely as follows:

THEOREM 1. *With the above notations,* $\Phi(1)N_\Phi(\underset{\sim}{x}) \le \|\underset{\sim}{x}\|_\Phi = \|\underset{\sim}{x}\|'_\Phi \le 2N_\Phi(\underset{\sim}{x})$, *and* $\{L^{\Phi}_{F_0}(\mu), N_\Phi(\cdot)\}$ *is a Banach space. Further if* $S^{\Phi}_{F_0} = S_{F_0} \cap L^{\Phi}_{F_0}(\mu)$ *and* $M^{\Phi}_{F_0}(\mu)$ *is the completion under* $N_\Phi(\cdot)$ *of the space* $S^{\Phi}_{F_0}$, *then* $M^{\Phi}_{F_0}(\mu) \subset \tilde{L}^{\Phi}_{F_0}(\mu) \subset L^{\Phi}_{F_0}(\mu)$ *with strict inclusions generally. There is equality here if* Φ *satisfies a growth condition such as* $\Delta_2 : \Phi(2u) \le C\Phi(u)$, $u \ge 0$, $0<C<\infty$.

Proof. The fundamental fields F_0 and F_0^* are fixed. To prove the middle equality, first it is noted that for $0<\varepsilon<1$, $\underset{\sim}{x}\in M_{F_0}(\Sigma)$, there is $\underset{\sim}{x}_\varepsilon^*\in M_{F_0^*}(\Sigma)$ such that

$$(1-\varepsilon)|\underset{\sim}{x}(\omega)| \le |\langle \underset{\sim}{x}(\omega),\underset{\sim}{x}_\varepsilon^*(\omega)\rangle| \ , \ 1-\varepsilon \le |\underset{\sim}{x}_\varepsilon^*(\omega)| \le 1+\varepsilon, \quad \omega\in\Omega \ . \tag{4}$$

Indeed, for the nontrivial case of $0 \ne \underset{\sim}{x}\in M_{F_0}(\Sigma)$, there exists an $\underset{\sim}{x}_0^*\in F^*$ such that $|\underset{\sim}{x}_0^*(\omega)| = 1$, and $\langle \underset{\sim}{x}(\omega),\underset{\sim}{x}_0^*(\omega)\rangle = |\underset{\sim}{x}(\omega)|$, by the Hahn-Banach theorem. Since $F_0^* \subset F^*$ is fundamental, there is an $\underset{\sim}{x}_\varepsilon^*\in F_0^*$ such that $|\underset{\sim}{x}_0^*(\omega) - \underset{\sim}{x}_\varepsilon^*(\omega)| < \varepsilon$, for all $\omega\in\Omega$ by condition (i). Hence $1-\varepsilon < |\underset{\sim}{x}_\varepsilon^*(\omega)| < 1+\varepsilon$, and moreover,

$$\begin{aligned}|\underset{\sim}{x}(\omega)| - \langle \underset{\sim}{x}(\omega),\underset{\sim}{x}_\varepsilon^*(\omega)\rangle &= \langle \underset{\sim}{x}(\omega),(\underset{\sim}{x}_0^*-\underset{\sim}{x}_\varepsilon^*)(\omega)\rangle \\ &\le |\underset{\sim}{x}_0^*(\omega)-\underset{\sim}{x}_\varepsilon^*(\omega)||\underset{\sim}{x}(\omega)| < \varepsilon|\underset{\sim}{x}(\omega)| \ .\end{aligned}$$

This implies (4).

With (4) and the definition of $\|\cdot\|_\Phi'$, consider for $0<\varepsilon<1$ and $\underset{\sim}{x}^*\in M_{F_0^*}(\Sigma)$, $\rho_\Psi(\underset{\sim}{x}^*) \le 1$,

$$\begin{aligned}\int_\Omega |\underset{\sim}{x}(\omega)||\underset{\sim}{x}^*(\omega)|\,d\mu(\omega) &\le \frac{1+\varepsilon}{1-\varepsilon}\int_\Omega |\underset{\sim}{x}^*(\omega)||\langle \underset{\sim}{x}(\omega),\frac{\underset{\sim}{x}_\varepsilon^*(\omega)}{1+\varepsilon}\rangle|\,d\mu(\omega) \ , \text{ by (4)},\\ &= \frac{1+\varepsilon}{1-\varepsilon}\int_\Omega |\langle \underset{\sim}{x}(\omega),\frac{|\underset{\sim}{x}^*(\omega)|}{1+\varepsilon}\underset{\sim}{x}_\varepsilon^*(\omega)\rangle|\,d\mu(\omega)\\ &\le \frac{1+\varepsilon}{1-\varepsilon}\sup\{\int_\Omega |\langle \underset{\sim}{x}(\omega),\underset{\sim}{y}^*(\omega)\rangle|\,d\mu : \rho_\Psi(\underset{\sim}{y}) \le 1\}\\ &= \frac{1+\varepsilon}{1-\varepsilon}\|\underset{\sim}{x}\|_\Phi \ .\end{aligned}$$

Taking suprema on the left side over $\{\underset{\sim}{x}^*\in M_{F_0^*}(\Sigma):\rho_\Psi(\underset{\sim}{x}^*) \le 1\}$ and then letting $\varepsilon \to 0$, one gets $\|\underset{\sim}{x}\|_\Phi' \le \|\underset{\sim}{x}\|_\Phi$ which together with the earlier observation proves the middle equality.

Let us now establish the other inequalities using the just established fact that $\|\cdot\|_\Phi = \|\cdot\|_\Phi'$. It suffices to consider the case that Φ is continuous, the other case (of discontinuous Φ) is easy and omitted. Let Φ' be the right derivative of Φ which exists everywhere. It is nonnegative and nondecreasing. Considering $0 \ne \underset{\sim}{x}\in M_{F_0}(\Sigma)$, let

$v(\omega) = \Phi'(|\underset{\sim}{x}(\omega)|/\|\underset{\sim}{x}\|_\Phi)$. Then $v(\cdot)$ is μ-measurable and $\rho_\Psi(v) \le 1$ (cf. [11], p. 171). Also there is equality in the inequality of (3) for this choice so that

$$(|\underset{\sim}{x}(\omega)|/\|\underset{\sim}{x}\|_\Phi)v(\omega) = \Phi(|\underset{\sim}{x}(\omega)|/\|\underset{\sim}{x}\|_\Phi) + \Psi(v(\omega)) \text{ , a.e.}(\mu) \text{ .} \tag{5}$$

Integrating, and using the preceding result, one gets

$$\int_\Omega \Phi\left(\frac{|\underset{\sim}{x}(\omega)|}{\|\underset{\sim}{x}\|_\Phi}\right) d\mu(\omega) \le \int_\Omega \frac{|\underset{\sim}{x}(\omega)|}{\|\underset{\sim}{x}\|_\Phi} v(\omega)\, d\mu(\omega) \le 1 \text{ .} \tag{6}$$

If $\alpha = \Phi(1) \le 1$, the convexity of Φ and (6) yield

$$\int_\Omega \Phi\left(\alpha\frac{|\underset{\sim}{x}(\omega)|}{\|\underset{\sim}{x}\|_\Phi}\right) d\mu(\omega) \le \alpha\int_\Omega \Phi\left(\frac{|\underset{\sim}{x}(\omega)|}{\|\underset{\sim}{x}\|_\Phi}\right) d\mu \le \alpha = \Phi(1) \text{ .}$$

Hence $\Phi(1)N_\Phi(\underset{\sim}{x}) \le \|\underset{\sim}{x}\|_\Phi$. For the last inequality, let $\underset{\sim}{y}^* \in M_{F_0^*}(\Sigma)$, $\rho_\Psi(\underset{\sim}{y}^*) \le 1$, and $\beta \ge N_\Phi(\underset{\sim}{x})$. Then using the inequality of (3), and (6):

$$\int_\Omega |\langle \underset{\sim}{x}(\omega), \underset{\sim}{y}^*(\omega)\rangle|\, d\mu(\omega) \le \beta\left[\int_\Omega \Phi\left(\left|\frac{\underset{\sim}{x}(\omega)}{\beta}\right|\right) d\mu(t) + \rho_\Psi(\underset{\sim}{y}^*)\right] \le 2\beta \text{ .}$$

Taking the suprema on $\{\underset{\sim}{x}^* \in M_{F_0^*}(\Sigma) : \rho_\Psi(\underset{\sim}{x}^*) \le 1\}$ and then letting $\beta \searrow N_\Phi(\underset{\sim}{x})$, the last inequality follows. Thus all the three norms are equivalent.

The fact that $\{L^\Phi_{F_0}(\mu), N_\Phi(\cdot)\}$ is a Banach space as well as the last inclusions and equalities follow by a simple modification of the argument of ([7_a], p. 81). The details will not be reproduced here. Thus the result holds as stated.

To analyze the structure of $L^\Phi_{F_0}(\mu)$, it is convenient to introduce:

DEFINITION: (i) Let Y be a Banach space and $B(F_0, Y)$ be the space of continuous linear mappings from the normed vector space F_0 into Y , with the uniform norm. Then $\nu : \Sigma \to B(F_0, Y)$ is *additive* if $A \in \Sigma$, $\underset{\sim}{x}, \underset{\sim}{y}$ in F_0 with $\underset{\sim}{x}\chi_A = \underset{\sim}{y}\chi_A \Rightarrow \nu(A)\underset{\sim}{x} = \nu(A)\underset{\sim}{y}$, and $\nu(A \cup B) = \nu(A) + \nu(B)$ for disjoint A, B in Σ . Further ν is *σ-additive* if $A_n \in \Sigma$, disjoint, implies $\nu(\bigcup_{n=1}^\infty A_n) = \sum_{n=1}^\infty \nu(A_n)$, the series converging in $B(F_0, Y)$. Then ν is also termed a *vector measure*.

(ii) An additive ν, as above, is of *(quasi-)* Φ*-bounded variation* relative to F_0 and μ on $A \in \Sigma$, if $I_\Phi(\nu:A) < \infty$ ($\bar{I}_\Phi(\nu:A) < \infty$), where with $\Sigma(A) = \{A \cap A_i : A_i \in \Sigma\}$,

$$I_\Phi(\nu:A) = \sup\{ \sum_{i=1}^{n} \Phi\left[\frac{\|\nu(A_i)\|}{\mu(A_i)}\right]\mu(A_i) : 0 < \mu(A_i) < \infty \ , \ A_i \in \Sigma(A), \text{ disjoint}\} \ , \tag{7}$$

and

$$\bar{I}_\Phi(\nu:A) = \sup\{\sup\left[\sum_{i=1}^{n} \Phi\left(\frac{|\nu(A_i)\underset{\sim}{x}|}{\mu(A_i)}\right)\mu(A_i) : 0 < \mu(A_i) < \infty, \ A_i \in \Sigma(A), \text{ disjoint}\right] : \|\underset{\sim}{x}\| \le 1\} \ . \tag{8}$$

One writes $I_\Phi(\nu)$ for $I_\Phi(\nu:\Omega)$ and similarly $\bar{I}_\Phi(\nu)$.

Clearly $\bar{I}_\Phi(\nu:A) \le I_\Phi(\nu:A) \le \infty$. It can be verified ([2]) that there is equality if $\mathcal{Y} = \mathbb{C}$ or $\Phi(u) = |u|$. Moreover, if the complementary function Ψ of Φ is continuous (so $\Psi(u) < \infty$ for $|u| < \infty$), then for $\nu:\Sigma \to B(F_0,\mathcal{Y})$, $\lim_{\mu(A)\to 0} \|\nu(A)\| = 0$ implying that ν is μ-continuous (as in [7_a]) whenever ν is additive.

For some work, F_0 should satisfy a *separation axiom* (of Godement):

(G): The fundamental field F_0 ($\subset F$) contains a denumerable set $\{\underset{\sim}{x}_n, n \ge 1\}$ such that $\{\underset{\sim}{x}_n(\omega), n \ge 1\} \subset X_\omega$ is dense for all $\omega \in \Omega$.

The functional $I_\Phi(\cdot,\cdot)$ has a key role to play in the theory of $L^\Phi_{F_0}(\mu)$-spaces, and it has the following structure under reasonable conditions:

THEOREM 2. *Let* $\mathcal{Y}$ *be a separable dual Banach space, the fundamental field* F_0 *also satisfy axiom* (G), *and* μ *be a localizable measure. (Haar measure has this property, when* Ω *is a locally compact group, but not necessarily σ-finite!) If* $\nu:\Sigma \to B(F_0,\mathcal{Y})$ *is a vector measure,* $I_\Phi(\nu) < \infty$, *and* Φ *is continuous, then there is a μ-unique operator field* $V = \{V(\omega):\omega \in \Omega\}$, *determined by* ν, $V(\omega) \in B(X_\omega,\mathcal{Y})$, *such that*

$$I_\Phi(\nu:A) = \int_A \Phi(\|V(\omega)\|)\,d\mu(\omega), \quad A\in\Sigma . \tag{9}$$

Hence $I_\Phi(\nu:\cdot):\Sigma \to \mathbb{R}^+$, *is a* μ*-continuous measure.*

Proof. Let $\Sigma_0 = \{A\in\Sigma:\mu(A) < \infty\}$. Then for each $A\in\Sigma_0$, $\Sigma(A) = \Sigma_0(A)$, and $I_\Phi(\nu:A) \le I_\Phi(\nu) < \infty$. This and ([10], Lemma 13) imply that ν on $\Sigma(A)$ is of bounded variation. If α is the variation of ν , then the fact that ν is μ-continuous (noted after (8)) implies that $\alpha:\Sigma(A) \to \mathbb{R}^+$ is a measure which is μ-continuous. Let $g = \frac{d\alpha}{d\mu}$ be the Radon-Nikodým derivative. Then the hypotheses on $\mathcal{V}, F_0$ and the fact that ν is of bounded variation on $\Sigma(A)$ together allow an application of the *vector* Radon-Nikodým theorem of Dinculeanu and Foiaş (1961) (cf. [2], §18), to yield a μ-unique operator field $V_A = \{V_A(\omega),\omega\in\Omega\}$ such that $\|V_A(\omega)\| = g(\omega)$ a.e.(μ) , and if $\|\underset{\sim}{x}\| \le 1$,

$$\int_B \underset{\sim}{x}(\omega)\,d\nu(\omega) = \int_B V_A(\omega)\underset{\sim}{x}(\omega)\,d\mu(\omega), \quad B\in\Sigma_0(A), \underset{\sim}{x}\in F_0 , \tag{10}$$

where the right integral is in the sense of Bochner, and the left side one is (its extension) in the sense of Bartle. A unified account of this integration theory is given in [2] for locally compact spaces; the abstract case is then an easy adjustment. Now by the support line property of a convex function (and $\alpha(A) < \infty$), it follows that $L^1(\Sigma(A),\alpha) \supset L^\Phi(\Sigma(A),\alpha)$ for the scalar spaces, and hence $\int_A f\,d\alpha = \int_A fg\,d\mu$, $f\in L^1(\Sigma(A),\alpha)$. In the scalar case, (9) is a consequence of [7_a] and the vector case needs additional arguments. This will be partly based on the work of [2], and the details are sketched as follows.

Let $0 \le f = \sum_{i=1}^n a_i\chi_{A_i}$, $A_i\in\Sigma(A)$, $A = \bigcup_{i=1}^n A_i$ (disjoint union) and $\varepsilon > 0$ be given. Choose $B_j^i\in\Sigma(A)$, a nontrivial partition of A_i , such that

$$a_i\alpha(A_i) < a_i \sum_{j=1}^{n_i} \|\nu(B_j^i)\| + \varepsilon/2n .$$

By definition of the (finite) variation of ν, one can find $\underset{\sim}{x}_j^i \in F_0$ such that $\|\underset{\sim}{x}_j^i\| \le 1$ and

$$\sum_{j=1}^{n_i} \|\nu(B_j^i)\| \le \sum_{j=1}^{n_i} |\nu(B_j^i)\underset{\sim}{x}_j^i| + \frac{\varepsilon}{2n}, \quad 1 \le i \le n .$$

Consequently,

$$\begin{aligned}\int_A f d\alpha - \varepsilon &= \sum_{i=1}^{n} a_i \alpha(A_i) - \varepsilon < \sum_{i=1}^{n} \sum_{j=1}^{n_i} a_i |\nu(B_j^i)\underset{\sim}{x}_j^i| \\ &\le k_1 k_2 \sum_{i=1}^{n} \sum_{j=1}^{n_i} \frac{\|\nu(B_j^i)\|}{k_1 \mu(B_j^i)} \cdot \frac{a_i}{k_2} \mu(B_j^i), \quad k_1 k_2 > 0 , \\ &\le k_1 k_2 \sum_{i=1}^{n} \sum_{j=1}^{n_i} \left[\Phi\left(\frac{\|\nu(B_j^i)\|}{k_1 \mu(B_j^i)}\right) + \Psi\left(\frac{a_i}{k_2}\right)\right] \mu(B_j^i) , \text{ by (3)}, \\ &\le k_1 k_2 \left[I_\Phi\left(\frac{\nu}{k_1}\right) + \rho_\Psi\left(\frac{f}{k_2}\right)\right] . \qquad (11)\end{aligned}$$

Letting $k_1 = N_\Phi(\nu) = \inf\{k > 0 : I_\Phi(\frac{\nu}{k}) \le \Phi(1)\}$, $k_2 = N_\Psi(f)$, it follows from (11), since $\Phi(1) + \Psi(1) = 1$, that

$$\int_A f d\alpha - \varepsilon < N_\Phi(\nu) N_\Psi(f) .$$

Hence

$$N_\Phi(\alpha : A) = \sup\{|\int_A f d\alpha| : N_\Psi(f) \le 1\} \le N_\Phi(\nu) < \infty . \qquad (12)$$

It follows from $\int_A f d\alpha = \int_A fg d\mu$, $f \in L^1(\Sigma(A), \alpha)$ and the inverse Hölder inequality, that $g \in L^\Phi(\Sigma(A), \mu)$. (Note that $L^\Psi(\Sigma(A), \mu) \subset L^1(\Sigma(A), \mu)$ also!) But the mapping $\alpha : B \mapsto \int_B g d\mu$, $B \in \Sigma(A)$ is an isometric isomorphism (cf. [10], Thm. 5) and hence with (12) one has, since $\|V_A(\cdot)\| = g(\cdot)$,

$$\int_A \Phi(\|V_A(\omega)\|) d\mu(\omega) = \int_A \Phi(g(\omega)) d\mu(\omega) = I_\Phi(\alpha : A) \le I_\Phi(\nu : A) . \qquad (13)$$

On the other hand, if $\|V_A(\cdot)\| \in L^\Phi(\Sigma(A), \mu)$, then by (10)

$$\|\nu(A)\| \le \sup\{\|\int_A \underset{\sim}{x} d\nu\| : \|\underset{\sim}{x}\| \le 1\} \le \int_A \|V_A(\omega)\| d\mu(\omega) .$$

So by Jensen's inequality and the monotonicity of Φ, one has

$$I_{\Phi}(\nu:A) = \sup\{\sum_{i=1}^{n} \Phi(\frac{\|\nu(B_i)\|}{\mu(B_i)})\mu(B_i):B_i\in\Sigma(A),\text{ disjoint}\}$$

$$\leq \sup\{\sum_{i=1}^{n} \Phi(\frac{1}{\mu(B_i)}\int_{B_i}\|V_A(\omega)\|d\mu(\omega))\mu(B_i):B_i\in\Sigma(A),\text{ disjoint}\}$$

$$\leq \int_A \Phi(\|V_A(\omega)\|)d\mu \ . \tag{14}$$

Thus (13) and (14) imply (9) on $\Sigma(A)$.

Since μ is localizable, V_A is μ-unique, it follows that $V_A = \tilde{V}_{\tilde{A}}$ on $A\cap\tilde{A}$ so that there exists a unique operator V such that $V(\omega) = V_A(\omega)$ for $\omega\in A\in\Sigma_0$. Moreover

$$I_{\Phi}(\nu) = \sup\{I_{\Phi}(\nu:A):A\in\Sigma_0\} = \int_{\Omega}\Phi(\|V(\omega)\|)d\mu(\omega) \ .$$

These computations are consequences of the properties of localizability (and obvious in the σ-finite case). In particular $I_{\Phi}(\nu:\cdot):\Sigma\to\mathbb{R}^+$ is a measure and is μ-continuous. This completes the proof.

The uniform convexity and smoothness conditions for the $L^{\Phi}_{F_0}(\mu)$-spaces are best studied through a representation of the dual of the relevant spaces. This will be considered in the following section since the representation theory has also independent interest.

III. DUALITY AND CONVEXITY OF $L^{\Phi}_{F_0}(\mu)$

The reflexivity and uniform convexity of the spaces are based on the following general result. From now on let μ be *localizable*.

THEOREM 3. *Let* $L^{\Phi}_{F_0}(\mu)$, $M^{\Phi}_{F_0}(\mu)$ *be the spaces as in Theorem 1, and* $\mathcal{Y}$ *be a separable dual Banach space. If* $T:M^{\Phi}_{F_0}(\mu)\to\mathcal{Y}$ *is a strongly continuous linear operator, as defined below, then there is a unique additive function* $\nu_T:\Sigma\to B(F_0,\mathcal{Y})$ *of* Φ*-bounded variation relative to* F_0,μ *such that*

$$T(\underset{\sim}{x}) = \int_{\Omega}\underset{\sim}{x}(\omega)d\nu_T, \qquad \underset{\sim}{x}\in M^{\Phi}_{F_0}(\mu) \ , \tag{15}$$

and

$$|||T||| = N_\Psi(\nu_T) \ , \tag{16}$$

where (Φ,Ψ) *is as in (3). Here* T *is strongly continuous, if* $|||T||| < \infty$ *where*

$$|||T||| = \sup\{ \sum_{i=1}^{n} \|T(x_i \chi_{A_i})\| : N_\Phi(\sum_{i=1}^{n} x_i \chi_{A_i}) \le 1, \ x_i \in F_0 \} \ .$$

If $T \in B(M^\Phi_{F_0}(\mu), \mathcal{Y})$, *then also (15) holds for* ν_T *of quasi* Φ*-bounded variation and (16) is replaced by*

$$\|T\| = \bar{N}_\Psi(\nu_T) \ , \tag{16'}$$

where $\|\cdot\|$ *is the operator (= uniform) norm and* $\bar{N}_\Psi$ *is defined as* N_Ψ *but with* $\bar{I}_\Psi$ *in place of* I_Ψ . *If further* Φ *is continuous,* F_0 *satisfies axiom* (G) *and* μ *is localizable, then there is a* μ*-unique operator field* $V_T = \{V_T(\omega), \omega \in \Omega\}$ *such that the following integral exists and equality holds:*

$$T(x) = \int_\Omega V_T(\omega) x(\omega)\, d\mu(\omega) , \qquad V_T(\omega) \in B(X_\omega, \mathcal{Y}) \ . \tag{17}$$

Moreover,

$$|||T||| = N_\Psi(\|V_T\|) \ , \quad \|T\| = \bar{N}_\Psi(\nu_T) \qquad (\le |||T|||). \tag{18}$$

In particular if Φ *satisfies the* Δ_2*-condition, then* $M^\Phi_{F_0}(\mu)$ *can be replaced by* $L^\Phi_{F_0}(\mu)$. *If* $\mathcal{Y} = \mathbb{C}$, $F_0^* \subset F^*$ *is a fundamental field of adjoint spaces and* Φ *is continuous, then for each* $\ell \in (M^\Phi_{F_0}(\mu))^*$, *there is a* μ*-unique* $x^*_\ell \in M_{F_0^*}(\Sigma)$ *such that*

$$\ell(x) = \int_\Omega \langle x(\omega), x^*_\ell(\omega) \rangle\, d\mu(\omega) , \quad x \in M^\Phi_{F_0}(\mu) \ . \tag{19}$$

When F_0^* *satisfies axiom* (G), *with* $\{(X_n, X_n^*), n \ge 1\}$ *forming the duality pairing, then* $x^*_\ell \in L^\Psi_{F_0^*}(\mu)$, *and*

$$\|\ell\| = N_\Psi(x^*_\ell) \ . \tag{20}$$

Proof. The argument is an extension of the classical case, and a condensed account will be given.

Case 1. Φ is a discontinuous Young function. Then $\Phi(u) \le c|u|$ for $|u| \le u_0 < \infty$, $= +\infty$ for $|u| > u_0 > 0$, where c is a constant. In this case it is seen that $M^{\Phi}_{F_0}(\mu) \subset L^{\infty}_{F_0}(\mu)$, the space of measurable essentially bounded vector fields. Then $\Psi(v) \le c'|v|$, $v \in \mathbb{R}$, and the Ψ-bounded variation becomes bounded variation on each $A \in \Sigma_0$, and $M^{\Phi}_{F_0}(\mu) = L^{\infty}_{F_0}(\Sigma(A),\mu)$, c' being another constant. In this case there is a unique $\nu_{T,A}$ satisfying (15) and (16) as a consequence of ([3], Thm. 6, p. 1522), and the general case on Σ follows from the uniqueness, as in the proof of Theorem 2.

Case 2. Φ is continuous. As above, first consider finite μ. If Ψ is discontinuous, then $L^{\Phi}_{F_0}(\mu) = L^{1}_{F_0}(\mu)$ and then the representation is a consequence of Dinculeanu's work ([2], §29). A different computation is needed however, when both (Φ,Ψ) are continuous, and this will be given in what follows.

Define ν_T by the equation $\nu_T(A)\underset{\sim}{x} = T(\underset{\sim}{x}\chi_A)$, $A \in \Sigma, \underset{\sim}{x} \in F_0$. Then it is easily verified that $\nu_T : \Sigma \to \mathcal{Y}$ is additive and vanishes on μ-null sets, and that (15) holds for step fields with this ν_T. If ν_T is shown to be of Ψ-bounded variation, then the result extends by continuity for all $\underset{\sim}{x} \in M^{\Phi}_{F_0}(\mu)$. Thus let $k = |||T|||$ and one needs to treat only the case, $0<k<\infty$. Consider a measurable partition $A_1,\ldots,A_n$ of Ω and $\underset{\sim}{x}_i \in F_0$, $\|\underset{\sim}{x}_i\| = 1$. If Ψ' is the right derivative of Ψ, $b_i = \Psi'_i(\|\nu_T(A_i)\underset{\sim}{x}_i\|/k\mu(A_i))$, let $f_0 = \sum_{i=1}^{n} b_i\chi_{A_i}$ ($\in L^{\Phi}(\mu)$). But by ([11], p. 175), if $0<h \in L^{\Phi}(\mu)$, and $g = \Psi'(h/N_{\Phi}(h))$, then $N_{\Psi}(g) = 1$, and there is equality in Hölder's inequality, so that

$$N_{\Phi}(h) = \int_{\Omega} hg\,d\mu \le \int_{\Omega} \Phi(h)\,d\mu + \int_{\Omega} \Psi(g)\,d\mu \text{ , by (3),}$$

$$\le \int_{\Omega} \Phi(h)\,d\mu + \Psi(1) \text{ .} \tag{21}$$

Taking $h = f_0$ here, since $\Psi(1) \le 1$, one has with (21)

$$\infty > \Psi(1) + \int_\Omega \Phi(f_0) \ge N_\Phi(|\sum_{i=1}^n b_i \underset{\sim}{x}_i \chi_{A_i}|) \text{ , since } \|\underset{\sim}{x}_i\| = 1 \text{ ,}$$

$$\ge \frac{1}{k} \sum_{i=1}^n \|T(b_i \underset{\sim}{x}_i \chi_{A_i})\| \text{ , by definition of } |||T||| \text{ ,}$$

$$= \sum_{i=1}^n b_i (\|\nu_T(A_i)\underset{\sim}{x}_i\| / k\mu(A_i))\mu(A_i) \text{ ,}$$

$$= \sum_{i=1}^n [\Phi(b_i) + \Psi(\frac{\|\nu_T(A_i)\underset{\sim}{x}_i\|}{k\mu(A_i)})]\mu(A_i) \text{ , since now there is equality in Young's inequality (3),}$$

$$= \int_\Omega \Phi(f_0)d\mu + \sum_{i=1}^n \Psi(\frac{\|\nu_T(A_i)\underset{\sim}{x}_i\|}{k\mu(A_i)})\mu(A_i) \text{ .} \tag{22}$$

The second term on the right side of (22) is thus bounded by $\Psi(1)$. Taking supremum over $\underset{\sim}{x}_i \in F_0$ of unit norm, and using the continuity of Ψ, one has

$$\sum_{i=1}^n \Psi(\frac{\|\nu_T(A_i)\|}{k\mu(A_i)})\mu(A_i) \le \Psi(1) \text{ .}$$

Now refining the partitions, this yields $I_\Psi(\nu_T/k) \le \Psi(1)$ so that $N_\Psi(\nu_T) \le k = |||T|||$.

For the opposite inequality, note that, since μ is finite, and $I_\Psi(\nu_T) < \infty$, ν_T has finite variation with α as its variation measure and Theorem 2 implies $I_\Psi(\nu_T) = I_\Psi(\alpha)$. Consider for a partition A_i ($\in \Sigma$), $1 \le i \le n$, of Ω, $\underset{\sim}{x}_i \in F_0$, $\|\underset{\sim}{x}_i\| = 1$, $\underset{\sim}{f} = \sum_{i=1}^n a_i \underset{\sim}{x}_i \chi_{A_i}$ ($\in L^\Phi_{F_0}(\mu)$), $a_i \in \mathbb{R}^+$,

$$\sum_{i=1}^n \|T(a_i \underset{\sim}{x}_i \chi_{A_i})\| = \sum_{i=1}^n a_i \|\int_\Omega \underset{\sim}{x}_i \chi_{A_i} d\nu_T\| \le \sum_{i=1}^n a_i \int_\Omega \chi_{A_i} d\alpha = \int_\Omega \|\underset{\sim}{f}(\omega)\| d\alpha \text{ ,}$$

$$\le N_\Phi(\underset{\sim}{f}) N_\Psi(\alpha) \text{ , by (12),}$$

$$\le N_\Phi(f) \cdot N_\Psi(\nu_T) \text{ , by (13).} \tag{23}$$

Taking supremum over $\{\underset{\sim}{f} : N_\Phi(f) \le 1\}$, (23) shows $|||T||| \le N_\Psi(\nu_T)$, so that $|||T||| = N_\Psi(\nu_T)$.

Since ν_T is clearly μ-continuous now, and if F_0 satisfies axiom (G), then ν_T can be represented as in (10); and by Theorem 2 both (17) and (18) hold. Finally, if μ is localizable (and Φ is continuous), then the results can be pieced together from each measurable set of μ-finite measure, using a standard procedure as in the scalar result (cf. [7_a]), because one has μ-uniqueness here also. Thus (15)-(18) hold in the general case as well.

If $\mathcal{Y} = \mathbb{C}$, then the preceding work implies that $\{V(\omega), \omega \in \Omega\}$ is a measurable vector field, and if $F_0^* \subset F^*$ also satisfies axiom (G), with matching duality pairing, then the formula,

$$\ell(\underset{\sim}{x}) = \int_\Omega \langle \underset{\sim}{x}(\omega), \underset{\sim}{x}_\ell^*(\omega) \rangle \, d\mu(\omega) ,$$

can be thought of as a continuous functional on $L^\Psi_{F_0^*}(\mu)$. In fact, the result applied to Φ, Ψ interchanged shows that $\underset{\sim}{x}_\ell^*$ can be chosen to be a measurable field so that by a simple modification of the argument of ([6], p. 109), $\underset{\sim}{x}_\ell^* \in L^\Psi_{F_0^*}(\mu)$. With this (19) and (20) are immediate consequences of (17) and (18).

Finally, the computations needed for the quasi-variation case are the same as (and slightly simpler than) the above work, and will not be repeated. Thus the result holds as stated.

Remark. There is no nice simplification of (18) for quasi-variation case when it differs from the Φ-bounded variation. Thus when $\mathcal{Y} \neq \mathbb{C}$, or $L^\Phi(\mu) \neq L^1(\mu)$, then $B(M^\Phi_{F_0}(\mu), \mathcal{Y}) \supsetneq L^\Psi_{F_0^*}(\mu)$, in general, where the latter space is isometrically identified in the former.

With the above result it is relatively easy to present conditions for reflexivity and uniform convexity of the $L^\Phi_{F_0}(\mu)$-spaces.

THEOREM 4. *(a) Let* (Φ, Ψ) *be a normalized pair of Young functions,* $F_0 \subset F$, $F_0^* \subset F^*$ *be fundamental fields, and* $L^\Phi_{F_0}(\mu), L^\Psi_{F_0^*}(\mu)$ *be the corresponding spaces of vector fields on a localizable space* (Ω, Σ, μ). *Suppose* *that*

(i) X_ω *is a reflexive Banach space for each* $\omega\in\Omega$,

(ii) F_0 *and* F_0^* *satisfy the axiom* (G), *with* $\{(X_n,X_n^*),n\geq 1\}$ *forming the duality pairing,*

(iii) $M^\Phi_{F_0}(\mu) = L^\Phi_{F_0}(\mu)$ *and* $M^\Psi_{F_0^*}(\mu) = L^\Psi_{F_0^*}(\mu)$ *(this is satisfied if both* Φ,Ψ *obey the* Δ_2*-condition).*

Then $L^\Phi_{F_0}(\mu)$ *is reflexive and* $(L^\Phi_{F_0}(\mu))^*$ *is isometrically isomorphic to* $L^\Psi_{F_0^*}(\mu)$.

(b) *A stronger conclusion obtains under the following stronger hypotheses:* (i) Φ' *is continuous and for each* $0<\varepsilon<1$ *there exist constants* $1<k_\varepsilon<c<\infty$ *such that*

$$\Phi'((1+\varepsilon)u) \geq k_\varepsilon\Phi'(u) \text{ , and } \Phi(2u) \leq c\Phi(u), \quad u\geq 0 , \tag{24}$$

(ii) *Each* X_ω, $\omega\in\Omega$, *is uniformly convex with a common modulus of convexity in the sense that given* $0<\eta\leq 2$, *there is* $1>\delta_\eta>0$ *such that* $|\underset{\sim}{x}(\omega)| = 1 = |\underset{\sim}{y}(\omega)|$, $|\underset{\sim}{x}(\omega)-\underset{\sim}{y}(\omega)| > \eta \Rightarrow |\underset{\sim}{x}(\omega)+\underset{\sim}{y}(\omega)| < 2(1-\delta_\eta)$, $\omega\in\Omega$ *and* $\underset{\sim}{x},\underset{\sim}{y}$ *in* F .

Then $L^\Phi_{F_0}(\mu)$ *is a uniformly convex space without assuming axiom* (G) *or anything on* μ *other than the finite subset property.*

In fact, the first part is a consequence of (19) and (20) of the preceding theorem so that $(L^\Phi_{F_0}(\mu))^*$ is isometrically isomorphic with $L^\Psi_{F_0^*}(\mu)$. Another application of the same result to $L^\Psi_{F_0^*}(\mu)$ yields that $(L^\Phi_{F_0}(\mu))^{**}$ is isometrically equivalent to $L^\Phi_{F_0}(\mu)$ since F_0^{**} can be identified with F_0 under the present hypothesis. The details are as in [7_b].

Regarding (b) the conditions in (i) imply the hypothesis of ([9], Thm. 5) so that the scalar space $L^\Phi(\mu)$ is uniformly convex. But X_ω, $\omega\in\Omega$, is uniformly convex with a common modulus of convexity, and so $L^\Phi_{F_0}(\mu)$ can be regarded as a "substitution space." It then follows by ([1], p. 749), that $L^{\Phi}_{F_0}(\mu)$ is uniformly convex (hence also reflexive).

By a general result in abstract analysis, the adjoint space of a uniformly convex space is (not necessarily uniformly convex but is) uniformly smooth. (The latter term is not defined here, but see [9].)

If each X_ω is a Hilbert space, then as is known, $\delta_\eta = 1 - (1 - \frac{\eta^2}{4})^{\frac{1}{2}}$ (independent of $\omega \in \Omega$) and if $\Phi(u) = \frac{|u|^p}{p}$, $1<p<\infty$, then (20) is satisfied and the hypothesis of part (b) is fulfilled. Hence the result reduces to Godement's ([5], p. 91) in this case. But the X_ω can be more general Banach spaces. For instance, each X_ω may be an Orlicz space $L^{\theta}(\lambda_\omega)$ on a probability space (S,A,λ_ω), $\omega \in \Omega$, and θ satisfies (24). Here $\lambda_\omega(\cdot)$: $A \to \mathbb{R}^+$ is a probability measure and $\lambda_{(\cdot)}(A):\Omega \to \mathbb{R}^+$ is a measurable function relative to Σ, for each $A \in A$. Evidently several other choices are possible.

REFERENCES

1. Day, M. M., "Uniform convexity - III," *Bull. Amer. Math. Soc. 49*(1943), 745-750.

2. Dinculeanu, N., *Integration on Locally Compact Spaces*, Noordhoff Int'l Publishers, Leyden, The Netherlands, 1974.

3. ______, "Linear operations on spaces of totally measurable functions," *Rev. Roum. Math. Pures Appl. 10*(1965), 1493-1524.

4. Dixmier, J. and A. Douady, "Champs continus d'espaces Hilbertiens et de C^*-algebras," *Bull. Soc. Math. France 91*(1963), 227-284.

5. Godement, R., "Sur la théorie des representations unitaires," *Ann. Math. 53*(2)(1951), 68-124.

6. Ionescu Tulcea, C. "Deux théorèmes concernant certains espaces de champs de vectors," *Bull. Soc. Math. France 79*(1955), 106-111.

7. Rao, M. M., "Linear functionals on Orlicz spaces," (a) *Nieuw Arch. Wisk. 12*(3)(1964), 77-98; (b) *Pacific J. Math. 25*(1968), 553-585.

8. ______, "Convolutions of vector fields - I," *Math. Z. 174*(1980), 63-79.

9. ______, "Smoothness of Orlicz spaces - II," *Indag. Math. 27*(1965), 681-690.

10. Uhl, Jr., J. J., "Orlicz spaces of finitely additive set functions," *Studia Math.* *29* (1967), 19-58.

11. Zygmund, A., *Trigonometric Series I,* 2d ed., Cambridge University Press, London, 1959.

ON THE CHOQUET'S THEORY AND VECTOR MEASURES

Elias Saab and Paulette Saab

The University of British Columbia, Vancouver, Canada

For a Choquet simplex K, it is shown that certain classes of bounded linear operators on $A(K)$, the space of all affine continuous real functions on K, behave in the same way as bounded linear operators on spaces of continuous functions on compact spaces. For example, if E is a Banach space every bounded linear operator $T: A(K) \to E$ can be represented by an E^{**}-valued finitely additive boundary vector measure G, T is compact if and only if G has a relatively compact range, T is absolutely summing if and only if G is of bounded variation, and T is nuclear if and only if G has a Bochner integrable Radon-Nikodym derivative with respect to its variation. Also, it is shown that if E does not contain a copy of c_o then every bounded linear operator is weakly compact.

I. INTRODUCTION

Let K be a compact convex subset of a locally convex Hausdorff topological vector space. The Choquet theorem in its geometrical form asserts that each element of K can be represented by a "boundary" probability measure on K. When the compact convex set K is metrizable, boundary scalar measures can easily be described as those measures on K that are supported by the G_δ set of extreme points of K. In what follows we shall be interested in Choquet simplexes, i.e., those compact convex sets for which uniqueness of the Choquet integral representation

ISBN 0-12-481820-X

theorem holds. Examples of Choquet simplexes are provided by the usual simplexes in the finite dimensional case. Other examples are provided by the compact convex sets of probability measure on compact Hausdorff spaces. For a detailed study of boundary measures in general and Choquet simplexes we refer the reader to [1], [2] or [4].

In this paper it will be shown that if K is a Choquet simplex and $A(K)$ is the space of all affine continuous real valued functions on K, then bounded linear operators on $A(K)$ behave in the same way as bounded linear operators on $C(\Omega)$ spaces. For example, it will be shown that if E is a Banach space, each bounded linear operator $T: A(K) \to E$ can be represented by a finitely additive "boundary" E^{**}-valued measure G; T is compact if and only if G has a relatively compact range; T is weakly compact if and only if G is regular; T is absolutely summing if and only if G is of bounded variation and T is nuclear if and only if G has a Bochner integrable Radon-Nikodym derivative with respect to the variation of G. All the above cited results generalize well known results for operators on spaces of continuous functions. This should be no surprise, for if Ω is a compact Hausdorff space one can view $C(\Omega)$ as the space of affine continuous functions on the Choquet simplex K of probability measures on Ω. Finally, Choquet's theory is used to prove another property that operators on $A(K)$ spaces share with their counterparts on $C(\Omega)$ spaces, namely it will be shown that if E does not contain a copy of c_o then every bounded linear operator on $A(K)$ is weakly compact.

II. PRELIMINARIES AND NOTATIONS

For a Banach space E we shall denote by E^* its topological dual. For a compact convex subset K of a locally convex Hausdorff space K we shall denote by Σ the σ-field of Borel subsets of K, by $C(K)$ the Banach space of all continuous real-valued functions, by $A(K)$ the subset of $C(K)$ consisting of affine functions on K, and by $M(K)$ the Banach space of all Borel regular scalar measures on K under the variation norm. If $\lambda \in M(K)$, $|\lambda|$ denotes its variation.

Definition 1. Let K be a compact convex set, E be a Banach space and let $G : \Sigma \to E^{**}$ be a finitely additive vector measure. The measure G is called a *boundary vector measure* if for each $x^* \in E^*$ the scalar measure $G(\cdot)x^* \in M(K)$ and its variation $|G(\cdot)x^*|$ is maximal in the Choquet ordering [2].

The study of operators on $A(K)$ spaces with K a Choquet simplex was initiated in [7]. In fact, it was shown that Choquet simplexes can be characterized by the property that for every Banach space E every bounded linear operator is represented by a boundary vector measure.

III. STRUCTURE OF OPERATORS ON $A(K)$:

The following representation theorem will be needed in the sequel.

THEOREM 2. *Let* K *be a Choquet simplex,* E *be a Banach space and let* $T : A(K) \to E$ *be a bounded linear operator. Then there exists a unique finitely additive vector measure* G *defined on* Σ *with values in* E^{**} *such that*

(i) *the measure* G *is a boundary vector measure;*

(ii) $x^*T(a) = \int a\, d(G(\cdot)x^*)$ *for each* $x^* \in E^*$;

(iii) *the mapping* $x^* \to G(\cdot)x^*|_{A(K)}$ *is weak*-to-weak* continuous;*

(iv) $\| T \| = \sup\{|G(\cdot)x^*|(K), \ \|x^*\| \leq 1\}$.

Conversely, any vector measure $G : \Sigma \to E^{**}$ *that satisfies* (i), (ii) *and* (iii) *defines a bounded linear operator from* $A(K)$ *to* E *that satisfies* (iv).

Proof. Let K be a Choquet simplex, E be a Banach space and let

$T : A(K) \to E$ be a bounded linear operator. Consider $T^* : E^* \to A(K)^*$ the adjoint linear operator of T. It follows from [8] that there exists an isometric linear selection map $S : A(K)^* \to M(K)$, i.e., S is linear and for each $\ell \in A(K)^*$, $S(\ell) = \ell$ on $A(K)$, $\| S(\ell) \| = \| \ell \|$ and $|S(\ell)|$ is maximal in the Choquet ordering [2]. Define now the vector measure $G : \Sigma \to E^{**}$ as follows: For each Borel set $B \in \Sigma$, let

$$G(B) = T^{**}(S^*(\phi_B)) ,$$

where ϕ_B is the element of $M(K)^*$ defined by $\phi_B(\lambda) = \lambda(B)$ for $\lambda \in M(K)$, and T^{**} (resp. S^*) is the adjoint linear operator of T^* (resp. of S). For each $x^* \in E^*$, it follows from the definition of G that

$$G(\cdot)x^* = S(T^*(x^*)) .$$

This proves (i), (ii) and (iii). Also since $|G(\cdot)x^*|(K) = \| T^*(x^*) \|$ for each $x^* \in E^*$, it follows that $\| T \| = \sup\{ |G(\cdot)x^*|(K) , \| x^* \| \leq 1\}$. The uniqueness of the boundary vector measure follows from condition (iii) and the fact that two boundary measures that agree on $A(K)$ must be identical, see [8]. The verification of the last statement of the theorem is routine.

As in the case of operators on spaces of continuous functions the above theorem has lots of applications. The following theorem characterizes compact operators on $A(K)$ spaces.

THEOREM 3. *Let* K *be a Choquet simplex, and* E *be a Banach space. A bounded linear operator* $T : A(K) \to E$ *is compact if and only if its representing boundary measure has a relatively compact range.*

Proof: Let $T : A(K) \to E$ be a bounded linear operator. Theorem 2 guarantees that T lifts to an operator $\tilde{T} : C(K) \to E^{**}$ where for $f \in C(K)$,

$$\tilde{T}(f) = \int f \, d\, G = T^{**}(S^*(f)) .$$

By definition of $\tilde{T}$, the operator T is compact if and only if $\tilde{T}$ takes its values in E and is compact. Moreover the measure G is also the measure representing the operator $\tilde{T}$ on $C(K)$, by [3] it follows that T is compact if and only if G has compact range. This completes the proof.

The proof of the following theorem is very similar to that of Theorem 3 therefore we shall omit it.

THEOREM 4. *Let* K *be a Choquet simplex, and let* E *be a Banach space. A bounded linear operator* $T : A(K) \to E$ *is weakly compact if and only if its boundary representing measure* G *is regular.*

Definition 5. Let X and Y be two Banach spaces. A bounded linear operator $T : X \to Y$ is called an *absolutely summing operator* if T maps weakly unconditionally Cauchy series into absolutely convergent series.

It is easy to check that an operator $T : X \to Y$ is absolutely summing if and only if there exists $\alpha \geqq 0$ such that for any finite set $x_1, x_2, \ldots, x_n$ in X the following inequality holds:

$$(*) \qquad \sum_{i=1}^{n} \| T x_i \| \leqq \alpha \sup\{ \sum_{i=1}^{n} |x^*(x_i)| : x^* \in X^*, \| x^* \| \leqq 1\} .$$

And the absolutely summing norm of T is

$$\| T \|_{as} = \inf\{\alpha \geqq 0 : (*) \text{ holds}\} .$$

Absolutely summing operators on spaces of continuous functions are those operators whose representing measures are of bounded variation. The following theorem shows that the same kind of theorem holds for absolutely summing operators on $A(K)$ spaces for Choquet simplexes.

THEOREM 6. *An operator* $T : A(K) \to E$ *is absolutely summing if and only if its representing boundary measure* G *is of bounded variation. Moreover in this case* $\| T \|_{as} = |G|(K)$.

Proof. Let $T : A(K) \to E$ be an absolutely summing operator. It follows from the so called "Principle of local reflexivity" [5], that T^{**} is also absolutely summing. Hence there exists $\alpha \geq 0$ such that for every finite cover $B_1, B_2, \ldots, B_n$ of K consisting of disjoint elements of Σ we have

$$\sum_{i=1}^{n} \| G(B_i) \| = \sum_{i=1}^{n} \| T^{**}(S^*(\phi_{B_i})) \|$$

$$\leq \alpha \sup\{ \sum_{i=1}^{n} | S^*(\phi_{B_i})(\ell) | : \ell \in A(K)^*, \| \ell \| \leq 1\}$$

$$\leq \alpha \sup\{ \sum_{i=1}^{n} | S(\ell)(B_i) | : \ell \in A(K)^*, \| \ell \| \leq 1\}$$

$$\leq \alpha \sup\{ |S(\ell)|(K) : \ell \in A(K)^*, \| \ell \| \leq 1\}$$

$$\leq \alpha .$$

Hence $|G|(B) \leq \alpha$, in fact this shows that $|G|(K) \leq \|T^{**}\|_{as} = \|T\|_{as}$.

Conversely, if G is of bounded variation then it is easy to check that the extension $\tilde{T}$ of T is absolutely summing. It follows that

$$\| T\|_{as} \leq \| \tilde{T} \|_{as} \leq |G|(K) .$$

Definition 7. Let X and Y be two Banach spaces, a bounded linear operator $T : X \to Y$ is called *nuclear* if there exists sequences $(x_n^*)_n$ in X^* and $(y_n)_n$ in Y such that $\sum_{n=1}^{\infty} \|x_n^*\| \ \|y_n\| < \infty$ and such that for each $x \in X$

$$(**) \qquad T(x) = \sum_{n=1}^{\infty} x_n^*(x) y_n .$$

If $T : X \to Y$ is a nuclear operator the nuclear norm of T is defined by

$$\| T\|_{nuc} = \inf\{ \sum_{n=1}^{\infty} \| x_n^*\| \ \| y_n\| \}$$

where the infimum is taken over all sequence $(x_n^*)_n$ and $(y_n)_n$ such that (**) holds.

THEOREM 8. *A bounded linear operator* $T : A(K) \to E$ *is a nuclear operator if and only if its representing boundary measure* G *has a Bochner integrable derivative* g *with respect to its variation* $|G|$. *In this case* $\|T\|_{nuc} = |G|(K) = \int_K \|g(\omega)\| \, d|G|(\omega)$.

Proof. If $T : A(K) \to E$ is a nuclear operator, then there exists sequences $(a_n^*)_n$ in $A(K)^*$, $(x_n)_n$ in E such that $T(a) = \sum_{n=1}^{\infty} a_n^*(a)x_n$ and $\sum_{n=1}^{\infty} \|a_n^*\| \, \|x_n\| < \infty$. Now, let G be the vector measure defined by

$$G(B) = \sum_{n=1}^{\infty} S(a_n^*)(B)x_n \quad \text{for each } B \in \Sigma ,$$

then G is well defined and it is easily checked that T and G agree on $A(K)$. Moreover, since for each $n \geqq 1$, $S(a_n^*)$ is a boundary measure, the measure $\sum_{n=1}^{\infty} |S(a_n^*)| \, \|x_n\|$ is also a boundary measure. Also since for each $B \in \Sigma$ one has

$$|G|(B) \leqq \sum_{n=1}^{\infty} |S(a_n^*)|(B) \|x_n\| ,$$

it follows by [2,27.7] that the vector measure G is a boundary vector measure of bounded variation. Hence by the uniqueness of the represent-ation theorem it follows that G is the unique boundary measure repre-senting the operator T and in fact this also shows that G does not depend on the choice of the sequences $(a_n^*)_n$ and $(x_n)_n$ and therefore $|G|(K) \leqq \|T\|_{nuc}$. This shows that the extension $\tilde{T}$ of T to $C(K)$ is nuclear and for each $f \in C(K)$ we have

$$\tilde{T}(f) = \sum_{n=1}^{\infty} S(a_n^*)(f)x_n .$$

Now by [3,p. 173] it follows that G, which also represents the operator $\tilde{T}$, has a Bochner integrable derivative with respect to $|G|$. The converse follows from the fact that if G has a Bochner integrable derivative then the extension $\tilde{T}$ of T will be nuclear and

$\|\tilde{T}\|_{nuc} \leq |G|(K)$, hence T is nuclear and $\|T\|_{nuc} \leq |G|(K)$.

COROLLARY 9. *If the Banach space* E *has the Radon-Nikodym property then for any Choquet simplex* K , *any absolutely summing operator* $T : A(K) \to E$ *is nuclear and* $\|T\|_{nuc} = \|T\|_{as}$.

Proof. Let K be a Choquet simplex, E be a Banach space and $T : A(K) \to E$ be an absolutely summing operator. If G is the representing boundary measure of T then $|G|(K) < \infty$ and T extends to an absolutely summing operator $\tilde{T}$ on $C(K)$, in particular $\tilde{T}$ is weakly compact and hence the measure G takes its values in E . Since E has the Radon-Nikodym property [3] there exists a Bochner integrable function $g : K \to E$ such that

$$G(B) = \int_B g(\omega)\, d|G|(\omega) \quad \text{For every } B \in \Sigma .$$

By [3, p. 174] T is nuclear, and by Theorem 8, $\|T\|_{nuc} = |G|(K) = \|T\|_{as}$.

In all the theorems we have proved so far, the use of the Choquet theory did not seem to be crucial except for the facts that we are working on Choquet simplexes and that the representing boundary vector measure is unique. The next Theorem makes a more decisive use of the Choquet theory and makes good use of the properties of boundary scalar measures.

THEOREM 10. *If* K *is a Choquet simplex, and if* E *is a Banach space not containing an isomorphic copy of* c_o , *then every bounded linear operator* $T : A(K) \to E$ *is weakly compact.*

Proof. Let $T : A(K) \to E$ be a bounded linear operator. By Theorem 2, T extends to a bounded linear operator $\tilde{T} : C(K) \to E^{**}$ with $\tilde{T}(f) = \int f\, dG$. Hence to complete the proof, it is enough to show that the extension $\tilde{T}$ takes its values in E . For this, note that by passing to an appropriate quotient map one can assume that the simplex K is metrizable. Now consider the unique boundary vector measure G representing the operator T. Let f be a continuous convex function on K, and let $\hat{f}(x) = \inf\{a(x), a > f, a \in A(K)\}$. The function $\hat{f}$ is of course upper semicontinuous and since K is a simplex, $\hat{f}$ is affine [2] or [1]. It follows by a result of Mokoboski [2] that there exists a decreasing sequence $a_n \in A(K)$ such that $a_n \downarrow \hat{f}$ pointwise. The existence of such a sequence is gauranteed by the metrizability of K.

Let $s_n = a_n - a_{n+1} \geqq 0$, then $\hat{f} = a_1 - \sum_{n=1}^{\infty} s_n$ pointwise. It follows that there exists an $M > 0$ such that for any $k \in K$ we have $\sum_{n=1}^{\infty} s_n(k) < M$ and therefore for every $x^* \in E^*$ one can deduce that

$$\sum_{n=1}^{\infty} \int s_n d|G(\cdot)x^*| < \infty .$$

Hence

$$\sum_{n=1}^{\infty} |x^*(T(s_n))| < \infty ;$$

it follows that the series $\sum_{n=1}^{\infty} T(s_n)$ is weakly unconditionally convergent in E, therefore it is unconditionally convergent because E does not contain an isomorphic copy of c_o. Hence $\sum_{n=1}^{\infty} T(s_n) \in E$. This shows that the Bartle integral of $\hat{f}$, $\int \hat{f}\, d\, G$ is an element of E. Moreover since $\hat{f} = f$ on the extreme points of K and since for every $x^* \in E^*, G(\cdot)x^*$ is a boundary measure, it can be shown that

$$\tilde{T}(f) = \int f\, d\, G = \int \hat{f}\, d\, G \in E .$$

Finally since the cone of continuous and convex functions on K is total in $C(K)$, it follows that $\tilde{T}$ as an operator on $C(K)$ takes its values in E. Apply now Pelcynzki's thoerem [6] to deduce that $\tilde{T}$ is weakly compact and therefore T is also weakly compact.

REFERENCES

[1] Alfsen, E. M., *Compact convex sets and boundary integrals*, Ergebnisse Math. Grenzgebiete, Band 57, Springer-Verlag, New York, 1971.

[2] Choquet, G., *Lectures on analysis, Vol. II*, W. A. Benjamin, New York, Amsterdam, 1969 (Mathematics Lectures Notes Series).

[3] Diestel, J., J. J. Uhl, Jr., *Vector Measures*, Math. Surveys No. 15, American Mathematical Society, Providence 1977.

[4] Phelps, R. R. , *Lectures on Choquet's theorem*, Van Nostrand, Princeton, N.J. 1966. MR 33 #1690.

[5] Lindenstrauss, J. and H.P.Rosenthal, "The L_p spaces," *Israel J. Math. 7*, (1969), 325-349.

[6] Pelcznski, A., "Projections in certain Banach spaces," *Studia Math. 19*(1960), 209-228.

[7] Saab, E., and P. Saab, "On operators on A(K,E)," to appear in *Bull. Sci. Math. France*.

[8] Saab, P., "The Choquet integral representation in the affine vector-valued case," *Aequationes Mathematicae 20*, (1980), 252-262.

A GENERAL RUDIN-CARLESON THEOREM FOR VECTOR-VALUED FUNCTIONS

Paulette Saab

The University of British Columbia, Vancouver, Canada

I. INTRODUCTION

In [3] Per Hag gave a general version of the Rudin-Carleson theorem for scalar-valued functions. Namely, he showed that if K is a closed subset of a compact Hausdorff space X, and A is a closed linear subspace of $C(X)$ such that for every scalar measure μ in the annihilator of A, the restriction μ_K of μ to K is also in the annihilator of A, then each element $a_o \in A_{|K}$ which is dominated by a positive lower semicontinuous function p, has an extension $a \in A$ that is also dominated by p.

The object of this paper is to obtain quickly a version of Hag's theorem for fector-valued functions and to use it to obtain an extension of Bishop's general Rudin-Carleson theorem [1] for scalar-valued functions. Other known results [2] are also deduced.

II. PRELIMINARIES AND NOTATIONS

All Banach spaces considered will be over the complex field. If Y is a compact Hausdorff space and E is a Banach space with dual E^* we shall denote by $C(Y,E)$ the Banach space of all continuous E-valued functions under the supremum norm. The symbol $M(Y,E^*)$ will stand for the space of all E^*-valued weak*-regular vector measures defined on the σ-field Σ of Borel subsets of Y and that are of bounded variation.

ISBN 0-12-481820-X

If $E = \mathbb{C}$ we shall simply write $C(Y)$ and $M(Y)$. If $\mu \in M(Y,E^*)$ and $e \in E$, $\langle e,\mu\rangle$ denotes the element of $M(Y)$ defined by

$$\langle e,\mu\rangle(B) = \mu(B)(e), \text{ for all } B \in \Sigma .$$

It is known [5] that the space $M(Y,E^*)$ is a Banach space under the variation norm and that $M(Y,E^*)$ is isometrically isomorphic to the dual of $C(Y,E)$. For $\mu \in M(Y,E^*)$ and $B \in \Sigma$, the restriction of μ to B is denoted by μ_B. If D is a closed linear subspace of $C(Y,E)$ we understand by $D^\perp$ the annihilator of D in $M(Y,E^*)$. If K is a closed subset of Y, the set of all restrictions of D to K will be denoted by $D_{|K}$.

III. REDUCTION TO THE SCALAR CASE

To extend the result of Per Hag [3] to Banach space-valued functions, let us observe that if X is a compact Hausdorff space, E is a Banach space with dual E^* and if $B(E^*)$ is the closed unit ball of E^* endowed with the weak*-topology, then it is easily checked that the mapping $I : C(X,E) \to C(X \times B(E^*))$ such that

$$I(\phi)(x,x^*) = x^*(\phi(x)),$$

for all $\phi \in C(X,E)$ and $(x,x^*) \in X \times B(E^*)$ embeds $C(X,E)$ isometrically in $C(X \times B(E^*))$.

This observation allows us to prove the key lemma of this paper.

LEMMA 1. *If* X *is a compact Hausdorff space,* K *is a closed subset of* X, E *is a Banach space, and* D *is a closed linear subspace of* $C(X,E)$, *then the following conditions are equivalent:*

(i) $\mu \in D^\perp \Longrightarrow \mu_K \in D^\perp$, *for all* $\mu \in M(X,E^*)$;

(ii) $\theta \in I(D)^\perp \Longrightarrow \theta_{K\times B(E^*)} \in I(D)^\perp$, *for all* $\theta \in M(X\times B(E^*))$.

Proof: (i) $\Longrightarrow$ (ii)

SECTION V

ANNOTATED BIBLIOGRAPHY

A BIBLIOGRAPHY ON APOSYNDESIS

E. E. Grace

Arizona State University, Tempe

A systematic attempt has been made to include all published papers and books that either state results about, or use, aposyndesis or a closely related concept. Ph.D. theses, preprints and work completed but not written up ("in progress") have been included where convenient. Second authors have been listed separately, with references to all primary listings of their joint papers. A systematic attempt was also made to list all of the reviews of these works appearing in *Mathematical Reviews* (MR) and *Zentralblatt für Mathematik und ihre Grenzgebiete* (Zbl). It has been impossible to annotate all entries. The choice of papers to be annotated was determined more by expediency than by logic.

"See [n]" in an annotation may mean either to see the nth paper or to see the annotation concerning that paper. In the annotations, a continuum is compact if it is not in a quotation.

Some open research questions are cited where appropriate in the text. However, no attempt is made to provide a complete source for such problems here. Instead, with the cooperation of Professor Howard Cook at the University of Houston, unanswered research questions have been systematically collected and entered in the (computerized) University of Houston Problem Book. It is hoped that new questions regarding aposyndesis and information about solutions of old problems will be sent to the Problem Book (University of Houston Problem Book, Mathematics Department, Cullen Blvd.,

ISBN 0-12-481820-X

Houston, TX 77004) as they arise so that it will continue to be an up-to-date source of information concerning research problems related to aposyndesis.

I wish to express my appreciation to the many people who contributed to the development of this bibliography, particularly to Rudy Gordh and Bob Hunter, who supplied information related to the subjects of their papers in this volume, and to my son, David, who did much of the library work.

1. Bellamy, D. P., "Continua for which the set function T is continuous," *Trans. Amer. Math. Soc.* 151(1970), 581-587. MR42#6791;Zbl 207 p. 530.

2. ______, "Aposyndesis in the remainder of Stone-Čech compactifications," *Bull. Acad. Polon. Sci. Sér. Sci. Math. Astronom. Phys.* 19(1971), 941-944. MR46#8176;Zbl 222#54028.

"...his main result is that $\beta(R^n)-R^n$ is an aposyndetic continuum." MR.

3. ______, "Set functions and continuous maps," these *Proceedings*.

4. Bellamy, D. P. and J. J. Charatonik, "The set function T and contractability of continua," *Bull. Acad. Polon. Sci. Sér. Sci. Math. Astronom. Phys.* 25(1977), 47-49. MR58#18370;Zbl 344 p. 54025.

"A sufficient condition of noncontractibility of Hausdorff continua is proved. It is of particular interest in the case of dendroids." See [25].

5. Bellamy, D. P. and H. S. Davis, "Continuum neighborhoods and filter-bases," *Proc. Amer. Math. Soc.* 27(1971), 371-374. MR43#2653;Zbl 206 p. 516.

6. Bellamy, D. P. and L. R. Rubin, "Indecomposable continua in Stone-Čech compactifications," *Proc. Amer. Math. Soc.* 39(1973), 427-432. MR 47#4219;Zbl 257#54035.

A lemma concerns the separation of irreducible, compact Hausdorff continua by sets T(x) .

7. Bennett, D. E., *Aposyndetic Continua and Some Characterizations of Dendrites*, Ph.D. Thesis, University of Kentucky, 1970.

8. ______, "Aposyndetic properties of unicoherent continua," *Pacific J. Math.* 37(1971), 585-589. MR46#4500;Zbl 198#559, 215#238.

9. ______, "A sufficient condition for countable set aposyndesis," *Proc. Amer. Math. Soc.* 32(1972), 578-584. MR45#2671;Zbl 228#54028.

10. Bennett, D. E., "A characterization of locally connectedness by means of the set function T ," *Fund. Math.* *86*(1974), 137-141. MR50#11189; Zbl 291#54035.

11. ______, "Strongly unicoherent continua," *Pacific J. Math.* *60*(1975), 1-5. MR52#4246.

"Main Theorem: If X is strongly unicoherent, and both aposyndetic and semi-locally connected at p , then X is hereditarily unicoherent at p ." MR.

12. ______, "Aposyndesis and unicoherence," these *Proceedings*.

13. Bennett, D. E. and J. B. Fugate, "Continua and their nonseparating subcontinua," *Dissertationes Math. (Rozprawy Mat.)* *149*(1977), 46 pp. MR56#9499.

"They prove...that if L is a subcontinuum of the continuum M which properly contains a terminal noncutting subcontinuum of M , then L is semi-locally-connected at K ." MR.

14. Bennett, Ralph, *On Some Classes of Noncontractible Dendroids*, Math. Institute of the Polish Academy of Sci., Mimeographed Paper, 1972 (unpublished).

"Proves" a theorem on noncontractibility of dendroids satisfying conditions on the T function, using a false result from the literature. The theorem was later proved as [4, Thm. 1].

15. Bing, R. H., "Some characterizations of arcs and simple closed curves," *Amer. J. Math.* *70*(1948), 497-506. MR10 p. 55;Zbl 41 p. 318.

"Theorem 10. A nondegenerate compact continuum is a simple closed curve if it is neither cut by any point nor separated by any one of its subcontinua."

The proof of this uses one of Jones's cutpoint theorems for nonaposyndetic continua [123]. The relationship, in the proof, between aposyndesis and the nonseparating property motivated the study of θ- and θ_n-continua in [45],[76],[193], etc.

16. Burgess, C. E., "Some theorems on n-homogeneous continua," *Proc. Amer. Math. Soc.* *5*(1954), 136-143. MR15 p. 814;Zbl 57 p. 151.

Conditions for a continuum to be indecomposable or to be indecomposable under index n are given in terms of aposyndesis.

17. ______, "Separation properties and n-indecomposable continua," *Duke Math. J.* *24*(1956), 595-600. MR18 p. 751;Zbl 72 p. 404.

Let $K = \{T(x) | x \in M\}$.

"Theorem 7. In order that, for a compact continuum M , the collection K should be finite, it is necessary and sufficient that there exist a positive integer n such that M is n-indecomposable."

18. ______, "Continua and various types of homogeneity," *Trans. Amer. Math. Soc.* *88*(1958), 366-374. MR20#1961;Zbl 86 p. 370.

Theorem. An aposyndetic, nearly homogeneous plane continuum is a simple closed curve or it separates the plane into infinitely many components.

Theorem. Decomposable, nearly 2-homogeneous, connected metric spaces are aposyndetic.

Theorem. Aposyndetic and hereditarily unicoherent metric continua are locally connected.

Here Burgess proved that 2-homogeneous continua are aposyndetic and asked whether they are locally connected (answered in the affirmative by G. S. Ungar).

19. Charatonik, J. J., "On irreducible smooth continua," *Proc. Int'l Symp. in Topology*, Budva (Yugoslavia), 1972 (1973), pp. 45-50. MR49#9818; Zbl 278#54033.

Aposyndesis is used in the study of irreducible smooth continua.

20. ______, "Problems and remarks on contractibility of curves," *General Topology and its Relations To Modern Anal. and Algebra, IV* (Proc. 4th Prague Topological Symp., Prague, 1976), Part B, pp. 72-76. Soc. Czechoslovak Mathematicians and Physicists, Prague, 1977. Zbl 373 #54029.

"Some...known conditions of noncontractibility of continua are expressed in terms of the set function T ." It also states, without proof, Thm. 2 of [21].

21. ______, "The set function T and homotopies," *Colloq. Math. 39*(1978), 271-274. Zbl 412#54035.

The subject of the title is studied in the context of compact, connected Hausdorff spaces.

22. Charatonik, J. J. and C. Eberhart, "On smooth dendroids," *Fund. Math. 67*(1970), 297-322. MR43#1129;Zbl 192 p. 600.

Several characterizations of smooth dendroids are given in terms of the aposyndetic set function T .

23. Charatonik, J. J. [4].

24. Czuba, S. T., "A concept of pointwise smooth dendroids" (Russian), *Uspehi Mat. Nauk 34*(1979), No. 6, 215-217.

25. ______, "The set function T and R-continuum," *Bull. Acad. Polon. Sci. Sér. Sci. Math. 27*(1979), 303-308. Zbl 424#54027.

"It is proved that if a dendroid X contains closed subsets A and B such that $T(A) \cap B = \emptyset = A \cap T(B)$ and $T(A) \cap T(B) \neq \emptyset$, then X contains an R-continuum. This is a generalization of the results of" [4].

26. ______, "On pointwise smooth dendroids," *Fund. Math.*, to appear.

"In this paper we study properties of pointwise smooth dendroids; in particular, we give some of their connections with the set function T and the concept of an R^3-continuum."

27. Czuba, S. T., "Open functions and pointwise smooth dendroids," *Bull. Acad. Polon. Sci. Sér. Math.*, to appear.

The set function T is used to prove that, for dendroids, pointwise smoothness is an invariant for open mappings.

28. Davis, H. S., "A note on connectedness im kleinen," *Proc. Amer. Math. Soc. 19*(1968), 1237-1241. MR40#8021;Zbl 181 p. 261.

"The author relates various properties (e.g. decomposability, connectedness im kleinen) with properties of the set function T ." MR.

Properties of T on weakly irreducible Hausdorff continua (= θ-continua) are also studied. See [46] and [196].

29. Davis, H. S. and P. H. Doyle, "Invertible continua," *Portugal. Math. 26*(1967), 487-491. MR41#1013;Zbl 186 p. 563.

Relationships between the set function T , a very weak form of aposyndesis (almost connectedness im kleinen), and invertibility are studied in Hausdorff continua.

30. Davis, H. S., D. P. Stadtlander and P. M. Swingle, "Properties of the set function T^n," *Portugal. Math. 21*(1962), 114-133. MR25#5501;Zbl 107 p. 401.

Here the set functions T^n are defined ($T^0(A) = T(A)$, $T^{n+1}(A) = T(T^n(A))$) and used in the study of indecomposable continua and generalizations.

31. ______, "Semigroups, continua and the set function T^n," *Duke Math. J. 29*(1962), 265-280. MR26#4325;Zbl 106 p. 20.

This continues the study of T^n begun in [30], now in the context of topological semigroups. Some results of Hunter [112] are generalized, and alternate proofs of some results of Koch and Wallace [138] are given.

32. Davis, H. S. and P. M. Swingle, "Extended topologies and iteration and recursion of set-functions," *Portugal. Math. 23*(1964), 103-129. MR36#5877;Zbl 134 p. 183.

The set function T is used as a model in a very general study of set functions in the spirit of Preston Hammer's extended topology.

33. Davis, H. S. [5].

34. Dickman, R. F., Jr., "On openness properties of mappings," *Portugal. Math. 26*(1967), 115-123. MR40#2016;Zbl 186 p. 562.

This relates the (semi-) local connectedness of the domain and the middle space of the monotone-light factorization of a map to certain mapping problems.

35. Dickman, R. F., R. L. Kelley, L. R. Rubin and P. M. Swingle, "Semigroups and clusters of indecomposability," *Fund. Math.* *56*(1964), 21-33. MR30#3452;Zbl 128 p. 410.

This contains a number of rather specialized results related to indecomposability and n-indecomposability.

36. Dickman, R. F., L. R. Rubin and P. M. Swingle, "Characterization of n-spheres by an excluded middle membrane principle," *Michigan Math. J.* *11*(1964), 53-59. MR28#4523;Zbl 116 p. 400.

37. ______, "Irreducible continua and generalization of hereditarily unicoherent continua by means of membranes," *J. Australian Math. Soc.* *5* (1965), 416-426. MR32#6424;Zbl 178 p. 260.

38. vanDouwen, E. K. and J. T. Goodykoontz, Jr., "Aposyndesis in hyperspaces and Čech-Stone remainders," these *Proceedings*.

39. Doyle, P. H., "Invertible spaces that are products," *Balkanica*, to appear.

Main result: $X\times[0,1] = Y$ is invertible and X is a metric continuum imply Y is contractible.

40. ______, "Product invertibility," Preprint.

If M is an invertible, metric continuum and $M = M_1\times[0,1]$, then, in each open subset U of M, there is a point p such that M is aposyndetic at p with respect to $M\backslash U$.

41. ______ [29].

42. Eberhart, C. [22].

43. Emeryk, A. and Z. Horbanowicz, "On atomic mappings," *Colloq. Math.* *27* (1973), 49-55. MR48#4987;Zbl 262#54033.

44. FitzGerald, R. W., "The cartesian product of nondegenerate compact continua is n-point aposyndetic," *Topology Conference, Arizona State Univ., 1967*, Ariz. State Univ., Tempe, 1968, pp. 324-326. MR38#5155; Zbl 217#482.

In [166] the title theorem is proved without the compactness condition.

45. ______, *Generalized Finite Graphs*, Ph.D. Thesis, Arizona State Univ., 1969.

Most of this was published as [46]. One theorem says (in effect) that θ-spaces are weakly irreducible [28].

46. ______, "Connected sets with a finite disconnection property," *Studies In Topology*, Academic Press, New York, 1975, pp. 139-173. MR51#1730; Zbl 306#54044.

This is the basic work on (Hausdorff) θ-spaces.

47. FitzGerald, R. W. and P. M. Swingle, "Core decompositions of continua," *Fund. Math.* *61*(1967), 33-50. MR36#7110;Zbl 179 p. 196.

For similar results, approached differently, see [150].

48. Fugate, J. B., "Irreducible continua," *Topology Conference, Arizona State Univ., 1967*, Ariz. State Univ., Tempe, 1968, pp. 100-103. Zbl 217 p. 196.

49. Fugate, J. B., G. R. Gordh, Jr. and Lewis Lum, "On arc-smooth continua," *Topology Proc.* *2*(1977), 645-656.

It is noted that every arc-smooth continuum is semi-aposyndetic.

50. ______, "Arc-smooth continua," *Trans. Amer. Math. Soc.*, to appear.

The aposyndetic set function T is modified so as to apply effectively to continua admitting arc-structures. The resulting set function is applied to obtain several characterizations of arc-smooth continua.

51. Fugate, J. B. [13].

52. Goodykoonz, J. T., Jr., "Aposyndetic properties of hyperspaces," *Pacific J. Math.* *47*(1973), 91-98. MR48#7192;Zbl 263#54027.

53. ______, "Aposyndesis and hyperspaces," *Proceedings: Conference on Metric Spaces, Generalized Metric Spaces and Continua* (honoring F. B. Jones), Guilford College, Greensboro, N. Carolina, 1980, pp. 129-136.

Expository. See [38].

54. ______ [38].

55. Gordh, G. R., Jr., "Monotone decompositions of irreducible Hausdorff continua," *Pacific J. Math.* *36*(1971), 647-658. MR43#6882;Zbl 197#490.

Several known results about the aposyndetic set functions K and T on irreducible metric continua are shown to be valid in the Hausdorff setting.

56. ______, "Concerning closed quasi-orders on hereditarily unicoherent continua," *Fund. Math.* *78*(1973), 62-73. MR48#1196;Zbl 258#54031.

Every semiaposyndetic continuum which is hereditarily unicoherent at some point is shown to be a dendroid. Several characterizations of smooth dendroids and smooth continua are given in terms of the aposyndetic set function T.

57. ______, "Aposyndesis in hereditarily unicoherent continua," these *Proceedings*.

58. Gordh, G. R., Jr., "Aposyndesis and the notion of smoothness in continua," these *Proceedings*.

59. Gordh, G. R., Jr. and C. B. Hughes, "On freely decomposable mappings of continua," *Glas. Mat. Ser. III 14(34)*(1979), 137-146. Zbl 411 #54011.

The notion of free decomposability for continua, which is equivalent to aposyndesis, is extended to mappings. The resulting class of mappings properly contains the monotone mappings and shares some of their properties.

60. Gordh, G. R., Jr. and Lewis Lum, "Monotone retracts and some characterizations of dendrites," *Proc. Amer. Math. Soc.* *59*(1976), 156-158. MR54#11296;Zbl 338#34068.

Theorem: The continuum X is a dendrite if and only if X is semiaposyndetic and for some point p in X, each subcontinuum containing p is a monotone retract of X.

61. ______, "On monotone retracts, accessibility, and smoothness in continua," *Topology Proc.* *1*(1976), 17-28. Zbl 389#54023.

The facts that smooth dendroids are semiaposyndetic, and aposyndetic dendroids are dendrites, are used in the study of monotone retractions.

62. ______, "Radially convex mappings and smoothness in continua," *Houston J. Math.* *4*(1978), 335-342. MR80e#54044.

The existence of certain radially convex mappings on an aposyndetic (resp., semiaposyndetic) continuum X, implies that X is a dendrite (resp., smooth dendroid).

63. Grace, E. E., *Certain Properties of Continua Related to Nonaposyndesis,* Ph.D. Thesis, University of N. Carolina, 1956.

64. ______, "Cut sets in totally nonaposyndetic continua," *Proc. Amer. Math. Soc.* *9*(1958), 98-104. MR20#1960;Zbl 138 p. 181.

65. ______, "Totally nonconnected im kleinen continua," *Proc. Amer. Math. Soc.* *9*(1958), 818-821. MR20#6079;Zbl 138 p. 181.

66. ______, "A totally nonaposyndetic, compact, Hausdorff space with no cut point," *Proc. Amer. Math. Soc.* *15*(1964), 281-283. MR29#595;Zbl 125 p. 399.

67. ______, "Cut points in totally non-semi-locally-connected continua," *Pacific J. Math.* *14*(1964), 1241-1244. MR30#4243;Zbl 139 p. 162.

See [178] for a strengthened result. See [214].

68. ______, "On local properties and G_δ sets," *Pacific J. Math.* *14*(1964), 1245-1248. MR30#4244;Zbl 151 p. 306.

See [213].

69. ______, "Certain questions related to the equivalence of local connectedness and connectedness im kleinen," *Colloq. Math.* *13*(1965), 211-216. MR31#4011;Zbl 135 p. 411.

70. ______, "On the existence of generalized cut points in strongly nonlocally connected continua," *Topology Conference, Ariz. State Univ., 1967*, Ariz. State Univ., Tempe, 1968, pp. 138-146. MR38#6548;Zbl 213 #497.

Mainly expository.

71. Grace, E. E., "Aposyndesis and weak cut points," *Proceedings: Conf. on Metric Spaces, Gen. Metric Spaces and Continua (honoring F. B. Jones)*, Guilford College, Greensboro, N. Carolina, 1980, pp. 151-166.

This is mostly expository but contains a generalization of part of [82].

72. ______, "Aposyndesis and weak cutting," these *Proceedings*.

73. ______, "A plane cut point theorem," in preparation.

This proves Theorem 12 of [105].

74. ______, "Monotone decompositions of θ-continua," in preparation.

This does for θ-continua what [76] does for θ_n-continua. See [195] or [196].

75. ______, "Semi-local-connectedness and cutting in nonlocally-connected, metric spaces," in preparation.

This generalizes part of [82].

76. Grace, E. E. and E. J. Vought, "Monotone decompositions of θ_n-continua," *Trans. Amer. Math. Soc.*, 263(1981), 261-270.

77. Grispolakis, J., A. Lelek and E. D. Tymchatyn, "Connected subsets of finitely Suslinian continua," *Colloq. Math.* 35(1976), 209-222. MR53 #11585;Zbl 326#54023.

It is shown that a semiaposyndetic metric continuum X is finitely Suslinian, if and only if every connected F_σ-subset of X is arc-wise connected or X satisfies any other of five similar conditions. But, see [212].

78. de Groot, J., "Connectedly generated spaces," *Proc. Int'l. Symp. on Topology and its Appl. (Herceg-Novi, 1968)*, pp. 171-175. *Savez Društava Mat. Fiz. i Astronom.*, Belgrade, 1969. MR43#1108;Zbl 209 p. 268.

Semi-locally-connected T_1-spaces are studied from an unusual point of view.

79. Hagopian, C. L., *A Classification of Nonlocally Connected Continua*, Ph.D. Thesis, Ariz. State Univ., 1968.

80. ______, "On nonaposyndesis and the existence of a certain generalized cut point," *Topology Conference, Ariz. State Univ., 1967*, Ariz. State Univ., Tempe, 1968, pp. 327-329. MR38#6549;Zbl 212#277.

81. ______, "On nonlocally connected continua," *Topology Conference, Ariz. State Univ., 1967*, Ariz. State Univ., Tempe, 1968, pp. 330-334. Zbl 212 p. 277.

Mutual aposyndesis was introduced here.

82. Hagopian, C. L., "Concerning semi-local-connectedness and cutting in nonlocally connected continua," *Pacific J. Math.* *30*(1969), 657-662. MR40#2007;Zbl 181 p. 510.

Generalizations occur in [71,75].

83. ______, "Mutual aposyndesis," *Proc. Amer. Math. Soc.* *23*(1969), 615-622. MR40#876;Zbl 184 p. 266.

84. ______, "Concerning arcwise connectedness and the existence of simple closed curves in plane continua," *Trans. Amer. Math. Soc.* *147*(1970), 389-402. MR40#8030;Zbl 191 p. 535.

For errata see [91].

85. ______, "On generalized forms of aposyndesis," *Pacific J. Math.* *34* (1970), 97-108. MR42#2439;Zbl 198 p. 558.

86. ______, "Arcwise connected plane continua," *Topology Conference, Emory Univ., 1970*, Emory Univ., Atlanta, 1970, pp. 41-44. MR49#7992;Zbl 247 #54034.

Mostly expository. Semiaposyndesis was introduced here.

87. ______, "A class of arcwise connected continua," *Proc. Amer. Math. Soc.* *30*(1971), 164-168. MR43#6883;Zbl 218 p. 363.

88. ______, "A cutpoint theorem for plane continua," *Duke Math J.* *38*(1971), 509-512. MR44#2204;Zbl 221#54029.

For errata see [94].

89. ______, "Arcwise connectedness of semiaposyndetic plane continua," *Trans. Amer. Math. Soc.* *158*(1971), 161-165. MR44#2205;Zbl 198 p. 559.

90. ______, "Arcwise connectivity of semi-aposyndetic plane continua," *Pacific J. Math.* *37*(1971), 683-686. MR46#6322; Zbl 218 p. 363.

91. ______, "Errata to 'Concerning arcwise connectedness and the existence of simple closed curves in plane continua,'" *Trans. Amer. Math. Soc.* *157*(1971), 507-509. MR43#1130;Zbl 216 p. 194.

Errata to [84].

92. ______, "Semiaposyndetic nonseparating plane continua are arcwise connected," *Bull. Amer. Math. Soc.* *77*(1971), 593-595. MR44#1004;Zbl 224 #54054.

93. ______, "The cyclic connectivity of plane continua," *Michigan Math. J.* *18*(1971), 401-407. MR45#9294;Zbl 224#54055.

94. ______, "Errata to 'A cutpoint theorem for plane continua,'" *Duke Math. J.* *39*(1972), 823. MR46#9947;Zbl 248#54039.

Errata to [88].

95. Hagopian, C. L., "The fixed point property and decomposable plane continua," *Proc. of the Univ. of Oklahoma Topology Conf., 1972*, Univ. of Okla., Norman, 1972, pp. 60-63. MR50#11192;Zbl 245#54045.

Mainly expository.

96. ______, "An arc theorem for plane continua," *Illinois J. Math. 17* (1973), 82-89. MR47#2562;Zbl 245#54034.

97. ______, "Characterizations of λ connected plane continua," *Pacific J. Math. 49*(1973), 371-375. MR50#5760;Zbl 281#54016.

" λ " should be replaced by " δ " in this and the next five papers in order to be consistent with earlier usage. See [106] for a definition of λ-connected.

98. ______, "Schlais' theorem extends to λ connected plane continua," *Proc. Amer. Math. Soc. 40*(1973), 265-267. MR#4744;Zbl 263#54016.

See [97].

99. ______, " λ connected plane continua," *Trans. Amer. Math. Soc. 191* (1974), 277-287. MR49#6186;Zbl 286#54023.

See [97].

100. ______, " λ connectivity and mappings onto a chainable indecomposable continuum," *Proc. Amer. Math. Soc. 45*(1974), 132-136. MR49#6185; Zbl 289#54022.

See [97].

101. ______, "Locally homeomorphic λ connected plane continua," *Pacific J. Math. 52*(1974), 403-404. MR52#4249;Zbl 291#54036.

See [97].

102. ______, " λ connectivity in the plane," *Studies in Topology*, Academic Press, New York, 1975, pp. 197-202. MR50#14695;Zbl 325#54013.

Expository. See [97].

103. ______, "Uniquely arcwise connected plane continua have the fixed-point property," *Trans. Amer. Math. Soc. 248*(1979), 85-104.

104. ______"Plane continua and the fixed point property," *Proceedings: Conf. on Metric Spaces, Gen. Metric Spaces and Continua (honoring F. B. Jones)*, Guilford College, Greensboro, N. Carolina, 1980, pp. 91-100.

Expository.

105. ______, "Aposyndesis in the plane," these *Proceedings*.

106. ______, "λ-connected products," these *Proceedings*.

107. ______, "Arcwise connectivity of continuum-chainable plane continua," preprint.

A failed analogy to semiaposyndetic plane continua is noted, and a related example by Oversteegen [157] is mentioned.

108. Hagopian, C. L. and L. E. Rogers, "Arcwise connectivity and continuum chainability," *Houston J. Math.*, to appear.

109. Horbanowicz, Z. [43].

110. Hughes, C. B., "Some remarks on freely decomposable mappings," *Topology Proc.* *2*(1977), 213-217. Zbl 407#54008.

"Theorem 3. If X is irreducible, Y is semi-locally connected, and $f:X \to Y$ is a FD mapping, then f is monotone. Consequently, if Y is nondegenerate, Y is an arc."

111. ______ [59].

112. Hunter, R. P., "On the semigroup structure of continua," *Trans. Amer. Math. Soc.* *93*(1959), 356-368. MR22#82;Zbl 97 p. 19.

Theorem. If S is a semigroup irreducible between two points with $S^2 = S$, then S/K is an arc (where K is the minimal ideal).

The crucial part of the proof of this, using the set function T, is generalized in [31, Thm. 16]. The latter part of the paper uses the sets T_p to give a rather detailed picture of one dimensional, compact, connected monoids. These are shown to be dendroids modulo K.

This is generalized in [31, 113, and 179].

113. ______, "Certain compact connected semigroups irreducible over a finite set" (Spanish), *Bol. Soc. Mat. Mexicana (2)6*(1961), 52-59. MR 25#2589.

Here the results of [112] are generalized to dendrites.

114. ______, "On a conjecture of Koch," *Proc. Amer. Math. Soc.* *12*(1961), 138-139. MR22#11070.

Here it is shown that the group of units of a compact, connected monoid cannot be a C-set. It follows that $T_p = \{p\}$ whenever p is a unit and that the monoid is 1-semi-locally-connected.

115. ______, "Sur la position des C-ensembles dans les demi-groupes," *Bull. Soc. Math. Belg.* *14*(1962), 190-195. MR25#1235;Zbl 105 p. 17.

Here it is shown that, modulo the minimal ideal, there are no C-sets.

116. ______, "On one dimensional semigroups," *Math. Ann.* *146*(1962), 383-396. MR25#557;Zbl 105 p. 18.

Some interesting examples involving nonaposyndesis are presented.

117. Hunter, R. P., "On the structure of homogroups with applications to the theory of compact connected semigroups," *Fund. Math.* 52(1963), 69-102. MR26#1864.

Some interesting examples involving nonaposyndesis are presented (without that being pointed out).

118. ______, "Aposyndesis in topological monoids," these *Proceedings*.

119. Jones, F. B., "Certain equivalences and subsets of a plane," *Duke Math J.* 5(1939), 133-145. Zbl 20 p. 409.

120. ______, "Concerning the boundary of a complementary domain of a continuous curve," *Bull. Amer. Math. Soc.* 45(1939), 428-435. Zbl 21 p. 161.

121. ______, "Aposyndetic continua and certain boundary problems," *Amer. J. Math.* 63(1941), 545-553. MR3 p. 59;Zbl 25 p. 240.

122. ______, "A characterization of a semi-locally-connected plane continuum," *Bull. Amer. Math. Soc.* 53(1947), 170-175. MR8 p. 397;Zbl 32 p. 315.

123. ______, "Concerning nonaposyndetic continua," *Amer. J. Math.* 70 (1948), 403-413. MR9 p. 606;Zbl 35 p. 109.

See [126]. Generalizations or analogous results occur in [64,67, 70,80,85,170,178, and 189]. Also see [66].

124. ______, "A note on homogeneous plane continua," *Bull. Amer. Math. Soc.* 55(1949), 113-114. MR10 p. 468;Zbl 35 p. 391.

Theorem. Any homogeneous, aposyndetic plane continuum is a simple closed curve.

125. ______, "Certain homogeneous unicoherent indecomposable continua," *Proc. Amer. Math. Soc.* 2(1951), 855-859. MR13 p. 573;Zbl 44 p. 380.

126. ______, "Concerning aposyndetic and nonaposyndetic continua," *Bull. Amer. Math. Soc.* 58(1952), 137-151. MR14 p. 71;Zbl 46 p. 403.

Expository. Together with [123], the principal source on aposyndesis.

127. ______, "On a certain type of homogeneous plane continuum," *Proc. Amer. Math. Soc.* 6(1955), 735-740. MR17 p. 180;Zbl 67 p. 405.

The decomposable, homogeneous, plane continua are characterized using an aposyndetic decomposition.

128. ______, "One-to-one continuous images of a line," *Fund. Math.* 67 (1970), 285-292. MR41#6180;Zbl 192 p. 601.

129. ______, "One-to-one continuous images of a line," *General Topology and its Relations to Modern Anal. and Algebra*, Proc. of the Kanpur Topological Conf., Academic Press, New York, 1971, pp. 157-160. Zbl 226#54031.

Expository.

130. Jones, F. B., "Aposyndesis revisited," *Proc. of the Univ. of Oklahoma Topology Conf., 1972*, Univ. of Okla., Norman, 1972, pp. 64-78. MR51 #1770;Zbl 244#54022.

Expository.

131. ______, "Aposyndetic continua," *Topics in Topology (Proc. Colloq., Keszthely, 1972)*, Colloq. Math. Soc. János Bolyai, Vol. 8, North-Holland, Amsterdam, 1974, pp. 437-447. MR51#1769;Zbl 316#54033.

Expository.

132. ______, "Aposyndesis, homogeneity and a bit more history," *Proceedings: Conf. on Metric Spaces, Gen. Metric Spaces and Continua (honoring F. B. Jones)*, Guilford College, Greensboro, N. Carolina, 1980, pp. 79-84.

Expository.

133. ______, "Aposyndesis," these *Proceedings*.

Expository.

134. ______ [197].

135. Kelley, R. L. [35].

136. Koch, R. J., "Note on weak cut points in clans," *Duke Math. J. 24* (1957), 611-616. MR19 p. 1064.

This is an early paper showing the equivalence of certain conjectures settled in [114] and proving the following.

Theorem. Let S be a homogeneous clan (with unit u) which is not a group. Then u is not a weak cut point, and S is 1-semi-locally connected.

137. ______, "A theorem a day," *Rev. Roumaine Math. Pures Appl. 17*(1972), 879-883.

This is "an account of" A. D. Wallace's contributions to topological semigroups and related areas." It mentions "the conceptually useful idea that the $T(p)$ sets...substitute for missing inverses."

138. Koch, R. J. and A. D. Wallace, "Admissibility of semigroup structures on continua," *Trans. Amer. Math. Soc. 88*(1958), 277-287. MR20#1729; Zbl 81 p. 255.

This is the first paper dealing with T_p in topological semigroups. (" T_p " was used instead of Jones's original notation " L_p " since L_p is the generating set for the principal left ideal generated by p.) Here certain basic facts are established, including the fact that $p \in I$ if T_p meets the ideal I.

139. Krasinkiewicz, J. and P. Minc (presented to the Auburn Univ. Topology Seminar).

They have proved that any atriodic, strictly nonmutually-aposyndetic [105], nondegenerate continuum can be mapped onto an indecomposable continuum. See [215].

140. Lelek, A. [77].

141. Lum, Lewis, "Weakly smooth continua," *Trans. Amer. Math. Soc. 214* (1975), 153-167. MR52#6679;Zbl 295#54044.

'Characterizations of weakly smooth continua in terms of (a) hyperspaces, (b) the weak cut point order, and (c) the aposyndetic set function T are obtained." MR

142. ______ [49,50,60,61 and 62].

143. Maćkowiak, T., "Some characterizations of smooth continua," *Fund. Math. 79*(1973), 173-186. MR47#9566;Zbl 259#54028.

The aposyndetic set function T is used in the study of smooth continua.

144. ______, "On smooth continua," *Fund. Math. 85*(1974), 79-95. MR51#1784; Zbl 282#54018.

A general notion of smoothness is defined and it is observed that smooth continua satisfy a rather strong aposyndetic condition.

145. ______, "Arcwise connected and hereditarily smooth continua," *Fund. Math. 92*(1976), 149-171. MR55#1324;Zbl 345#54034.

A known aposyndetic property of smooth continua is used in studying hereditarily smooth continua.

146. ______, "On some characterizations of dendroids and weakly monotone mappings," *Bull. Acad. Polon. Sci. Sér. Sci. Math. Astronom. Phys. 24*(1976), 177-182. MR53#14451;Zbl 341#54010.

"Theorem 2 is proved by the use of a general result to the effect that every arcwise connected continuum satisfies an aposyndesis-like condition." MR

147. McAllister, B. L., "Cyclic elements in topology, a history," *Amer. Math. Monthly 73*(1966), 337-350. MR34#780.

Semi-local-connectedness is mentioned in passing.

148. McAuley, L. F., *On the Aposyndetic Decomposition of Continua*, Ph.D. Thesis, Univ. of N. Carolina, 1954.

This work was published as [149].

149. ______, "On decomposition of continua into aposyndetic continua," *Trans. Amer. Math. Soc. 81*(1956), 74-91. MR19 p. 158;Zbl 73 p. 396.

See [150] for the culminating work.

150. McAuley, L. F., "An atomic decomposition of continua into aposyndetic continua," *Trans. Amer. Math. Soc. 88*(1958), 1-11. MR23#A1353; Zbl 151 p. 307.

A generalization of Whyburn's nonseparated collections is used to define a "minimal" decomposition into a aposyndetic continuum. For similar results based on the set function T , see [47].

151. ______, "Local cyclic connectedness. I. Cyclic subelements of cyclic elements," *Notices Amer. Math. Soc. 5*(1958), 698 (abstract).

"A cyclic subelement theory...is developed...for aposyndetic continua."

152. McWaters, M. M. and J. H. Reed, "Semi-locally connected spaces and unicoherence," *Bull. Acad. Polon. Sci. Sér. Math. Astronom. Phys. 20* (1972), 657-661. MR47#1018;Zbl 238#54036.

153. Minc, P. [139 and 215].

154. Moore, R. L., *Foundations of Point Set Theory*, rev. ed., Amer. Math. Soc. Colloq. Publ. 13, Amer. Math. Soc., Providence, R.I., 1962. MR 27#709;Zbl 5 p. 54.

Jones's work on aposyndesis is mentioned.

155. Nadler, S. B., Jr., *Hyperspaces of Sets---A Text With Research Questions*, Marcel Dekker, New York, 1978.

The following known results about hyperspaces are stated: (a) If X is a Hausdorff continuum, then C(X) and 2^X are each aposyndetic. (b) Aposyndesis (and some related properties such as semiaposyndesis) are Whitney properties.

156. Oversteegen, L. G., "Open retractions and locally confluent mappings of certain continua," *Houston J. of Math. 6*(1980), 113-125. Zbl 407 #54010.

Semi-local-connectedness and a weak form of aposyndesis (that could also be considered a weak form of semi-local-connectedness) are used in the study of irreducible, locally confluent mappings and of open mappings on hereditarily unicoherent continua. Some related examples are given.

157. ______, "A continuum chainable aposyndetic plane continuum," preprint.

An example is given of such a continuum that is not arcwise connected.

158. Petrus, Ann, "Whitney maps and Whitney properties of C(X) ," *Topology Proc. 1*(1977), 147-172.

Various forms of aposyndesis are shown to be Whitney properties.

159. Rakowski, Z. M., "Smooth Hausdorff continua," *Bull. Acad. Polon. Sci. Sér. Sci. Math. Astronom. Phys. 25*(1977), 563-566. MR57#4127;Zbl 361 #54017.

160. Reed, J. H. [152].

161. Rogers, J. T., Jr., "Dimension of hyperspaces," *Bull. Acad. Polon. Sci. Sér. Sci. Math. Astronom. Phys.* *20*(1972), 177-179. MR51#6762; Zbl 226#54029.

162. ______, "Completely regular mappings and homogeneous, aposyndetic continua," *Canad. J. Math.*, to appear.

163. ______, "Applied aposyndesis---the work of F. B. Jones on homogeneous continua," *Proceedings: Conf. on Metric Spaces, Gen. Metric Spaces and Continua (honoring F. B. Jones)*, Guilford College, Greensboro, N. Carolina, 1980, pp. 85-90.

Expository.

164. ______, "Aposyndesis and homogeneity," these *Proceedings*.

Expository.

165. Rogers, L. E., *n-Mutual Aposyndesis*, Ph.D. Thesis, Univ. of Calif., Riverside, 1970.

166. ______, "Concerning n-mutual aposyndesis in products of continua," *Trans. Amer. Math. Soc.* *162*(1971), 239-251. MR45#2676;Zbl 224#54056.

167. ______, "Mutually aposyndetic products of chainable continua," *Pacific J. Math.* *37*(1971), 805-812. MR51#1774;Zbl 198 p. 280.

168. ______, "Aposyndesis and the separation axioms," *Proc. of the Univ. of Okla. Topology Conf., 1972*, Univ. of Okla., Norman, 1972, pp. 264-273. MR51#1766;Zbl 245#5432.

169. ______, "Continua in which only semi-aposyndetic subcontinua separate," *Pacific J. Math.* *43*(1972), 493-502. MR47#7713;Zbl 223#54023.

170. ______, "Non-n-mutually aposyndetic continua," *Proc. Amer. Math. Soc.* *42*(1974), 595-601. MR49#1490;Zbl 276#54034.

171. ______, "Products with nonlinear finite-set-aposyndetic continua," *Colloq. Math.* *32*(1975), 199-206, 309. MR52#1646;Zbl 314#54037.

172. ______, "Aposyndesis in product spaces," these *Proceedings*.

Expository.

173. ______ [108].

174. Rosasco, John, "A note on Jones' function K ," *Proc. Amer. Math. Soc.* *49*(1975), 501-504. MR51#4188;Zbl 303#54012.

175. Rubin, L. R. [6,35,36, and 37].

176. Schlais, H. E., *Non-aposyndesis and Indecomposability*, Ph.D. Thesis, Ariz. State Univ., 1971.

This research was published as [177].

177. Schlais, H. E., "Non-aposyndesis and non-hereditary decomposability," *Pacific J. Math.* *45*(1973), 643-652. MR47#5845;Zbl 274#54021.

Let $K(x) = \{y | x \in T(y)\}$. Main Theorem. "If M is a compact continuum and for some $x \in M$, $\mathrm{Int}(K(x)) \neq \emptyset$, then M is not hereditarily decomposable."

178. Shirley, E. D., "Semi-local-connectedness and cut points in metric continua," *Proc. Amer. Math. Soc.* *31*(1972), 291-296. MR44#3294;Zbl 229#54032.

179. Stadtlander, D. P., "Actions with topologically restricted state spaces," *Duke Math. J.* *37*(1970), 199-206. MR41#4507;Zbl 199 p. 339.

Here the results of [112] are extended to semigroup actions.

180. ______ [30 and 31].

181. Stratton, H. H., "On continua which resemble simple closed curves," *Fund. Math.* *68*(1970), 121-128. MR41#7639;Zbl 198 p. 279.

182. Swingle, P. M. [30-32, 35-37, and 47].

183. Thomas, E. S., Jr., "Monotone decompositions of irreducible continua," *Rozprawy Math.* *50*(1966), 74pp. MR33#4907;Zbl 142 p. 212.

Theorem. Let M be an irreducible metric continuum. Then M admits a monotone, upper semi-continuous decomposition, each element of which has void interior, with the decomposition space an arc, if and only if $[T(z)]^0 = \emptyset$, for each z in M .

184. Tymchatyn, E. D. [77].

185. VandenBoss, E. L., *Set Functions and Local Connectivity*, Ph.D. Thesis, Michigan State Univ., 1970.

This is a study of a set function Y that can be defined in terms of domain aposyndesis [69] in the way T is defined in terms of aposyndesis. The relationships of Y to T and to monotone maps are considered in the context of Hausdorff spaces.

186. Vought, E. J., *Stronger Forms of Aposyndetic Continua*, Ph.D. Thesis, Univ. of Calif., Riverside, 1967.

187. ______, "A classification scheme and characterization of certain curves," *Colloq. Math.* *20*(1969), 91-98. MR38#6550;Zbl 186 p. 563.

188. ______, "Concerning continua not separated by any nonaposyndetic subcontinuum," *Pacific J. Math.* *31*(1969), 257-262. MR41#2633;Zbl 184 p. 481.

See [169] for generalizations.

189. ______, "n-Aposyndetic continua and cutting theorems," *Trans. Amer. Math. Soc.* *140*(1969), 127-135. MR39#3462;Zbl 182 p. 567.

190. ______, "Strongly semi-aposyndetic continua are hereditarily locally connected," *Proc. Amer. Math. Soc.* *33*(1972), 619-622. MR50#14700;Zbl 246#54041.

191. Vought, E. J., "Monotone decompositions of continua into generalized arcs and simple closed curves," *Fund. Math. 80*(1973), 213-220. MR48 #5029;Zbl 265#54035.

192. ______, "Monotone decompositions into trees of Hausdorff continua irreducible about a finite subset," *Pacific J. Math. 54*(1974), 253-261. MR51#13999;Zbl 302#54031.

193. ______, "Monotone decompositions of continua not separated by any subcontinua," *Trans. Amer. Math. Soc. 192*(1974), 67-78. MR49#6189; Zbl 289#54020.

See [74] and [76] for generalizations.

194. ______, "Monotone decompositions of Hausdorff continua," *Proc. Amer. Math. Soc. 56*(1976), 371-376. MR53#14440;Zbl 332#54027.

195. ______, "Structure of θ and θ_n-continua," *Proceedings: Conf. on Metric Spaces, Gen. Metric Spaces and Continua (honoring F. B. Jones)*, Guilford College, Greensboro, N. Carolina, 1980, pp. 121-128.

Expository.

196. ______, "Monotone decompositions of continua," these *Proceedings*.

197. Vought, E. J. and F. B. Jones, "Stronger forms of aposyndetic continua," *Topology Conf., Ariz. State Univ., 1967*, Ariz. State Univ., Tempe, 1968, pp. 170-173. MR39#925;Zbl 212 p. 276.

198. Vought, E. J. [76].

199. Wallace, A. D. [138].

200. Ward, L. E., "A general fixed point theorem," *Colloq. Math. 15*(1966), 244-251. MR34#806;Zbl 151 p. 309.

Here it is shown that in an aposyndetic, Hausdorff continuum the cutpoint partial ordering has a closed graph and other nice properties.

201. ______, "Recent developments in dendritic spaces and related topics," *Studies in Topology*, Academic Press, New York, 1975, pp. 601-647. MR 50#14709;Zbl 318#54036.

Theorem. A dendritic space is rim compact if and only if it is aposyndetic.

202. ______, "Partially ordered spaces and the structure of continua," *Proceedings: Conf. on Metric Spaces, Gen. Metric Spaces and Continua (honoring F. B. Jones)*, Guilford College, Greensboro, N. Carolina, 1980, pp. 137-150.

Expository. This mentions the relevance of [200, annotation] to Whyburn's cyclic element theory.

203. Whyburn, G. T., "The cyclic and higher connectivity of locally connected spaces," *Amer. J. Math. 53*(1931), 427-442.

Semi-local-connectedness was introduced here but called local divisibility.

204. Whyburn, G. T., "Semi-locally connected sets," *Amer. J. Math. 61* (1939), 733-749. MR1 p. 31; Zbl 21 p. 359.

205. ______, *Analytic Topology*, Amer. Math. Soc. Colloq. Publ. 28, Amer. Math. Soc., Providence, R. I., 1942. MR4 p. 86;Zbl 36 p. 124.

206. Wilder, R. L., "Generalized closed manifolds in n-space," *Annals of Math. 35*(1934), 876-903. Zbl 10 p. 181.

207. ______, "Sets which satisfy certain avoidability conditions," *Časopis Pěst. Mat. 67*(1938), 185-198. Zbl 18 p. 427.

"...types of avoidability...are employed...to obtain...results concerning the relations of closed sets to their complements in euclidean n-space."

208. ______, "Property S_n," *Amer. J. Math. 61*(1939), 823-832. MR1 p. 45; Zbl 22 p. 413.

209. ______, *Topology of Manifolds*, rev. ed., Amer. Math. Soc. Colloq. Publ. 32, Amer. Math. Soc., Providence, R. I., 1963. MR10 p. 614. MR 32#440;Zbl 39 p. 396. Zbl 117 p. 162.

"...an almost 0-avoidable point of a continuum M is a point at which M is semi-locally-connected..., and conversely."

210. Wiser, H. C., "Decompositions and homogeneity of continua on a 2-manifold," *Pacific J. Math. 12*(1962), 1145-1162. MR26#6938;Zbl 114 p. 390.

"Theorem 10. If a nondegenerate proper subcontinuum of a 2-manifold is aposyndetic and homogeneous, then it must be a simple closed curve."

Also, aposyndesis is used in proving decomposition theorems, including one on decomposition of a hereditarily finitely multicoherent continuum into a simple closed curve.

211. ______, "Near-homogeneity on 2-manifolds," *Amer. Math. Monthly 74* (1967), 423-426. MR35#994;Zbl 148 p. 169.

"Theorem 3. If a nondegenerate proper subcontinuum X of a 2-manifold is aposyndetic and continuously near-homogeneous then X is a simple closed curve."

ADDENDUM

212. Abo-Zeid, E., "On σ-connected spaces," *Colloq. Math.* *40*(1978), 85-90.

The semiaposyndesis condition in the result in [77] is shown to be unnecessary.

213. Grace, E.E., *Local Properties in II_z Spaces*, Technical Report No. 51, Ariz. State Univ., 1980.

[68] specializes these results to complete, metric spaces.

214. Grace, E.E., *Cut Points in Totally Nonsemi-locally-connected II_z Continua*, Technical Report No. 52, Ariz. State Univ., 1980.

[67] specializes these results to locally compact, metric spaces.

215. Krasinkiewicz, J. and P. Minc, "Dendroids and their endpoints," *Fund. Math.* *99*(1978), 227-244.

"We...prove that in every dendroid X the minimal arcwise connected set spanning the set of all endpoints of X at which X is semi-locally connected is a dense subset of X. Moreover, the set of endpoints at which X is semi-locally connected is a G_δ-set in X and the remaining endpoints form a subset of the first category in X."

216. ______, "Continua with contable number off Arc-components", *Fund. Math.* *102* (1979), 119-127. MR80#54041a.

217. ______, "Continua and their open sub-sets with connected complements", *Fund. Math.* *102* (1979), 129-136. MR80#54041b.

PARTICIPANTS

Leonard A. Asimow
David P. Bellamy
R. H. Bing
Karol Borsuk
Beverly Brechner
C. Edmund Burgess
Bruce L. Chalmers
John E. de Pillis
Nicolae Dinculeanu
W. M. Dugger
James Dugundji
Eldon Dyer
Edward G. Effros
Joseph Brauch Fugate
B. D. Garrett
George R. Gordh, Jr.
Edward E. Grace
Neil E. Gretsky
Francois Guenard
Charles L. Hagopian
Roger W. Hansell
Robert W. Heath
Robert P. Hunter
Robert C. James
F. Burton Jones
I. Juhasz
John L. Kelley
Lewis Lum
Garr S. Lystad
Byron L. McAllister
Louis F. McAuley
Donald A. Martin
Austin C. Melton
Mark Michael
Edwin E. Moise
Deane Montgomery
Jan Mycielski
Issac Namioka
Peter J. Nyikos
John W. Petro
Roman Pol
S. Purisch
M. Rajagopalan
M. M. Rao
Michael D. Reagan
Prabir Roy
Mary E. Rudin
David E. Rush
Elias Saab
Paulette Saab
Albert R. Stralka
Franklin D. Tall
Howard G. Tucker
Eric K. van Douwen
Eldon J. Vought
John J. Walsh
Lewis E. Ward, Jr.
David C. Wilson
C. T. Yang